AF388655

Trends in GPR and other NDTs for Transport Infrastructure Assessment

Trends in GPR and other NDTs for Transport Infrastructure Assessment

Editors

Mercedes Solla
Vega Pérez-Gracia
Simona Fontul

MDPI • Basel • Beijing • Wuhan • Barcelona • Belgrade • Manchester • Tokyo • Cluj • Tianjin

Editors

Mercedes Solla	Vega Pérez-Gracia	Simona Fontul
CINTECX,	Universitat Politecnica de Catalunya,	Transportation Department,
Universidade de Vigo,	GIES Research Group	National Laboratory for
GeoTECH Research Group	Spain	Civil Engineering
Spain		Portugal

Editorial Office
MDPI
St. Alban-Anlage 66
4052 Basel, Switzerland

This is a reprint of articles from the Special Issue published online in the open access journal *Remote Sensing* (ISSN 2072-4292) (available at: https://www.mdpi.com/journal/remotesensing/special_issues/GPR_NDTs).

For citation purposes, cite each article independently as indicated on the article page online and as indicated below:

LastName, A.A.; LastName, B.B.; LastName, C.C. Article Title. *Journal Name* **Year**, *Volume Number*, Page Range.

ISBN 978-3-0365-2123-7 (Hbk)
ISBN 978-3-0365-2124-4 (PDF)

Cover image courtesy of Mercedes Solla.

Contents

About the Editors

Mercedes Solla

received her Ph.D. degree from the Universidade de Vigo, Spain, in 2010. She is an environmental engineer, currently working at the Department of Natural Resources and Environmental Engineering (Universidade de Vigo) with a "Ramón y Cajal" contract (RYC2019-026604-I/AEI/10.13039/501100011033). Her studies include the use of Ground Penetrating Radar (GPR) for subsurface prospection, mainly devoted to studying civil engineering and cultural heritage structures. As a researcher in more than 25 European and Spanish projects, she has authored more than 250 publications in peer-refereed journals, book chapters and conference proceedings, with 55 scientific papers on prestigious international journals included in the JCR database.

Vega Pérez-Gracia

received her Ph.D. degree from the Universtat Politècnica de Catalunya, Spain, in 2001. She has a degree in physicist specialized in geophysics and a master in Earthquake Engineering and Structural Dynamics, currently working at the Department of Strength of Materials and Structural Engineering (Universitat Politècnica de Catalunya) as professor. Her studies include the use of GPR in cultural heritage, buildings and other civil structures, as well as for subsurface prospection. She participated in more than 14 European and Spanish projects, and she has authored more than 260 publications in peer-refereed journals, book chapters and conference proceedings, with 48 scientific papers on prestigious international journals included in the JCR database.

Simona Fontul

received her Ph.D. degree from the Coimbra University, Portugal, in 2005. She is a civil engineer, currently working as Assistant Researcher at the Transportation Department of National Laboratory for Civil Engineering (LNEC). Her studies include functional and structural evaluation of roads, runways and railways, as well as the use of GPR for infrastructure assessment. She is also invited Assistant Professor at Nova University of Lisbon. She has authored more than 150 publications in peer-refereed journals, book chapters and conference proceedings, besides reports of applied research at LNEC.

Preface to "Trends in GPR and other NDTs for Transport Infrastructure Assessment"

The effective and timely assessment of structural health conditions becomes crucial to assure the safety of the transportation system and time-saver protocols and reduce excessive repair and maintenance costs. Ground-Penetrating Radar (GPR) is widely recognized as one of the most powerful and versatile nondestructive testing (NDT) methods for routine subsurface inspections. Although it has been demonstrated that GPR has significantly benefited the procedures for inspection and successfully solved some of the limitations of traditional methods, it still presents some constraints. To maximize the benefit obtained through the use of GPR, it is important to be aware of these constraints and to be able to develop new systems and adapt the methodologies for testing and analysis of the results for decision-making procedures. Recent trends also show an increasing interest in the combination of NDT methods for high-resolution diagnosis.

This book contains reviews and recent advances of general interest on the use of GPR to assess transport infrastructure. It is composed of 13 papers, compiled into the following four sections:

- GPR application on transport infrastructure.

This first section introduces a review of GPR applications on transport infrastructure, focusing on pavements, railways, retaining walls, bridges, and tunnels. This review discusses the best practices, troubleshooting, and future perspectives of the method for infrastructure inspection.

- Roads and airports.

This section comprises seven papers devoted to studying pavement structure with GPR and other complementary methods such as a pavement density profiler (PDP) and emerging Persistent Scatterers Interferometry (PSI). A new methodology for pavement thickness analysis is presented based on three-dimensional (3D) GPR models. The use of more advanced deep learning techniques is also addressed in this section with two different goals: first, to reduce incoherence measurement noise through deep-learning-based super-resolution (SR) image enhancement, and second, to automatically detect internal defects in asphalt pavement.

- Railways.

The paper in this section focuses on new perspectives in the decision-making process in performing mechanized ballast cleaning based on GPR assessment.

-    Concrete bridges.

The last section of this book deals with GPR applications for the condition monitoring of concrete bridge structures. It is composed of four papers. The first paper presents a novel approach using GPR to estimate the variability of material properties over time and space for the structural health assessment of concrete bridges. The second provides an integrated geophysical investigation, Water-Penetrating Radar (WPR) supplemented by sonar, for the underwater inspection of bridge-foundations-related scour and erodible-scour-based infill. The third paper evaluates the combination of three different NDT methods (GPR, impact–echo, and metal magnetic memory) to investigate the best practice to analyze the reinforcement conditions in concrete beams. Finally, the last paper develops a new 3D modeling approach based on the Complex Refractive Index Model (CRIM) to estimate the bulk permittivity of concrete.

Mercedes Solla, Vega Pérez-Gracia, Simona Fontul
Editors

Review

A Review of GPR Application on Transport Infrastructures: Troubleshooting and Best Practices

Mercedes Solla [1,*], Vega Pérez-Gracia [2] and Simona Fontul [3,4]

[1] CINTECX, GeoTECH research group, Universidade de Vigo, 36310 Vigo, Spain
[2] Department of Strength of Materials and Structural Engineering, Universitat Politècnica de Catalunya, Campus Diagonal Besós, Barcelona East School of Engineering, EEBE, Av. Eduard Maristany, 16, 08019 Barcelona, Spain; vega.perez@upc.edu
[3] Department of Transportation, National Laboratory for Civil Engineering—LNEC, 1700-066 Lisbon, Portugal; simona@lnec.pt
[4] Civil Engineering Department, NOVA School of Science and Technology, 2829-516 Caparica, Portugal
* Correspondence: merchisolla@uvigo.es

Abstract: The non-destructive testing and diagnosis of transport infrastructures is essential because of the need to protect these facilities for mobility, and for economic and social development. The effective and timely assessment of structural health conditions becomes crucial in order to assure the safety of the transportation system and time saver protocols, as well as to reduce excessive repair and maintenance costs. Ground penetrating radar (GPR) is one of the most recommended non-destructive methods for routine subsurface inspections. This paper focuses on the on-site use of GPR applied to transport infrastructures, namely pavements, railways, retaining walls, bridges and tunnels. The methodologies, advantages and disadvantages, along with up-to-date research results on GPR in infrastructure inspection are presented herein. Hence, through the review of the published literature, the potential of using GPR is demonstrated, while the main limitations of the method are discussed and some practical recommendations are made.

Keywords: ground penetrating radar; inspection; pavements; railways; bridges; tunnels; retaining walls; methodologies; limitations; benefits

Citation: Solla, M.; Pérez-Gracia, V.; Fontul, S. A Review of GPR Application on Transport Infrastructures: Troubleshooting and Best Practices. *Remote Sens.* **2021**, *13*, 672. https://doi.org/10.3390/rs13040672

Academic Editor: Roberto Orosei

Received: 19 January 2021
Accepted: 9 February 2021
Published: 13 February 2021

Publisher's Note: MDPI stays neutral with regard to jurisdictional claims in published maps and institutional affiliations.

1. Introduction

The deterioration and distress mechanisms that are active under the surface cannot be assessed with high accuracy using only traditional methods such as visual inspection, hammer sounding, chain dragging, and not even by destructive coring and test pits. Alternative methods are therefore required for a more complete inspection. Among other geophysical methods, ground penetrating radar (GPR) is widely recognized as one of the most powerful and versatile non-destructive testing (NDT) methods for routine subsurface inspections. GPR is a geophysical method that allows for the analysis of the propagation capacity of electromagnetic waves through media with different dielectric constants. A transmitting antenna emits an electromagnetic signal into the ground, which is partly reflected at the interface between two different media with sufficient dielectric contrast and partly transmitted into deeper layers. Then, the reflections produced are recorded from the receiving antenna, which is either in a separate antenna box or in the same antenna box as the transmitter. The strength (amplitude) of the reflected fields is proportional to the change in the magnitude of the dielectric constant. As the antenna is moved along the ground surface, a two-dimensional image (known as a radargram or B-scan) is obtained, which is an XZ graphic representation of the detected reflections. The x-axis represents the antenna displacement along the survey line, and the z-axis represents the two-way travel time of the pulse emitted and received. If the time required to propagate to a reflector and back is measured, and the velocity of the signal propagation in the medium is known, the depth

1

of the reflector can therefore be determined. By using three-dimensional data acquisition and imaging techniques, the reflected signals recorded by the GPR can be transformed into 3D imaging, which allows for a better perception of the underground structures. A detailed description of the methodology along with a deep theoretical background can be found in [1–3]. GPR is a portable and fast data acquisition technique that provides dense and accurate data, with a much higher resolution compared with competing geophysical approaches.

The major strengths of GPR are that the method allows for the collection of data at high speed, continuously, using mobile acquisition units and in a non-contact mode. Furthermore, it enables one to keep the prospected transport infrastructure area in-service during the GPR assessment, thus reducing costs and discomfort to users. Conversely, the main limitations occur in the presence of high-conductivity materials (such as wet clay) and in heterogeneous conditions causing signal attenuation and complex scattering phenomena, respectively. Another drawback is that the interpretation of radargrams is generally non-intuitive and requires considerable expertise to properly process and understand the measurements.

A few international organizations have promoted recommendations to properly use GPR in transport infrastructures. The American Society for Testing and Materials (ASTM) has produced a standard guide, ASTM D6432-19, that summarizes the equipment and field procedures for GPR subsurface investigation, as well as processing methods used to interpret GPR data [4]. The British Highway Agency has published technical specifications (DMRB 3.1.7 and DMRB 7.3.2) regarding the use of GPR for testing highway structures (concrete and masonry bridges) [5] and for pavement assessment [6], respectively. In Europe, the EuroGPR has produced guidelines for pavement structural surveys [7] including limitations of GPR use on pavements, survey assessment, survey specification, location referencing, data quality and backup, and data reporting. In Italy, the Comitato Elettotecnico Italiano (CEI) published the Italian standard (CEI 306–8) for performing preliminary surveys with GPR before laying underground utilities and infrastructures [8].

The use of GPR in civil engineering applications began to appear in the mid-70s and early 80s, mainly focused on the detection of hidden utilities [9–12] and pavement inspection [13,14]. From these first studies, the number of applications and developments of the methodology increased remarkably, being nowadays successfully and widely employed for a great variety of tasks. A comprehensive overview on the civil engineering applications of GPR can be found in [15–17].

GPR surveys in pavements (rigid or flexible) are frequently applied as a complementary technique for the assessment of roads and airport pavements. GPR images provide valuable information about the continuity of layers and their thickness, which could be combined with falling weight deflectometer (FWD) tests. In addition, GPR inspection allows for the analysis of the pavement layer's condition and locating defects, which is crucial in preserving the structure and its service life. In road assessment, the main analyses are focused on measuring layer thickness and detecting voids or delamination. Other, more recent studies are dedicated to the detection of water presence, mainly in the base and subgrade. Quality studies of the material density, water infiltration and cracking are also some examples of the applications that have been recently tested. Most of those researches are based on the analysis of the GPR images through the two-way travel time conversion into depths and the reflected wave amplitude analysis. However, during the last few years, several tests were prepared to determine additional parameters by studying frequencies.

For railway assessment, the main applications consist of ballast layer thickness measurement and detection of changes in track structure and substructure. During the last decade, a huge research effort has been carried out to improve the systematic application of GPR for ballast condition assessment, such as ballast fouling and moisture content.

In the study of retaining walls, GPR focuses on the detection of rebar in concrete retaining walls, detecting joints and dowels, and the location of damage and water infiltration. In some cases, the GPR analysis objective is the detection of the contact between the wall and the natural soil or between the wall and further restorations. Other applications are the detection and the analysis of the foundations. Those walls support high loads and the assessment of their state is important in maintaining or repairing parts of the structure.

Regarding masonry bridges, the main GPR applications comprehend the detection of unknown geometries such as hidden arches and ancient profiles, restorations and reconstructions in stonework, moisture content in masonry, bridge foundation, voids and cracks in masonry, filling distribution, and thickness of ashlars (e.g., voussoirs). Concerning concrete bridges, GPR evaluation encompasses the diagnosis of bridge–deck thickness, mapping rebar and tendon ducts, moisture content, corrosion assessment, and mapping delamination.

Finally, in tunneling inspection, the applications of GPR include the assessment of concrete segments and backfill grouting thickness, reinforcement, damages (e.g., cavities) in lining, moisture content and corrosion.

This paper presents a review of published works in the frame of GPR application in the evaluation of transport infrastructures, including roads, highways, airport runways, railways, retaining walls, concrete and masonry bridges, and tunnels. Although it has been demonstrated that GPR has significantly benefited the procedures for inspection and successfully solved some of the limitations of traditional methods, it also has some constraints. In order to maximize the benefit obtained through the use of GPR, it is important to be aware of these constraints and to be able to adapt the methodologies for testing and for analysis of the results to the job's specific conditions. Thus, through the review of the methodologies shown by other authors in the literature, the present paper highlights some of these best practices and recommendations (type of antennas, setting parameters, acquisition mode, processing, etc.). Moreover, some recommendations are made in terms of best practices aiming to improve the application of GPR and to avoid the misinterpretation of results.

As Supplementary Materials, different tables are addressed with a compilation of interesting published works carrying out the on-site survey of transport infrastructures. Those present a synthesis on GPR equipment, inspection methodologies, and set parameters for data acquisition and data processing, as well as the most relevant results obtained and the drawbacks identified.

2. GPR Equipment for Transport Infrastructures Surveying

There are several GPR manufacturers and commercial equipment available, and some experimental prototypes also exist. Different GPR systems will have different capabilities according to the type of antennas and their frequency, which affect the operating speed, the resolution, the penetration and the sampling rate [17,18]. The frequency and depth of penetration are related, with higher frequency pulses achieving lower penetration, but better resolution. Impulse GPR systems are the most widely used, with two main groups of GPR antennas, dipole and horn antennas, and with frequencies nowadays ranging from 10 MHz to 6 GHz. Currently, the most commonly used technology is the time-domain impulse radar. Additionally, several investigations point to the step-frequency radar as a potential technology for broad resolution range, although these systems do not allow real-time visualization of data during acquisition.

It should be noted that there are technical standards regulating the electromagnetic emissions of GPR equipment. The following are the main standards in Europe, USA and Canada: European Telecommunications Standard Institute (ETSI) EN 203 066-1, ETSI EN 203 066-2, ETSI EN 203 489-32 and ETSI EG 202 730, USA regulations on UWB-GPR: Part 15 of Federal Communications Commission (FCC) Regulations, and Industry Canada Radio Standards Specification RSS-220 (Issue 1) "Devices Using Ultra-Wideband (UWB) Technology.

2.1. GPR with Horn Antennas

GPR horn antennas were specifically designed for use in transport infrastructures evaluation, since they can operate at traffic speed. In the last ten years, this type of equipment has evolved from prototype status to routine use in pavement evaluation studies.

Horn antennas have frequencies ranging from 1 to 2.5 GHz, corresponding to penetration depths in the order of 1 m to 0.4 m, respectively. The minimum layer thickness that can be detected is about 50 mm, for 1 GHz antennas and 25 mm, for the higher frequencies.

The antennas are "air-coupled", and normally they work mounted on a mobile vehicle and are suspended at a certain distance from the surface (Figure 1a,b), typically ranging from 0.4 to 0.6 m. They perform measurements at traffic speeds (up to 80–120 km/h) without any interference with traffic, and therefore they are suitable for the evaluation of in-service pavements without major disturbance to road users.

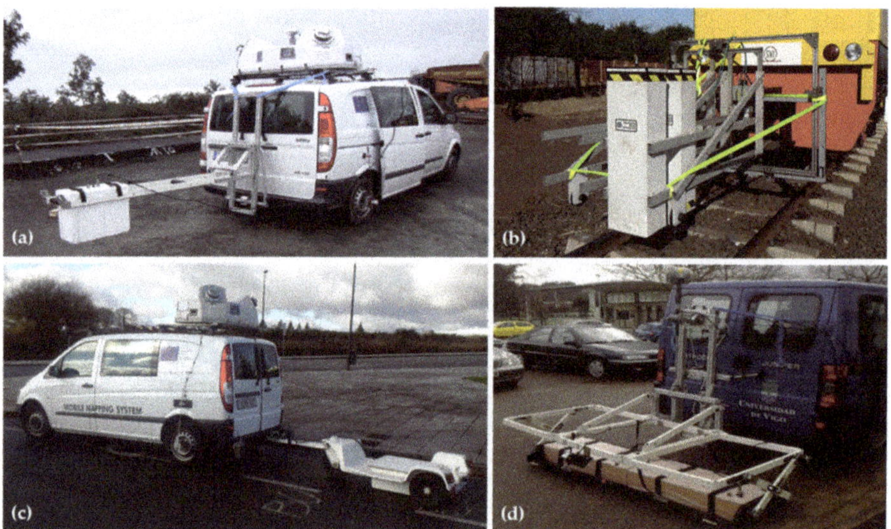

Figure 1. Different ground penetrating radar (GPR) antennas: (**a**) air-coupled (horn) antennas, (**b**) multi-antenna system (horn), (**c**) ground-coupled (dipole) antennas in a trolley, and (**d**) array multi-channel (ground-coupled) antennas.

2.2. GPR with Dipole Antennas

Dipole antennas were primarily developed for use in geological survey, normally ground-coupled (Figure 1c). They have frequencies ranging between 10 MHz and 6 GHz. For transport infrastructure applications, the best results are obtained with antennas from 400 MHz to 2.5 GHz central frequency. In general, the higher the frequency, the lower the penetration depth and the higher the resolution is. For example, 1.5 GHz dipole antennas will give a penetration depth of 0.50 m, while the 400 MHz will give a penetration of 2.00 m [19].

Dipole antennas were mainly developed for use in contact with the surface, or suspended just above it (2–5 cm), and they are suitable for testing at maximum speeds of 20–30 km/h. In this condition, the radar signal is "ground-coupled". Ground coupling introduces a stronger signal into the pavement, and therefore these antennas are normally employed for detailed studies over limited areas, as they allow one to obtain higher resolution [20].

2.3. GPR with Antennas Array Multi-Channel

GPR array multi-channel systems consist of a large number of closely-spaced antennas recording at the same time. Different multi-channel prototypes have recently been provided with different configurations, and they can include both air- and ground-coupled antennas (Figure 1b,d, respectively). Commonly, such multi-static systems are composed of 4–16 couples of transmitting and receiving channels mounted in a parallel broadside configuration with a cross-line trace spacing of 4–12 cm, depending on the manufacturer.

The main advantage is that they enable faster data collection by increasing the extension of the investigated area per time unit, and they make it easier for the operator to produce 3D images.

Generally, in transport infrastructure inspection, the antennas are mounted on a mobile vehicle to minimize traffic disruption. Mobile GPR is positioned connected to an external real-time kinematic (RTK) global navigation satellite system (GNSS) for trace tagging (georeferenced data) or to a distance measurement indicator (DMI) to control the distance trace-interval and to measure the travelled distance. The system also uses a computer navigation guided system to correctly follow profile direction and keep a constant overlap among parallel profiles without any physical marker on the ground surface.

3. Overview on GPR in Roads and Airports

Mobility and communications are the axis of our society's development. Road transport and planes move population and merchandising, being crucial for the development of industry and the cities. The roads and airports are the infrastructures that support this high level of activity. The safety and the efficiency of the transport and communication routes require detailed maintenance plans. The plans must include detailed inspection of infrastructures and development of maintenance strategies based on this information. Damage due to age, traffic and weather highly affects this type of infrastructure and, without a careful conservation and maintenance strategies, can lead to significant financial losses and can cause accidents. Roads and airports are two types of pavement structures designed to support different kinds of vehicles. Generally, the pavement consists of several layers: the surface course (asphalt or concrete layer), the base course (usually an asphalt base layer) and the subbase course (an unbound aggregate layer). This structure is supported by the natural subsoil (subgrade layer). The number of layers and their thickness depend on the type of pavement, being different in the case of roads, airports and industrial and harbor platforms (slabs). The airport pavement layers' thicknesses are always higher than in the roads. Depending on their structure, it is also possible to distinguish between rigid, flexible and semi-rigid pavements. In the first case, the surface layer is a reinforced concrete slab (although in some cases it is overlaid with an asphalt course, becoming a semi-rigid pavement) laid over a lean concrete layer. In the case of flexible pavements, the surface course is an asphalt layer. Differences in the structure of pavements yield on differences in the pavement behavior. Deformation in flexible pavements due to normal traffic loads is generally recoverable, whereas degradation due to excessive loads in rigid pavements are permanent. Changes in temperature also induce stress in rigid pavement and significantly affect the asphalt layers' moduli that decrease with an increase in temperature, due to the viscoelastic characteristics of this material. However, the strength in the case of flexible pavements depends also on the strength and quality of the subgrade soil, whereas in the case of rigid pavements, this only depends on the concrete layer design. Therefore, rigid pavements are usually used in zones with soils that present an inadequate support condition, in bridges or to cross voids or failure zones.

The requirement in many cases of surveying without interrupting the use of roads and airports forces the use of non-destructive tests that, at the same time, can be carried out without obstructing or disturbing the traffic [21,22]. GPR assessment in flexible pavement rehabilitation is recommended in the American Association of State Highway and Transportation Officials (AASTHO) guides [23], in combination with the use of FWD to obtain the thickness and the strength of each layer. In addition, GPR evaluation is frequently used to complement the studies about the structural conditions of pavements, detecting defects that affect their service life and preservation. The study of defects in pavements by means of GPR began at the end of the 20th century. The first studies allowed one to determine the contact between the asphalt and base layers, being difficult to distinguish thinner layers and details [24]. However, the studies about the relation between the GPR images and the material parameters [25–27] allowed for the application of the technique in the assessment, focused on particular aspects such as the material density for quality controls [28], control of layer thickness [29–31], detection of voids [32], measuring the depth to damage, and the detection of contact between rehabilitated (new) and old pavement [33]. Figure 2 presents an example of GPR data obtained in the assessment of roads with the objective of detecting possible voids in the ground, under the structure of the pavement.

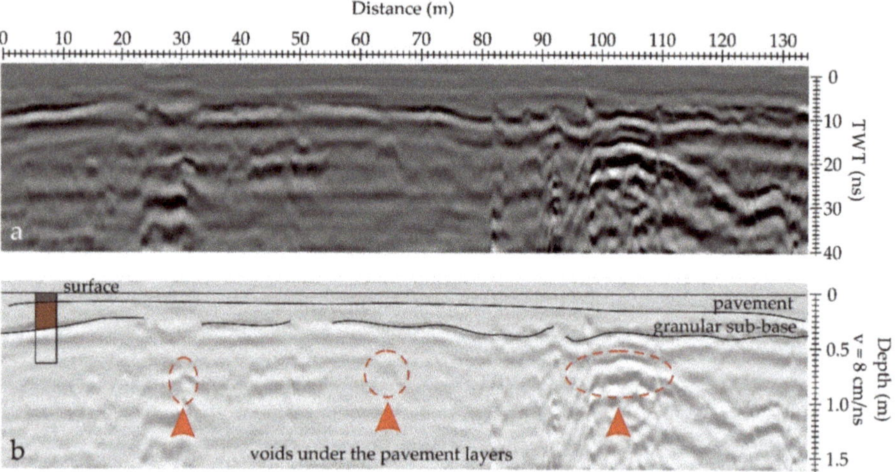

Figure 2. Detection of voids under the structure of the pavement with a 400 MHz center frequency antenna. Red arrows indicate the location of the anomalies associated to voids under the granular subbase, but the different layers of the pavement cannot be distinguished. (**a**) Radargram and (**b**) data interpretation.

More recent applications were focused on the damage and water infiltration detection in order to analyze the bearing capacity of the pavements, highly affected by the water content underneath the asphalt [34,35]. They also include the detection of pavement layers [36–38], the analysis of material density [39], cracking analysis [40–42], water detection [43–45], and the analysis of drainage structures under pavements [46]. Other recent studies have analyzed the effect of moisture in GPR amplitudes and frequencies [47,48] or delamination [47,49].

3.1. Roads

GPR application in roads is mainly focused on the assessment of flexible pavements as a complementary survey combined with more traditional methods: visual inspection, drilling and sampling, deflections and dynamic response monitoring systems. In some cases, GPR is also combined with infrared thermography, laser scanning and accelerometers. The application of this geophysical method in the assessment of pavements is regulated and recommended in the AASTHO guides [23]. Moreover, the guide ASTM D4748-98 [50] presents the procedures for the inspection of the upper layers of both bituminous and concrete pavements, using a short-pulse GPR. The methods included in this international standard are focused on the thickness evaluation of pavement layers. The report presents the functioning principles of a short-pulse radar and the test equipment configuration. Therefore, the guide includes technical topics such as calibration and standardization, procedures, calculation and reliability of the results. The document also emphasizes the importance of safety issues in GPR survey activities, since the apparatus involves potential microwave radiation hazard. However, the different characteristics of countries have driven the European Nordic countries to develop the Mara Nord Project, in order to provide recommendations and guidelines for the use of GPR in road construction quality control [51]. This project started in 2009 and involved several research units such as universities, national transport administrations, road rehabilitation and consultant companies, and GPR manufacturers from Finland, Sweden and Norway. The goal of this project was to define common criteria for the equipment standards and for the results requirements.

The European Cooperation in Science and Technology (COST) Action TU1208 has published some tests and recommendations for suitable GPR system performance compliance, obtained by scientists from Belgium (Belgium Road Research Center), Czech Republic (University of Pardubice), Portugal (National Laboratory for Civil Engineering), and Serbia (Faculty of Technical Sciences, Novi Sad) [52]. The D6087-08 standard, emitted by the ASTM [53], describes four procedures for the calibration of GPR systems equipped with air-coupled antennas. After a critical analysis of those procedures, four improved tests were proposed, which can be carried out to evaluate the signal-to-noise ratio, short-term stability, linearity in the time axis, and long-term stability of the GPR signal.

GPR is also applied in the assessment of rigid pavements, with one of the main objectives being the identification of metal bars and layers thickness [54], and measuring the rebar cover depth [55]. The assessment to detect damage such as cracking and voids is another application in rigid pavements [41,42,56]. These studies are mainly focused on the detection of damage areas, but in some cases the origin of the damage is also investigated, including the effect of roots in pavements [57] and the impact of weather and load conditions. Figure 3 presents an example of GPR data acquired in a rigid pavement, showing the location of the rebar and the cover depth, besides the different layers and cracking.

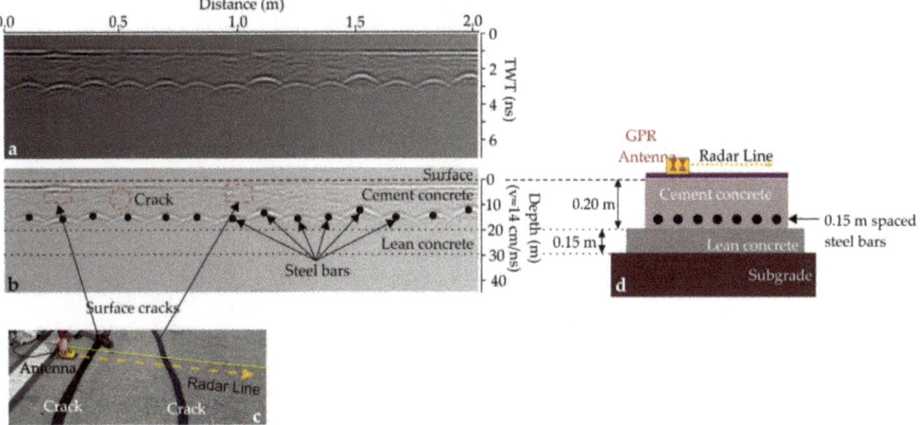

Figure 3. Results obtained in the study of rigid pavement. (**a**) Radargram. (**b**) Data interpretation highlighting the existence of cracks also visible at pavement surface (**c**) and cracks in the deeper layers, not yet visible. Moreover, image (**b**) shows the location of the rebar, the cover depth and the thickness of the different layers, obtaining a final model of the structure (**d**) (adapted from Rasol et al. [56]).

To summarize, GPR applications in pavement are mainly focused on:

- Measuring layer thickness [30,36,38,58–63].
- Detection of cracks and voids under the pavement, including the analysis of possible causes (weather, loads, roots, etc.) [40,56,57,64–68].
- Quality control of asphalt, mainly in the case of new pavement [69–72].
- Moisture and changes of water content detection [34,43,48,73,74].

Table S1, in Supplementary Materials, presents a compilation of the relevant on-site survey of roads.

3.2. Airports

The GPR assessment of airport infrastructure has the same objectives as road inspections. Most of them are focused on the detection of anomalies that could indicate damage in the airport runways, taxiways and apron pavements. The main difference between road and airport pavements is the layer thickness that in the case of an airport can be about 0.50 m, while in highways and roads is about 0.30 m [75]. The structure in airport pavements is composed of asphalt, concrete or reinforced concrete slabs. Moreover, the pavement of airport runways has a greater durability due to the higher content of asphalt bitumen. This infrastructure is designed to support the load due to the weight of the airplanes (between 70 Tn and 420 Tn). The highest loads are supported at the top (headland) of the runway where the airplanes accelerate when starting the take-off and stop after landing.

Despite the great loads supported by this infrastructure and its importance in communication and transport, the examples of GPR assessment of airports that can be found in the literature are less than those found on roads. In most of those examples, the objective is determining a procedure for the location of rebar and recognition of cracks and voids (in and under the concrete slab) [76–80]. Cracks wider than 1 mm are also detectable [81]. Other studies are focused on the detection of pavement structures and layers, detecting the dowels, joints and expansion joints [77] and different layers thickness [75,82–84]. GPR is also used to evaluate the quality of grouting treatments [78] applied to stabilize damaged pavements. GPR is also used in the assessment of the asphalt layers' interfaces to detect debonding [75,85]. Other studies were focused on the analysis of airports under special conditions: (i) in the case of airports built over the sea, in order to detect the contact between the filled layers and the natural soil under the sea [86]; and (ii) in the study of airports in frozen areas, in order to detect the depth to the permafrost layer [87].

To summarize, the main applications in airport pavements are focused on:

- Measuring layer thickness [75,77,82–84].
- Detection of cracks and voids in pavement [76,79,81,88].
- Detection of thin cracks [81].
- Quality control of restoration treatments [78].
- Thickness of filling layers in special airports [86].
- Detection of debonding [75,85].
- Depth to permafrost layer [87].
- Structures in the pavement [77].
- Detection of rebar in reinforced concrete pavements [80].

Table S2, in Supplementary Materials, presents a compilation of the relevant on-site survey of airports.

4. Overview on GPR in Railways

The railway (railroad in USA English) is a transport infrastructure energetically efficient and environmentally friendly, and it is gaining more and more importance for passengers and freight transport. The railway network use is increasing and, consequently, not only the deterioration under traffic load is higher, but also the time window available for monitoring and maintenance actions is becoming more restricted. The maintenance has to be well planned, due to the impact that it has on the railway users' comfort and safety, and to enable it, a proper inspection is essential. In this context of efficient and expedient monitoring, NDT such as GPR represents a powerful tool to assess the track condition. The information gathered with GPR is generally used together with other track inspection results, such as track geometry and both rail shape and rail integrity measurements. While the track geometry and rail condition data give an indication of track deterioration, GPR survey provides the identification of the real cause of deterioration, such as changes in structure, settlements, ballast fouling and drainage problems. The information provided by GPR is therefore crucial for the definition of efficient maintenance action and depth of the intervention, mainly if the deterioration is due to substructure defects. Some examples of GPR application to railway assessment are shown in Figure 4.

4.1. Ballasted Railways (Superstructure and Substructure)

One of the first studies of GPR application to railway was performed in late 90s by Sussmann [89]. Several studies were developed since then, some of them referred to herein. Nevertheless, there are only a few guides and standards for GPR application to railways [90].

On railways, a major limitation to the systematic application of GPR is the presence of the rails, as it is a metal, and consequently an ideal reflector that can shadow the information gathered in the subsurface. To overcome this, dedicated antennas and software were developed for railway assessment [91,92], which are discussed later in this paper.

The main applications of GPR for railway infrastructure evaluation are:

- Layer thickness measurement, mainly ballast [92–96].
- Investigation of embankment defects and moisture [92,94].
- Location of ballast pockets, subsidence and trapped water areas [92,93,95,97].
- Frost susceptible areas [98,99].
- Fouled ballast assessment [100–104].

In order to support the knowledge and to calibrate the on-site surveys, several laboratory and real scale tests were performed, such as:

- Track substructure assessment, layer thickness (experimental test sections) [105].
- Ballast condition characterization, fouling and fragmentation (experimental test sections) [101,106].
- Fouled ballast assessment (laboratory tests) [101,107–110].
- The influence of concrete sleepers on the assessment of ballast condition (laboratory tests) [111].

4.2. Ballastless Railways

The objectives of GPR application on a ballastless track are similar to other concrete reinforced structures, such as rigid pavements and bridge decks. The main applications are:

- Layer thickness measurement [112].
- Rebar detection [112].
- Mud detection under the slabs [113].
- Debonding between sleepers and slabs and between layers [112].

Table S3, in Supplementary Materials, presents a compilation of the relevant on-site survey of railways.

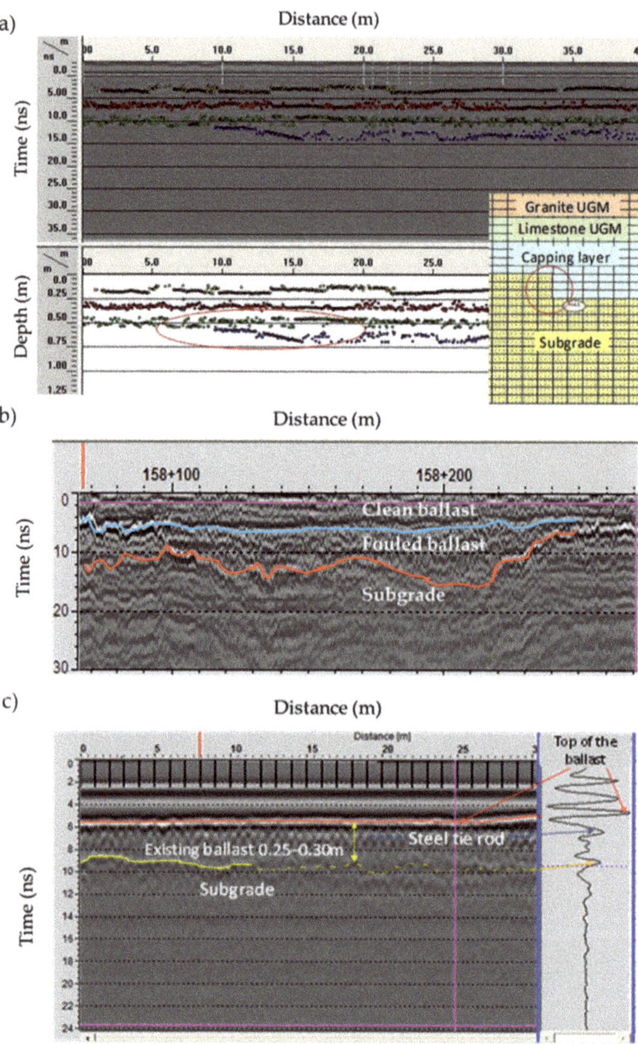

Figure 4. Example of information obtained by GPR surveying on a ballasted track (1.0 GHz air-coupled antenna): (**a**) subballast thickness and continuity in the Alcácer Bypass (Portugal) showing an increase in capping layer thickness (red ellipse); (**b**) an in-service national railway line (Portugal) displaying the interfaces between the clean ballast/fouled ballast/subgrade where the settlement location is evident due to weak subgrade (adapted from Fontul et al. [95]); (**c**) GPR record of an in-service railway line showing the metal reflection of a steel tie rod. (UGM means Unbound Granular Material).

5. Overview on GPR in Retaining Walls

Retaining walls are structures prepared to support the soil laterally, in order to avoid landslides. Those structures retain the soil behind them and are designed to support high pressures due to the tones of the materials. The pressures, in most cases, are increased as a consequence of soil water content. This type of structure allows one to define different levels of soil, and in most cases are indispensable constructions in roads, bridges, tunnels or railways. Depending on the construction, there are different types of retaining walls. Some of the most usual types are:

- Gravity retaining walls, which are massive and require a significant gravity load to be stable under the soil horizontal pressures. These structures can be built with concrete, masonry, stone or precast concrete boxes filled with coarse material structures.
- Cantilever retaining walls that are usually built using concrete and reinforced with pre-stress concrete. These walls are composed of the stem and a base slab as a foundation. This is the most usual type.
- Anchored retaining walls, composed of the stem and cables anchored with concrete in the ground.
- Piled retaining walls, built with adjacent piles, making a wall.

As a consequence of aging, degradation of materials, changes in the loads (due to variations in the water content or in consequence of the construction of other structures) or changes in the temperature and moisture, the walls can suffer different levels of damage, leading even to the collapse of the structure. The detection of the pathologies reduces the possibility of the structure failure. The assessment of those walls requires the study of the structure and the analysis of the interaction between the structure and the soil. The inspection of the structures for retained soil requires the assessment of the wall that determines the decisions about the structure (repair, maintenance, change or frequency of future inspections). The assessment involves the visual inspection and simple sonic and thermographic tests. Depending on the results, more invasive tests could be needed.

Non-destructive inspection techniques are currently applied to obtain valuable information about the state of the wall. GPR is a promising technique in this field. However, the application of GPR to vertical structures still requires development. The US Department of Transportation prepared a guide for retaining wall inspection [114], including GPR as a possible technique that requires further analysis and improvement. The assessment of those structures requires the analysis of the wall conditions (damage, moisture, rebar), the study of the soil (mainly water content) and the evaluation of the supporting structures (cables and anchors) and the foundations. Therefore, the GPR applications in retaining walls focus on the detection of the rebar in concrete retaining walls [115], detecting joints and dowels and obtaining high quality 3D models [116]. The studies define the limits of the method, showing a clear first layer of rebar and a low-quality image of the second rebar layer [117]. Figure 5 presents two examples of the reinforced wall inspections. In both cases, the rebar is detected and several anomalies highlight the existence of possible damage.

Only a few applications in masonry retaining walls can be found in the literature. Those applications are based on detecting the contact with the natural soil, estimating the thickness of the structure, and changes due to old repair works and the inner structure [118], including the foundations [119]. Although in many cases, GPR data is acquired with the antenna on the surface of the wall, in some cases, the analysis of the wall acquiring data from its top part (Figure 6) could provide information about the number of stone rows and the contact between ashlars [120]. Figure 6 shows the image and the interpretation of the GPR data. In the study of those walls, the velocity could be estimated from the hyperbolas produced at the contacts between stones.

In some cases, the retaining walls are part of subterranean structures, as in the case of some Mycenaean tombs [121]. Those structures are composed of irregular stones, in two or more layers. Figure 7 is an example of the GPR images that can be obtained in this type of retaining wall. The different stone layers could be detected, and the irregular pattern of anomalies in the GPR images corresponds to the uneven stones. In addition, changes in the amplitudes could inform about the variations of physical of chemical properties of the ground or the structure, highlighting zones that could be damaged or that must be restored.

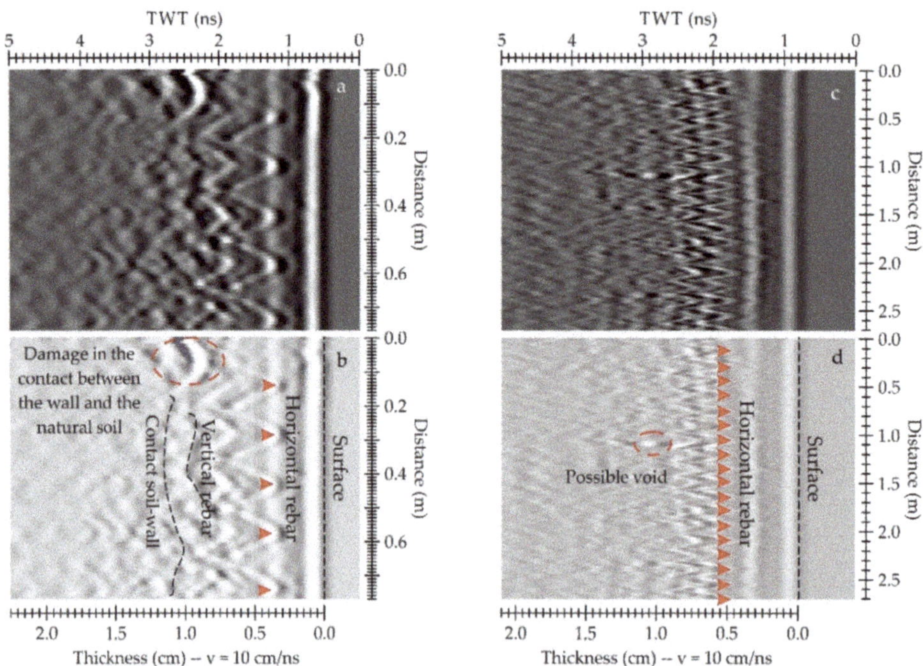

Figure 5. Two examples of GPR images (**a** and **c**) obtained in reinforced retaining wall assessments with a 2.3 GHz antenna. Both cases focus on the detection of the rebar and the location of zones with possible damage (**b** and **d**).

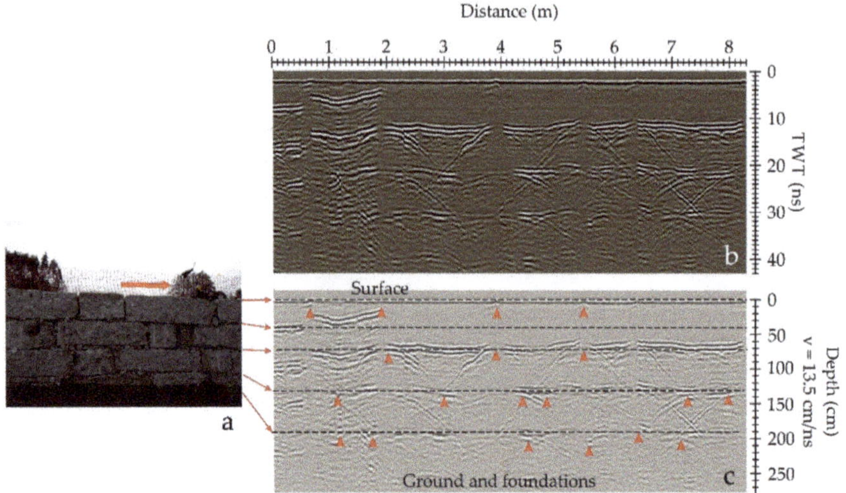

Figure 6. GPR assessment of masonry walls composed by granite ashlars. (**a**) The data were acquired from the upper part of the retaining wall. (**b**) GPR image. (**c**) Interpreted radargram, showing the contact between the different rows of ashlars and the anomalies produced by the contact between stones (red arrows) (adapted from Solla et al. [120]).

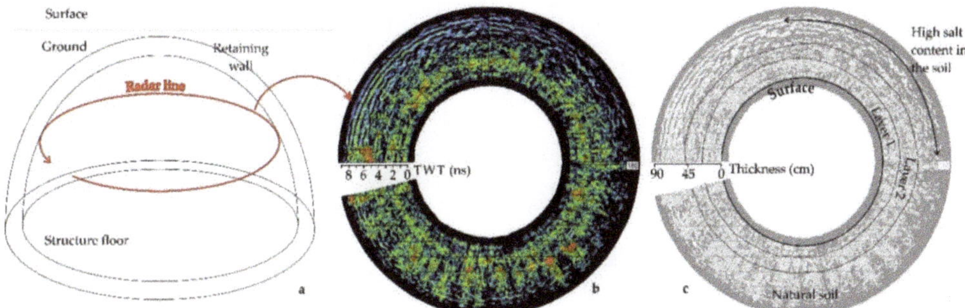

Figure 7. GPR images from the study of the retaining walls in a subterranean tomb (adapted from Santos-Assunçao et al. [121]). (**a**) Scheme of the structure and location of the profile line (in red). (**b**) Processed GPR image. (**c**) Data interpretation. The wall is composed of two layers of uneven stones. The contact with the natural soil is detected and the changes in the amplitude denote changes in the soil salt content, which could affect the preservation of the structure.

A particular type of retaining wall is used as a support system in deep excavations in the case of unstable soils or shallow water table. Those walls are built with concrete in different construction phases and in different sections inside the ground. The soil is excavated until the depth of the constructed section and, after that, the next section of the wall is built. This process continues until the excavation is complete. Prior to the excavation, the assessment of the already constructed wall section increases the safety of the process. Borehole GPR is a reliable method in this quality control process. The GPR images provide information about the integrity of the underground retaining wall, showing zones that must be repaired before excavation, such as damage or voids in the ground in contact with the structure that could lead to the collapse during the excavation procedure. Figure 8 shows an example of a borehole GPR image in the retaining wall during an excavation. The objective is the detection of both wall surfaces, the possible defects and voids in the concrete and in the ground behind the wall, and zones with possible changes in water content.

To summarize, the main applications in retaining walls depend on the type of the structure. However, the detection of the contact between the natural soil and the wall in order to identify possible voids, and changes in humidity or defects on the surface of the wall, is a common application in all types of walls. In masonry walls, the common objectives of surveying are the analysis of layers, the contact between blocks, and the assessment of foundations. Conversely, in concrete walls the main goal focuses on the detection of the rebar and possible defects that occurred during the construction of the wall.

Table S4, in Supplementary Materials, presents a compilation of the relevant on-site survey of retaining walls.

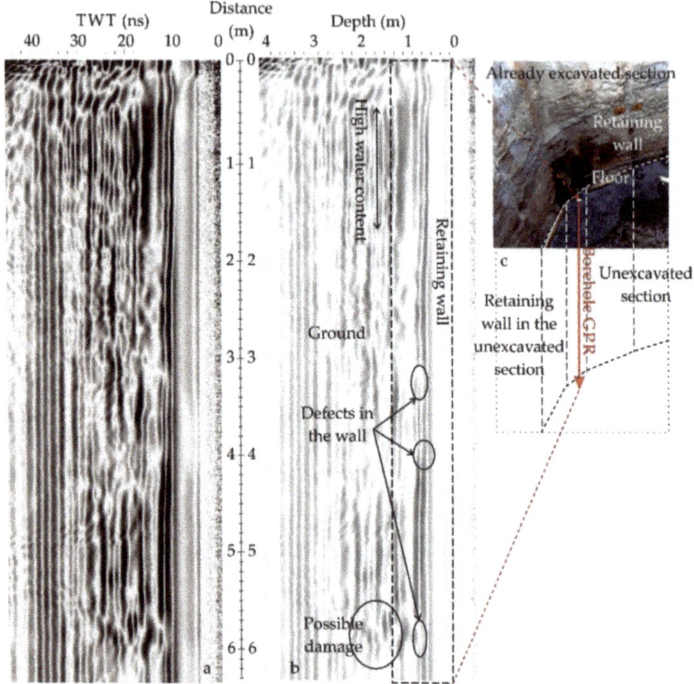

Figure 8. Borehole GPR images. (**a**) Radargram. (**b**) Data interpretation. (**c**) The excavation and the position of the GPR borehole. The images allow one to determine the contact between the wall and the ground, as well as zones that must be repaired or reinforced before the excavation.

6. Overview on GPR in Bridges

Bridges are considered vital to human life because they connect different lands and facilitate economic, industrial and social development. However, a great majority of these structures, in operation within the transportation system, require desperate repair and maintenance strategies. The condition monitoring of bridges focuses on detecting material degradation and structural pathologies aiming to accurately assess the safety and service life of existing structures. Conventionally, diagnosis is accomplished by visual inspection and assessment can therefore be difficult as there is structural design and damage of the structure inaccessible and hidden from view. Indeed, the GPR method is one of the non-destructive techniques most used in internal bridge inspections [122–126], due to its high practicality in the field in a relatively quick time and with the minimum intervention without affecting the structure's integrity. Internal damage detection can prevent the unpredictable and premature collapse of structures.

This section provides an overview of the main GPR applications on the condition monitoring of both masonry and concrete bridge structures.

6.1. Stone Masonry Arch Bridges

A large number of the existing masonry arch bridges still in use within the transportation network are ancient structures remaining from Roman and Mediaeval periods. In addition to their age, and environmental conditions, the stability of these bridges (both road and railway bridges) is questionable because of the actual loading, significantly higher than the estimate for their original design [127]. Consequently, they require periodic inspections of the changes in the structural condition, aiming to develop effective preventive and maintenance tasks.

The use of NDT technology becomes essential for the health assessment of these former structures in order to preserve their historical character and structural integrity. In this context, the GPR method has been demonstrated as effective for the diagnosis of masonry arch bridges, in which the detection of the following characteristics was possible:

- Unknown geometries remaining in the interior of the bridge such as hidden arches and ancient profiles (shape) of the structure [128–130].
- Evidences of restorations and/or reconstructions in stonework [130–133].
- Existence of cavities and fractures/cracking in masonry [131,134–136].
- Moisture in masonry [134,137,138].
- Bridge foundations [137,139–141].
- Filling distribution in masonry [130,136,139–143].
- Thickness of ashlars (pavement, ring arch, spandrel walls, etc.) [124,130,139,140,142–146].

There are different studies combining GPR with structural assessment on masonry arch bridges. The knowledge of the internal geometry and material characterization of the bridge allows engineers to perform more accurate numerical simulations and detailed assessment of the bridge's structural load capacity and safety [136,139,140,142,143,145–151].

Figure 9 presents some GPR data obtained in three different masonry arch bridges, which demonstrate the feasibility of the technique for the documentation of previously unknown information from the interior of the structure.

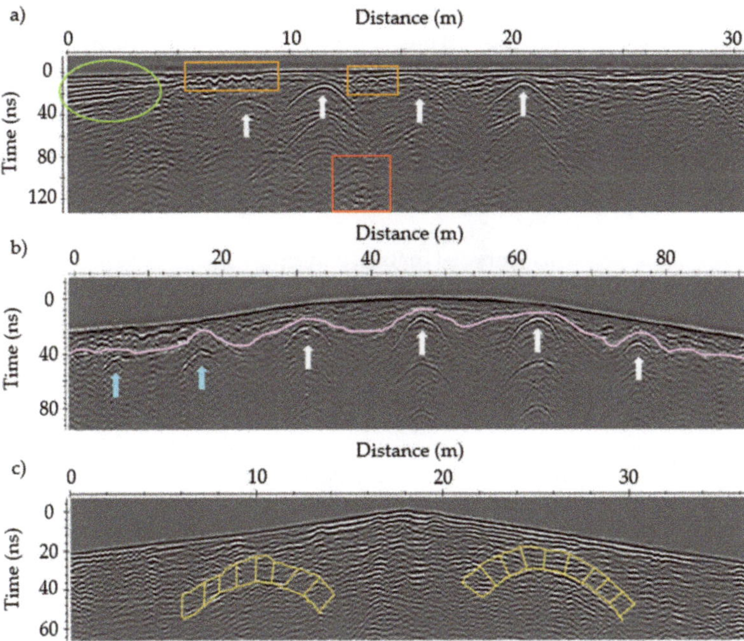

Figure 9. Examples of information obtained by GPR surveying on stone masonry arch bridges: (a) Traba bridge (Noia, Spain) showing an ancient profile (green ellipse), evidence of reinforced concrete used for restoration (orange rectangles) and presence of a void in a pier (red square); (b) Monforte bridge (Monforte de Lemos, Spain) displaying the interface between the original and the new backfill used for restoration (pink line) and the existence of two hidden arches (cyan arrows); (c) Vilanova bridge (Allariz, Spain) reproducing the voussoirs of the arch-ring (yellow lines).

6.2. Concrete Bridges

Bridge deck condition assessment is the most important part of concrete bridge health diagnostics and maintenance. Reinforced concrete bridges have steel bars, or rebar, which are embedded within the deck for structural strength. The corrosion of internal reinforcing bars is a major cause of concrete bridge deck deterioration and the most significant contributor to bridge deficiencies (with the formation of cracks and delamination). These deficiencies cause severe damage even on the top pavement layers, often having a crucial effect on driving safety and even structural failure. In order to prevent failure at early stages, extend the service life of concrete bridge decks, and reduce maintenance costs, proper inspection must be carried out periodically. GPR is seen predominantly as an effective NDT technique for deterioration assessment. The ASTM has a standard procedure, the ASTM D6087-08, that defines a methodology for using GPR to evaluate the condition of concrete bridge decks [53]. It is also important to mention the guidelines published by the European project—Mara Nord—for the use of GPR in bridge deck surveys [152], which describes GPR equipment, survey planning and performance, data processing and interpretation, and the reporting and delivery of results.

GPR has been successfully used in a wide range of applications in concrete bridge inspection, including:

- Estimation of concrete cover depth [153–155].
- Mapping reinforcing bars (deck and beams) [153,154,156–162].
- Location of cable ducts and other utilities such as deck joints or drain grate [154,158,160,163].
- Damage detection on concrete (corrosion, cracking, spall, delamination, etc.) [157,158, 160,161,164–175].
- Moisture detection and water content estimation [155,157,158,176].

When dealing with a GPR bridge inspection, a large amount of data are produced and extensive manual processing is required to extract useful information, in which the manual identification and localization of rebar is time-consuming and labor intensive. In such a context, some authors have developed algorithms for the automatic detection of rebar [157,173,174,176–182] and cracks/delamination [165,177,178].

Additionally, as in the case of stone masonry arch bridges, there are different studies that use GPR data to create structural models aiming to analyze the load-carrying capacity of concrete bridges [159,183].

Table S5, in Supplementary Materials, presents a compilation of the relevant on-site survey of masonry and concrete bridges.

7. Overview on GPR in Tunneling

Tunnels are essential civil engineering structures that constitute an important part of roadways, passages, sewer systems, utility networks and railways. There are various categories of tunnel construction, including bored, drill and blast, horizontal directional drilling, and pipe jacking/microtunnels. Moreover, the bored tunnel can be divided into two subcategories, controlled bored tunnels and tunnel boring machine (TBM) tunneling. Shield tunneling is the typical method used for railways, which is considered a more stable method than mining.

Tunnel distresses might not be visible, and it is therefore necessary to conduct regular inspections during its service life in order to ensure safety and long-term viability. Various processes and subsequent failures may cause damage to tunnel structural integrity; those that occur during construction (ground settlement) and after opening to service (aging, physical and chemical agents, etc.). Undoubtedly, the frequency of periodic inspection should be increased when the tunnel has become aged. If potential and hidden deficiencies of the tunnel are detected in advance, and treated in time, many accidents may be avoided, whereas excessive repair and maintenance costs can be saved. Firstly, the geological and hydro-geologic conditions in which tunneling occur are essential factors in the structural integrity. Secondly, more and more problems appear in the lining with aging, such as water leakage, cracks, spalling and voids behind the lining, which greatly affect the safety of the tunnel and the users. The thickness distribution of the grouting layer behind lining segments is a crucial factor to control the quality of the tunnel and it directly affects the overall bearing capacity and structural adhesion of the tunnel. Overall, based on the International Tunneling and Underground Space Association (ITA/AITES) [184], the stresses and settlements that appear at the surface of the lining are indicators of an inadequate TBM method on shield tunnels: (i) over-excavation, (ii) the ploughing/heading effect and steering, (iii) lining deformation and insufficient grouting and (iv) swelling/consolidation of the surrounding ground.

Conventional methods for tunnel assessment are destructive and have great limitations and uncertainties in obtaining a representative model of the overall structure. Nowadays, the stability of tunnels is being assessed more and more often through geophysical surveys. In fact, geophysical NDT technology is meant to obtain a rapid analysis of rock shearing and concrete lining quality. The ASTM produced a standard guide, ASTM D6429-99, to provide assistance in selecting the most appropriate geophysical method for a range of geologic, geotechnical, hydrologic, and environmental investigations [185]. Hence, the GPR method has demonstrated its reliability in quality testing, since it provides continuous measurements of grouting distribution. Some examples of GPR application to tunneling inspection are shown in Figure 10.

This section provides an overview of the main GPR applications on the condition monitoring of tunnel structures. Currently, the most common applications include the following objectives:

- Thickness of concrete segment/lining [186–199].
- Thickness of the backfill grouting layer [187,189,193,196,198,199].
- Damages in concrete lining and grouting layer [186,188,192,194,196,200,201].
- Damages (e.g., cracks/fissures, fractures and voids) behind tunnel linings [188,195, 197,200–205].
- Moisture/water content [186,188,195,203,204,206].
- Depth and location of reinforcement (rebar) [189,192,194,195,201,207,208].
- Inspection of other reinforced concrete structures (e.g., steel arch and shotcrete layer) [188,197,204,206–208].
- Location of immersion joints [207,208].
- Identification of depth and presence of insulation material [208].

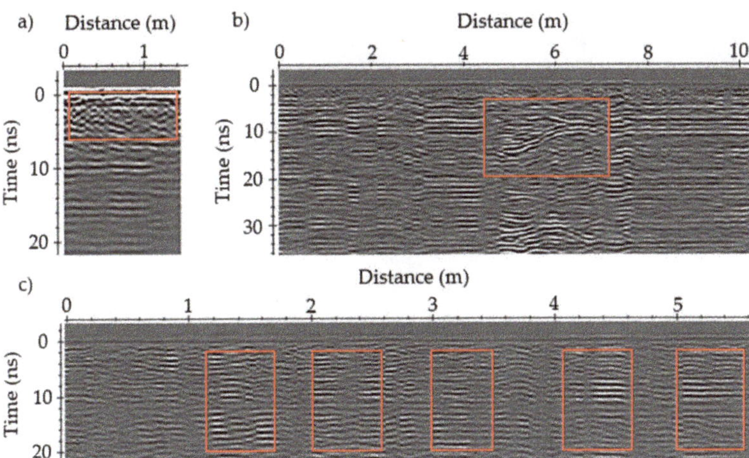

Figure 10. Results obtained from a GPR tunneling inspection: (**a**) shotcrete layer, (**b**) cavity and (**c**) embedded steel arch (adapted from Prego et al. [204]).

Table S6, in Supplementary Materials, presents a compilation of relevant on-site survey of tunnels.

8. Discussion Table on Limitations and Best Practices

8.1. Roads and Airports

GPR surveys in pavements are most likely some of the most common and developed inspections. Generally, the main objectives are focused on the detection of layer thickness and damage in the pavement layers or under the road structure. Therefore, the application is included in several regulations [7,50,53,149]. The standard ASTM D4748 [50] defines the GPR test method for the determination of pavement layer thickness, advising of two potential complex scenarios (extremely wet pavement or saturated pavements, and high iron content in the aggregates) at which attenuation could affect the data interpretation. The ASTM D6087-08 [53] develops the test methods to assess the asphalt covering concrete decks with GPR. Both standards cover flexible and rigid pavements. However, these standards are only recommendations, and the application depends on the different countries' regulations. In Europe, there are no specific common guidelines about the application of GPR in pavement surveys, even though some proposals are developed in different countries [209]. Examples of those guidelines are in the Mara Nord Project [51] and in the British [5,6] and Belgian [210] regulations. Many authors declare that an optimal pavement assessment requires a GPR calibration process. The studies demonstrate that the correct calibration causes an extreme decrease in the error in the estimation of the pavement thickness. The analysis of the uncertainty of GPR data interpretation in the study of pavements demonstrates that the calibrated data is closer to real data than non-calibrated data [211], being the results and the interpretation affected by the calibration process. The divergence between GPR data and real data is produced by several factors, such as: (i) the distance between the antenna and the pavement surface; (ii) the pavement age; (iii) the heterogeneity of the medium (changes in water content, presence of cracks, changes in density, etc.); (iv) the wave attenuation; (v) the scatters at the pavement surface (depending on its roughness); (vi) water, leaves or sand presence at pavement surface; (vii) changes in the time-zero; (viii) clutter; and (ix) changes in the data acquisition velocity [75,84,212]. The analysis of different calibration algorithms demonstrates that all of them diminish the systematic error during the data acquisition [212], and the tests allow one to evaluate the operation and the stability of the GPR system [213]. In some cases, calibration consists of comparing GPR data with cores, obtaining radar-wave velocities by comparing the thickness of each layer with the time to each continuous anomaly in the B-scans. In other cases, it consists of determining the dielectric permittivity of the surface layer comparing the amplitude of the reflected wave at the surface with the amplitude of the wave reflected on a metallic plate. However, in other cases, the dielectric permittivity of the layers can be determined with common midpoint (CMP) or wide-angle reflection and refraction (WARR) measurements. In the case of rigid pavements, the wave velocity in the concrete layers can be estimated considering the hyperbolic anomalies produced on GPR data as a consequence of reflections on the rebar [80]. An error less than 3% is observed in the thickness concrete layer estimation using this method [55].

Some of the authors use ground-coupled antennas [47,48,214]. In some cases, when surveying is dedicated to limited sections of the roads, the antenna is manually moved or mounted on a cart survey. Nevertheless, for large-scale surveying, the ground-coupled antenna is generally mounted on a vehicle that uses special devices and moves at a traffic speed. These devices are usually platforms or trailers that support the antenna at a certain distance from the surface of the road or the airport platform. The platforms are built with wood [84,215], plastics [21] or fiberglass [75] in order to reduce as much as possible clutter in the radar data. However, in some cases the survey is carried out using a platform in contact with the pavement, connected to the vehicle. The ground-coupled antenna is placed on that platform, assuring the maximum contact between the antenna and the pavement surface [216], and the maximum transmission of the wave into the medium. This type of data acquisition reduces the velocity of the vehicle during the survey, with the speed between 16 km/h [216] and 25 km/h [217], but increases the penetration depth, compared with a survey using the same frequencies but with the antenna suspended at a certain distance from the pavement surface. In general, surveys with ground-coupled antennas are designed to place the antenna as close as possible to the pavement surface. In most cases, pavements are surveyed with air-launched antennas [217]. These devices are suspended at a distance between 150 mm to 500 mm from the surface of the pavement [17,75,80,84,216]. Those devices provide clean GPR images, allowing surveys at a traffic speed (up to about 90 to 120 km/h) [75,84,218], but the penetration depth is less than in the case of ground-coupled antennas in contact with the surface of the medium. The comparison between the data obtained with each one of those antennas epitomizes the change in the direct and reflected wave, depending on the distance between the antenna and the surface of the pavement [75,219].

Some of the most recent assessments use an array of antennas, combining in many cases ground-coupled and air-launched antennas [216,220]. In [63] the use of a single transmitter and five receiver Vivaldi antennas is proposed, obtaining results in tests and pavement surveys with an error less than 10%, which is in the same level of other methods, including drilling. Other studies use a step-frequency array system [221] detecting shallow and deeper features. Dérobert et al. [222] demonstrate that a combination of UWB antennas, ranging from 500 MHz to 6 GHz central frequency, detected the second course in the base of the pavement structure and thin layers of about 2.5 cm. In addition, controlled tests based on the detection of pavement layers and damage conclude that GPR is a useful tool that could be implemented as a quality control tool during the construction of new pavements or in the assessment of already existing structures, indicating that survey speed during the data acquisition velocity must be different, depending on the antenna [218].

Even though in many of the applications the study is based on the analysis of anomalies in B-scans [82,86,87], in some cases those analyses are combined with other analyses from amplitudes and data obtained in CMP or WARR surveys [85], detection of reflections patterns in A-scans and B-scans [76], and velocity analysis [81]. In addition, many applications propose the use of arrays and multi-frequency antennas [77,81,83,85].

8.1.1. Soil Subgrade Assessment and the Detection of Bedrock

Some of the first studies in pavements were focused on the analysis of the soil subgrade. This layer is highly sensitive to the environmental changes, being also affected by the traffic loads [223]. Saarenketo [69] points to three applications:

- Analysis of new road alignments, studying the ground materials and the depth to the bedrock and to the water table, reaching in some cases a depth of about 5 m.
- Assessment of existing roads in order to detect filling materials [86], voids or weak zones under the pavement [40,56,64–68].
- Quality control of restorations [69–72].

One of the most important difficulties in the assessment of the subgrade is the wave attenuation. Two possible causes are the clay content in the soils and the existence of a significant water content [102,223], which recommends the use of different center frequency antennas to improve the data interpretation. In general, a combination of several assessment techniques is the most useful tool to detect the anomalous zones in the subgrade, being applicable in the quality control of roads under construction [223]. The tests presented by [223] are a clear example of the use of combined methodologies and types of antennas to enhance the final results. They conclude that air-coupled antennas are more appropriate in detecting interfaces between layers and with the subgrade; whereas the ground-coupled antennas are more appropriate in determining anomalous zones that could be associated with damage.

Related to the detection of the subgrade, there are two main types of studies: (i) the analysis of the ground under the pavement and (ii) the estimation of the depth to the bedrock. The location of the bedrock is essential, mainly in the design of road lines. In these applications, the attenuation causes an extreme decrease in the signal-to-noise ratio at higher depths, being difficult to distinguish clearly the contact with the bedrock or the existence of fractures and stratification in the ground under the pavement. These geological features could affect the stability of the road. Moreover, the data accuracy in the data interpretation depends on the precision in GPR wave velocity estimation [224], being a complex task in heterogeneous media. The studies presented in [224] propose the 450 MHz ground-coupled antenna as the most appropriate system, because it offers higher resolution than antennas between 100 MHz to 200 MHz central frequency. This result is a consequence of the GPR wave attenuation, allowing for the detection of discontinuities until a maximum depth of 5 m in all cases, based on the observation of different textures in the B-scans. The authors recommend the combination of GPR with electric resistivity tomography (ERT) in the case of: (i) bedrock at depths higher than 5 m and (ii) soils with superficial conductive clays. Those clays produce an extreme attenuation of the GPR signals. Other studies demonstrate that GPR is effective in determining the asphalt layers, being moderately successful in the detection of base thickness and shallow stratigraphy and is not applicable in the location of the near-surface bedrock [225]. However, other studies demonstrate that a 600 MHz center frequency antenna allows the detection of the bedrock contact and also changes in the compaction of the subgrade, locating areas under the pavement with poor soil compaction [226]. Low compaction of the subgrade soil could be the cause of road degradation, and GPR assessment could be a support in the strategies for road restorations. Saarenketo and Scullion [69] warn about the weak anomalies that could be associated with the bedrock contact depending on the dielectric properties of the pavement and the soil and the rock; hence concluding that the anomalies will probably be weak and non-continuous reflections.

8.1.2. Pavement Layer Thickness Analysis

Pavement is usually assessed with GPR in combination with FWD [223,226–228]. In these studies, one of the most common analysis is the determination of layer thickness [30,36,38,58–63,75]. This analysis requires high-frequency antennas, in all cases with a central frequency higher than 800 MHz. For example, [36] uses a 2 GHz antenna, while [229] reports the application of a 1.5 GHz ground-coupled antenna for the layer thickness determination, and [48] compares the results obtained with two antennas with 800 MHz and 900 MHz central frequency. The detection of the contact between layers is observed in the B-scans as a continuous anomaly produced by the reflection of the radar-wave at the interface between two media with different dielectric permittivity. The studies point to the easier detection of boundaries in the case of asphalt layers than in the case of concrete layers [17,84,228]. The results indicate that the error in the layer thickness determination is higher in the case of thinner layers and deteriorated structures. Errors less than 10% are reported in several works referring to new pavements [36,63,75]. Another analysis observes an error of about 12% in the surface asphalt layer with a common data processing method, but about 3% when the deconvolution is applied to separate the anomalies [60]. The analysis developed by [230] obtains the following results: (i) with air-coupled antennas of 1.8 GHz central frequency and the CMP (common mid-point) survey, the error in thinner asphalt layers ranges between 6% and 18%; and (ii) with ground-coupled antennas of 1.5 GHz central frequency and the CO (common offset) mode, the error ranges between 1% and 26%. This study highlights the importance of the data processing for the proper picking of anomalies associated with reflections in the base of thin layers. It is noticeable that both methods introduce the same error (an average value of 7%) in the case of thicker layers. This study also notices the influence of dust or sand on the surface of the pavement for the analysis of the dielectric permittivity using ground-coupled antennas, although the error depends on the thickness and the depth to the layer. In [216] an average error of 6.8% was reported for pavement thickness layers ranging from 280 to 350 mm, while for pavement thickness layers ranging from 100 mm to 200 mm the error was 3.8%. Other authors also conclude that the error in the bituminous layer differs from the error in the granular layer. In [231] an average error of 7.5% was obtained for asphalt surface layers ranging from 5 mm to 500 mm thick, but this error increased until it was 12% in the case of granular base layers ranging from 150 mm to 330 mm thick. A higher error was also obtained in [230] for thinner layers, observing a correlation between the thickness of the layer and the error in the case of CO surveys. However, no correlation was observed in the case of CMP surveys.

Additionally, the error depends on the quality and homogeneity of the layers. Al-Qadi et al. [232] observe lower errors in the case of plain concrete layers; whereas the studies in old pavements [216] and in new pavements [217], show an important change in the error comparing both studies in new and old pavements.

The error diminishes in the case of a proper calibration using several cores [36,214,223], although the accuracy depends on the number of cores. The results obtained without calibration present higher errors and demonstrate that the GPR used as a unique NDT assessment of the pavement is only useful for a quick and rough evaluation, being nonetheless a satisfactory technique.

8.1.3. Damage Assessment: Detection of Voids and Cracks

In most surveys, the objective is determining a procedure for the recognition of cracks and under-slab voids, as well as the rebar location in reinforced concrete slabs [76–81,88], being mostly used the CO acquisition mode and determining anomalies in B-scans. However, Yi et al. [81] observed that thin cracks are not detected using the common reflection analysis, and propose a 3D CMP using an array of antennas in order to determine slight changes in the radar-wave velocity that could be associated with zones affected by cracks wider than 1 mm.

On-site measurements compared with cores highlight that clear images of cracks in the B-scans correspond to large fissures (several centimeters wide), in many cases filled with foreign material [42,56]. Early fissures are not visible in GPR images because the anomaly is masked with the anomalies produced in the grains and heterogeneities of the layers [66]. However, zones affected by many smaller early cracks produce in some cases blurred images in GPR B-scans. This effect can be used to select possible zones with early damage. The effect of internal small cracks is also visible in the frequency spectrum. The heterogeneity due to the existence of cracks, mainly in the subgrade, increases the energy dispersion and subsequently attenuates the signal, generating an irregular image [81] with vague boundaries [66]. The radar-wave velocity in damaged areas differs slightly from the radar-wave velocity in undamaged asphalt [81]. In addition, the amplitude of the frequency spectrum decreases, which denotes the existence of possible anomalous zones [47].

Several tests report that 1 GHz center frequency antenna [40,41,65], or even larger frequency antennas [66], are valuable systems to characterize vertical pavement cracks. However, lower frequency antennas, for example 250 MHz, are in many cases more effective for detecting cracks because the larger wavelength is less affected by the heterogeneities and the aggregates grading of the layers [65]. In consequence, most pavements are assessed with antennas presenting a range of frequencies. For example, the studies in [68] explore depths between 50 mm to 160 mm using antennas with central frequency between 2.5 GHz and 700 MHz; [233] obtains details of the top structure and the subgrade soil with 1.5 GHz and 400 MHz center frequency antennas, respectively; [234] explores the bituminous layer with a 1.6 GHz center frequency antenna and the underneath structure (base, subbase and subgrade) with a 600 MHz center frequency antenna; [235] combines 800 MHz and 500 MHz center frequency antennas to detect subsidence in a port area, showing visible deterioration in paving blocks and loss of filling; and [221] uses an array of antennas with central frequencies ranging between 50 MHz and 3050 MHz.

Many works are devoted to laboratory tests and controlled measurements in order to determine the limits in the detection of cracks. The laboratory tests described by [56] in specimens representing rigid pavements are focused on detecting the bottom of the cracks depending on its width, length, and filling material. The results conclude that it is possible to detect 2 mm wide cracks, and the presence of material filling the crack increases its detectability, being possible to determine the bottom of the cracks: (i) wider than 5 mm and filled with saturated sand; and (ii) wider than 15 mm and filled with dry sand. The detection also depends on the depth of the crack and its position in the pavement structure. The controlled field studies described in [65] also present the detection of the bottom of the cracks in asphalt pavements, concluding that the stronger anomaly is obtained when the bottom of the crack intersects the bottom of the asphalt pavement. The results obtained in the laboratory experiments presented in [67] highlight that the stronger response depends on the contrast between adjacent layers in the case of cracks crossing the entire upper layer. The analysis presented in [40] determines an error of about 5.5% in the estimation of the length of the cracks when the bottom is detected. In those experimental tests, additional computational methods (e.g., finite-difference time-domain (FDTD) modelling) were applied to compare synthetic data with laboratory and controlled measurements.

To summarize, different authors point to the applicability of GPR in the detection of cracks and voids inside different layers of the pavement structure. The method provides clear images in the case of large cracks, although small cracks (until 2 mm) can be detected in the upper layers. The existence of zones with a large number of thin or early cracks can be also detected because of the changes in the radar-wave velocity with respect to the velocity in undamaged pavement. In addition, the bottom of the cracks can be also detected in the case of large and superficial cracks or in the case of cracks filled with water or saturated sand. The dielectric permittivity of those materials differs strongly from the dielectric permittivity of the pavement layer materials. More heterogeneous layers, presenting different grading, make the detection of cracks or voids more difficult. Cracks crossing one entire layer are more detectable, especially in the case of a large contrast in their dielectric permittivity.

8.1.4. Damage Assessment: Debonding

The loss of adhesion between adjacent pavement layers produces a change in the behavior of the pavement, which can contribute to its early failure, producing permanent deformations or superficial cracks and adversely affecting its performance. This problem is considerably severe in the case of airport pavements that undergo high shear stresses during the landing and take-off of the planes. The early detection of debonding is therefore crucial for pavement assessment. The application of GPR in the detection of debonding is usually combined with FWD analysis [75,85], with consistent results validating each other. In addition, the analysis of the GPR signal frequency indicates that the spectrum is sensitive to the structural condition of the pavement, correlating the results with the standardized deflection tests [47] in roads: when comparing damaged zones with undamaged zones, the spectrum bandwidth reduces while the amplitude and the central frequency decreases. However, the clear detection of debonding is only possible in the case of water presence in the debonded zone. Another analysis focused on the assessment of airport pavements [85] refers that the identification of such damage is difficult because it is usually produced at the shallow zone of the pavement structure (within the first several centimeters). However, GPR lateral wave in the CMP data acquisition is highly sensitive to the material properties, and the amplitude maps have proved to be efficient in determining debonding zones in which the high amplitude corresponds with the damaged zones detected by acoustic sounding methods. Methodologies based on linear prediction and supported by vector regression have been proposed and tested using a 10 GHz bandwidth antenna in an accelerated pavement testing facility [236], providing promising results in the detection of thin debonding. Another proposal was the application of a support vector machine to the A-scans, allowing for the detection of horizontally stratified debonding and geotextiles between the two upper pavement layers [237].

8.1.5. Damage Assessment: Moisture

The analysis of changes in water content in the different pavement layers of the substrate was analyzed in a laboratory [238], in controlled field tests [43], and in specific zones of roads [48,73,74]. Some results epitomize the ability of GPR to reveal variations in the water content in the pavement layers beneath the asphalt layer. Comparing GPR data with direct measurements of gravimetric water content reveals the ability of GPR to detect changes in moisture and also to assess the functionality of the drainage layers [34]. The results presented in [74] compare 1 GHz and 2 GHz center frequency antennas, concluding that, although data was similar in both cases, the higher penetration depth with the 1 GHz center frequency antenna makes it more appropriate for detecting zones affected by moisture. In addition, the existence of high water content in fractures [42] and debonding [47,48] between layers is a particular condition for the best detection of those damages. The representation of A-scan datasets in time-domain and in frequency-domain, obtained in wet and dry parts of the pavements, illustrates the great difference in amplitude due to the difference in dielectric permittivity contrast. In many assessments, the moisture damage is detected by observing the amplitude of the maximum peak in the time-domain. However, Zhang et al. [45] propose the use of more properties of the dataset to detect wet zones, considering statistical properties that are affected by moisture, selected with linear discriminant analysis. The comparison of results from GPR tests and passive microwave radiometry demonstrates that GPR is able to detect water content at higher depths, enabling the detection of water table in the soil under the structure [237]. Nevertheless, the identification in the B-scans of the anomaly associated with water table can require different seasonal measurements in order to determine the permanent and variable reflectors, which allows one to distinguish between natural layers or pavement structures, and water table.

The central frequency of the antennas recommended for the detection of moisture ranges from 900 MHz (to analyze moisture in the ground under the pavement [239]) to higher frequencies such as 2.5 GHz (to determine the existence of water in porous layers of the pavement structure [45], between those layers or inside cracks and voids).

The frost susceptibility is another issue to be treated in pavements placed in zones with extreme temperature changes. It is related to water content and drainage of the subgrade. In roads affected by frozen water, a strong reflection is produced at the interface between frozen and non-frozen sections when the subgrade presents high dielectric permittivity and subsequent high frost susceptibility [69].

8.1.6. Quality Control of New Structures: Asphalt Air Void Content and Segregation

Quality control is important in the construction of new pavements and in the restoration of already existing roads and airfields [78,240,241]. The analysis requires the improvement of signal-to-noise ratio of the GPR scans. Bianchini Ciampoli et al. [240] present and discuss a processing flow to improve the GPR images during the quality control of structures. In these applications, a previous calibration is required to obtain accurate results [217].

Quality control involves the thickness control during pavement construction [217], the quality analysis of the layer bonding [209,242], the porosity of the materials [71] and the analysis of thin layers overlaid on old and rehabilitated pavements [217].

The existence of thin layers over the pavements, added as a rehabilitation of damaged zones, produces an overlap between the wave reflected on the surface and the wave reflected on the contact between the thin layer and the old pavement, making the detection of the anomaly and the thickness estimation of the rehabilitated layer difficult. The existence of those thin layers considerably degrades the accuracy of the GPR interpretation of the layer's thickness and, therefore, accurate signal processing can be required. GPR surveys provide precise results in the estimation of the layer's thickness when used to assess the quality control of flexible pavements during their construction, allowing traffic speed velocities during the data acquisition (up to 90 km/h with air-coupled antennas and 25 km/h with ground-coupled antennas [217]). Nevertheless, the use of air-coupled antennas is highly recommended [217].

The analysis of quality in the pavement layer bonding is based on the water accumulated in the debonded zones. GPR is an efficient tool because of its sensitivity to changes in water content. Furthermore, it has been observed that GPR is also sensitive to changes in the quality of the material in the case of new pavements [242]. These changes are perceived mainly as a consequence of changes in the travel wave velocity. However, it is difficult to discern between the real cause of changes in the wave travel time: changes in the material conditions or variations in the layer thickness.

The void content in the asphalt pavement is related to the dielectric permittivity values that are obtained when calibrating the wave velocity by comparing GPR data to cores [72]. Dielectric permittivity ranges between 4.5 and 6.5 in dry aggregates while the asphalt permittivity varies between 2.6 and 2.8 [71]. Saarenketo [71] introduces an empirical relationship between the air voids content and the measured dielectric values from GPR data, with an accuracy of about 90%. The uncertainty in these kind of studies can be reduced by comparing GPR data to cores [28]. The analysis presented in [71] also detects the damage in pavements as a consequence of segregation, which is produced in small zones of the shallow layers of pavements with low density materials. This segregation is also associated with an excess of coarse aggregates. The proposed analysis is based on the estimation of the surface dielectric permittivity using GPR data, in which extremely low values could be associated with segregated zones [69]. Computational methods corroborate the occurrence of changes in the GPR signal when variation exists in the pavement density. The A-scan's amplitude increases when the asphalt pavement density increases because of the higher dielectric permittivity, also showing an augment in the amplitude of the frequency range [243].

8.1.7. Rebar Detection and Corrosion

In rigid and reinforced pavements, a usual assessment also involves the detection of the rebar [42], [56], [80] and the cover depth to the rebar [55], obtaining an average error of about 2.6% when comparing GPR results with cores. The estimation of the cover depth is based on the analysis of hyperbolas produced by reflections on the rebar. The analysis of the rebar location is, in most cases, based on B-scans acquired along the pavement surface. However, in many cases the 2D horizontal analysis can provide accurate and effective images of the rebar distribution inside the rigid pavement. Those images can be obtained from a grid of B-scans obtained in profile lines separated by 10 cm [244]. Combining GPR with other techniques such as electromagnetic induction, it is possible to determine the position of the rebar, the cover thickness, and the rebar diameter [245]. The main limitation occurs in the case of a densely-meshed rebar, requiring antennas with narrower transmission energy cones. The proposal of applying deep learning techniques to B-scans suggests accurate results in the classification of the rebar and other buried objects in pavements after building a data library for pavement objects. The technique is based on the recognition of changes in the GPR data amplitude and the anomalies associated with interfaces. The results presented in [246] allowed to automatically detect both the hyperbolas and the interfaces.

Another aspect in the rebar assessment of rigid pavement is the analysis of corrosion. Some studies have analyzed the GPR response due to corrosion of the rebar in reinforced structures: the damage obscures the B-scans and the amplitude of the hyperbolas produced by a corroded rebar decreases extremely when compared with the amplitude of the hyperbolas obtained from the reflection by a non-damaged rebar [247,248].

In addition, the existence of water with high salt content facilitates the corrosion of those elements. GPR B-scans show an important attenuation in the zones affected by high salt content water, which allows for the detection of possible problematic zones [247]. Detailed studies indicate that the reflected wave amplitude is sensitive to mapping corrosion in the rebar while the direct wave frequency amplitude is sensitive to chloride content [249], where the dispersion of the energy acts as an indicator of the cracks most likely produced in the concrete around rebar in later stages of corrosion process [248]. Other studies correlate damage in the surface of rigid pavements with the corrosion of the rebar [168].

8.1.8. Recent Developments and Applications

The most recent developments in the pavement assessment are focused on four main purposes: (i) controlled experiments, (ii) deep learning and automatic detection of damage and layers, (iii) antenna positioning, and (iv) design of new devices.

Some of those more controlled experiments are realized in testing sites on roads, comparing GPR results to other complementary tests and cores. Other tests are developed in a laboratory, using specimens prepared to simulate damage and structures. Many tests are prepared to analyze the effect of different conditions on the GPR images and to monitor damages. The experimental tests address the evaluation of different problems. In this way, for example, Rasol et al. [56] present the effects of size and filling of cracks on the GPR images, and Baltrušaitis et al. [250] use the GPR results in order to estimate the quality of materials by determining the bulk density of bituminous mixture, and [251] determines the bitumen content in different mixtures. In all cases, the measured parameter is the dielectric permittivity, and mathematical models are compared to cores and experimental results.

Regarding laboratory tests and field surveys, some developments applied neural networks in data analysis, deep learning techniques and automatic detection of damage by means of post-processing techniques. For example, Zhang et al. [45] propose a training convolutional neural network structure for moisture damage detection from GPR B-scans, which works by using an incremental random sampling method to automatically select the most appropriate plot scale at each radargram. Other authors, such as Kim et al. [246], propose the reconstruction of GPR data to detect underground objects under pavements in urban areas using deep learning techniques, concluding that the precision of the results depends directly on the quality and number of field training data. In this work, four features were studied: (i) pavement layer interfaces, (ii) the existence of a manhole cover in the surface, (iii) the detection of buried targets that produce hyperbolas, and (iv) subsoil background. More complex elements in the study could require a larger number of training data. Larger databases are compiled and used in the study of pavement distress using deep learning methods, showing that the frequency of the antenna also affects the results because of the detail loss [252]. In [253] a review of deep learning applications in GPR is presented, comparing the results and grouping the works depending on the type of data used. The conclusion is that the methods using A-scans [254,255] present slightly better results than those using B-scans [252,256] or C-scans [257,258]. However, the best spatial information is obtained from C-scans, and the most recent works [257] are focused on 3D GPR data. Notwithstanding, this processing requires more complex architecture and a large volume of datasets. The optimal solution to achieve a proper development in the deep learning based on C-scans is to create a big GPR dataset, sharing data from research around the world [253].

The efficiency of the monitoring, at high speed, turns the GPR into an important tool when combined with new loading tests devices, such as FWD and, more recently, with TSD (traffic speed deflectometer) [259–262]. For studies at the network level, in some countries the use of GPR is already considered mandatory for layer thickness assessment due to its efficiency to support pavement management system decisions [263]. The joint interpretation of GPR with load tests, such as FWD and TSD, has also led to changes performed on the GPR system in order to better process the data collected. For this purpose, a dedicated GPR, a noise-modulated ground penetrating radar (NM-GPR), was developed in Australia [259]. This GPR variant uses coded signal modulation and efficient receiver hardware to enable traffic speed multi-channel performance and consists of hardware receivers combined with an array of ground-coupled antennas and custom damping systems in a trailer format to collect 3D data while travelling at speeds of up to 100 km/h. The NM-GPR is used for the assessment of calibrated road layer depth measurements and quantitative moisture mapping. Geospatial views, developed to align and join the data gathered with NM-GPR, TSD and FWD, represent an important tool for understanding these complementary data together and in the context of the surrounding environment [259].

Other new developments are focused on finding antenna setups (positioning systems) that can be used in different scenarios. In that way, the studies by Yuan and Cai [264] propose a mechanism to automatically adapt the antenna trajectories. The methodology consists of comparing synthetic data obtained for different angles between the antenna and the target with field data, thus contributing to automation in the antenna positioning. The self-detection of the antenna position is possible using fixed reflectors on the surface [265] or using a geodetic network in order to georeference the position of each A-scan and the anomalies in the radargrams [266]. The work by Gabrys and Ortyl [267] compare several georeferencing systems in GPR surveys for the detection of layers and pipes under the pavement in C-scans. The position was determined using a GPS synchronized with the UTC time and a total station. The results demonstrate that the higher accuracy in the anomaly position is obtained with the synchronized GPS, although in some cases the system does not provide the correct antenna trajectories. This is caused because this system is based on signals from satellites and in some circumstances, as in dense cities, the horizon visibility is limited. In those cases, the most efficient position is obtained with the total station.

Finally, the design of new devices for GPR surveys also includes the positioning of the antenna with robotic systems. In the field of road inspections, a few more developed prototypes have been built with the objective of early detection and repairing damage, using cameras for visual inspections. Wu et al. [268] design and test (in controlled experiments) a six-channel antenna mounted on a robotic system, and [269] proposes and constructs a robotic system with 3D printing, with the scope of using the prototype both for assessing damage and repairing early cracks. In addition, some works suggest pavement surveys by using drones [270]. The UAVs (unmanned aerial vehicles) are flexible platforms that could support different types of remote sensing sensors [271] and, in the inspection of roads, have a low impact in the traffic while allowing one to obtain images from several lanes at the same time and build 3D images [272].

8.2. Railways Inspection

Although GPR is applied in several railways for monitoring, as it provides a rapid continuous measuring, its application presents limitations that have to be considered. The GPR setup has to be chosen according to the scope of the study. On the one hand, GPR assessment can be applied at different stages of railway life cycle, during construction of new lines, for in-service monitoring or for rehabilitation purposes. On the other hand, GPR can be used for distinct diagnosis goals, such as detecting settlements of deeper layers, ballast quality evaluation, debonding and crack detection. The specific requirements for an efficient application of GPR on railway assessments and the main limitations, together with advice in overcoming them, are referred herein.

Designated inspection vehicles are used in railways to assess the railway condition, through inclinometers, digital videography, laser scanning and, most recently in Europe, USA and Australia, the GPR. This inspection enables a diagnosis of railway elements related to traffic safety and comfort, such as track geometry, rail profile, and sleeper and fastening defects. Based on this assessment, maintenance works are planned when the track condition reaches a pre-defined level, known as the "intervention level" [107,273]. The main maintenance intervention on the railway is tamping, performed when the settlement of the rail is occurring, and consists of adding a ballast below the sleepers to raise the rail level [274]. This is an action that corrects the track geometry but does not solve the deterioration causes, which are mainly due to substructure settlements [17,96,97,107], that can be only detected by GPR, thus exemplifies its importance.

GPR had been initially been applied to railway diagnosis in the 90s, at research level, using ground-coupled antennas [84,89,275] and, some years later, using high-frequency air-coupled horn antennas [97,276,277] that enable non-contact tests at higher speeds. Some key limitations delayed the GPR application to railways. One is the nature of track elements, ballast and sleepers, that are rough contact surfaces for the ground-coupled antennas.

detection of water content changes during rainfall and, as a novel approach, for the location of soft clay and peat deposits that can cause accelerated ground settlements under dynamic amplification caused by high-speed trains (over 200 km/h). This research made extensive recommendations for conducting GPR investigations on railway lines, which included the application of optimal system configuration consisting of ground-coupled and air-coupled antennas. More details can be found in specific Deliverables (D), namely on GPR applications referred to above in D1.2, the joint interpretation of GPR with track geometry in D2.4, a novel damage detection methodology developed for bridge assessment in D 2.2 and the assessment of earthworks in D2.3 [301].

At the same time, the Federal Railway Administration (USA) has also financed a project aiming to develop a device for rail scan and guidelines for GPR implementation on rail assessment [90]. A GPR system was installed on a survey vehicle for field-testing and the integration of GPR results with track stiffness and geometry was also addressed. Several tests and calibrations were performed for over 130 km and improvements were made, mainly to the signal quality of shoulder antennas. In addition, dynamic cone penetrometer (DCP) tests, test pits and laboratory tests were performed to validate the results for fouling and moisture assessment. The system installed consists of three 2 GHz horn antennas and one 400 MHz ground-coupled antenna that can measure at a maximum speed of 120 km/h. The specificities of the approach are: i) automated processing software that handle large volumes of data for rapid reporting and ii) the integration of GPR data with the track geometry and gage restraint measurement system (GRMS) data generated by the T-20 car [90]. Based on this extensive research, GPR indexes were defined depending on four parameters, namely ballast fouling, ballast depth, layer roughness and moisture detection. Those indexes were then combined with track roughness assessed through track geometry, and a track substructure quality index (TSQI) matrix was developed and implemented. Nevertheless, the study highlights the need for additional research: i) on the ballast fouling index sensitivity to type of fines and ii) to improve the shoulder antennas' results, namely the signal penetration and reflection characteristics as they are highly influenced by the edge of the tie.

It is clear that there is a general interest in GPR application for railway studies, given the importance of this test for the detection of real causes of track deterioration and to support the maintenance management decisions. Nevertheless, in spite of the studies developed so far there is still the need for future research to improve the testing methodology and data processing in order to make the GPR application more user friendly and time efficient.

In the case of ballastless railway tracks, GPR application is similar to other concrete reinforced structures, such as pavements and bridge decks. The main challenge is to select the frequency to be applied in order to measure the slab thickness and the base layer thickness and condition, as generally the slab is laid over a lean concrete layer. The central frequency of the antenna has to be lower in order to detect the interface between the lean concrete and soil, as the GPR wave dissipation in concrete is higher than in asphalt or in aggregates. Similar to ballasted railways, antennas with different central frequencies can be used for different purposes. In this case, the lower frequencies for layer thickness, and void and mud detection under the slabs [113], and higher frequencies for detecting debonding between sleepers and slab, cracks at the interface, and the rebar location [112]. In the case of ballastless railways, GPR is generally applied together with other NDT methods, such as the impact-echo method (IEM) and ultrasonic testing.

It can be concluded that GPR represents an important tool for railway assessment in order to support the asset management of this transport infrastructure.

8.3. Retaining Walls

GPR assessment of retaining walls is focused on many cases in the detection of the contact between the wall and the natural soil, in the search of voids or changes in water content. Other objectives depend on the type of wall.

In the case of most of the cantilever walls and concrete retaining walls, one of the objectives is the detection of the rebar [115,116]. In those cases, the application is similar to other usual inspections of concrete structures [302], and antennas with central frequencies higher than 900 MHz are normally used because the presence of metal mesh makes difficult the signal penetration to deeper targets, in the case of high density of rebar. However, several layers of reinforcement could be mapped and represented using 3D images of GPR slices [116], with the rebar position and structure perfectly detected and even including other reinforcement structures such as dowels [116,302]. The main difficulty for an accurate 3D representation of the vertical and horizontal rebar structure is the control of the exact position of the antenna [302]. Tests in different types of structures indicate that the accuracy of the antenna position in order to obtain precise 3D images is within 1 cm for m^2 [303]. Hugenschmidt and Kalogeropoulus [116] propose a 1 cm separation between the profile lines, and develop and use a survey apparatus based on guides for the antenna and an electric motor to move up and down the antenna on the wall, covering the entire surface with precision. Another important step in 3D imaging is the processing that requires in all cases an appropriate migration [116,303]. Hugenschmidt et al. [117] test different processing strategies, comparing results from the most classical 2D filter sequence and 3D processing with an inverse scattering followed by data fusion. The results show that the most classical processing allows for the detection of the top layer of the rebar, which decreasesing the resolution extremely, and focuses on the second layer of vertical bars, although the 3D processing followed by fusion showed the vertical and horizontal bars in the top layer of the rebar. The most complete map of both rebar layers was obtained with the inverse scattering and data fusion. However, the second rebar layer is defined with a lower quality image. The authors suggest that the influence of the concrete, acting as a low-pass filter, and the screen effect of the top rebar layer mask the deeper targets, highlighting the limits of dataset. Other authors also suggest the application of inverse scattering techniques in the assessment of reinforced concrete walls focused on the rebar location [304,305].

Therefore, concerning the detection of inner damage or the contact between the wall and the natural soil, the result depends on the rebar density. Hugenschmidt and Kalogeropoulus [116] and Hugenschmidt and Mastrangelo [302] test the results comparing three different center frequency antennas: 1.5 GHz, 900 MHz and 400 MHz. The results epitomize that, although the resolution decreases as the central frequency decreases, as expected, the lower frequency does not allow greater penetration into the medium. Hence, the inspected depth is the same in all cases as a consequence of the screen effect of the metallic rebar.

GPR assessment is also a promising tool in the case of quality control of retaining walls in deep excavations in unstable soils. In those cases, borehole GPR could be used before the excavation in order to check the integrity of the structure, detecting zones with possible defects, and increasing the safety of the dig. There are few applications of borehole GPR in the literature, most of them being in tunnels during the excavation [306] and in mining and geological explorations. The highest difficulty in this survey is determining the azimuthal position of the targets, mainly in the case of single borehole GPR [307].

The objectives in the assessment of stone and masonry retaining walls include the detection of foundations [119,120], the study of the structure [119], analyzing the contact between stones [120] and determining layers in the wall [121]. In addition, different damages or possible causes of damage are also examined. The most common analysis focuses on the study of voids [308], cracks and discontinuity between stones [120]. In some cases, the inspection focuses on the detection of changes in the mineral composition of the ground that could affect the wall. This is the case in the analysis of changes in the soil salt content [121] or sulphate and chloride content [309]. Changes in the aggregate size also produces changes in the GPR images, showing a large scattering [310]. This effect has been detected in the ground studies [311,312]. Those inspections require the detection of both surfaces of the wall, as it is important to detect the contact between the ground and the wall. Therefore, the central frequency of the antennas used depends on the aspects under interest. Beben et al. [115] utilize dual center frequency antennas of 200 MHz and 600 MHz to analyze the wall and the foundations and, depending on the terrain conditions, an antenna with a central frequency of 80 MHz. Solla et al. [120] inspect a stone retaining wall with a 1 GHz center frequency antenna, with survey lines in the top of the wall to detect the different stone layers and the contact between the wall and the ground, and the foundations. Santos-Asunçao et al. [121] use an 800 MHz center frequency antenna to detect a buried structure, and the 1 GHz and 2.3 GHz center frequency antennas in the assessment of the state of the buried walls, obtaining 3D images and GPR slices. Wiggenhauser et al. [310] check the inner structure of walls using two antennas with lower central frequencies of 200 MHz and 400 MHz. In addition, several tests in laboratory [313] or in field surveys [115] recommend the use of a high-frequency microwave synthetic aperture radar between 1 GHz and 6 GHz in order to obtain B-scans or to generate 2D slice images.

Wiggenhauser et al. [310] highlight the difficulty of the GPR and other NDT in the inspection of masonry retaining walls as a consequence of the limited access to the surfaces and their different stages of deterioration, whereas Santos-Assunçao et al. [121] remark the problem in accurate data acquisition due to irregular stone surfaces and also the importance of the precise position of the antenna, proposing a laser guide marking for the profile lines.

In general, although there are promising results in the assessment of retaining walls, two of the main difficulties are the uncertainty because the different targets could produce similar anomalies and the accurate position of the antenna, because in many cases data are gathered on hard-to-reach surfaces or over irregular surfaces. Different authors propose methodologies and apparatus to facilitate accurate data collection in these difficult conditions. Another inconvenience that affects the assessment of concrete reinforced walls is the limit in penetration depth due to the screening effect of the rebar and the low-pass filter effect of concrete. Although there is the application of accurate data processing, the detection of the contact between the wall and the natural soil is not possible no matter the frequency used. In the case of masonry walls composed of irregular stones and irregular surfaces, ringing noise usually appears in the images that makes the detection of pathologies (voids and cracks) difficult. In addition, there is not a standard procedure or guideline in order to apply GPR surveys to the different types of walls, excepting the guide proposed by the US Department of Transportation [114] and technical reports that include the GPR between the NDT methods for retaining wall inspection [309] or define GPR procedures [308] applied to different case studies. Finally, GPR wall inspection is usually combined with a preceding visual inspection and other tests such as laser scanning [120], infrared thermography, acoustic methods and X-Ray fluorescence material identification through samples [309].

8.4. Bridges Health Assessment

Surveying stone masonry or concrete bridges differ in the GPR equipment and in data acquisition methods used. Thus, the discussion on these types of structures reported herein is presented separately.

8.4.1. Stone Masonry Arch Bridges

The application of GPR aims to reach the bridge foundations while analyzing the shallower filling and structural elements. The 250 MHz and 500 MHz center frequency antennas are therefore more commonly selected as the most suitable due to their optimal compromise between penetration and resolution. Ground-coupled antenna configurations, usually single antennas mounted on a trolley or manually on a survey cart, are used in this type of bridge surveying. Alani et al. [132], Fauchard et al. [135], Pérez-Gracia et al. [140], Solla et al. [128–131,137,144], Lubowiecka et al. [142] and Arêde et al. [145] employed a combination of low (200–250 MHz) and medium (500–600 MHz) center frequency antennas to evaluate the stone masonry health assessment. Additionally, higher frequencies were used in the 800 MHz to 2 GHz frequency range to estimate the thickness of masonry walls and arches [124,132,134–136,138,140,141,143].

Longitudinal profile lines crossing the bridge in the common offset (CO) mode are recommended, with the antenna polarization being orthogonal to the direction of acquisition. Generally, the GPR data acquisition is based on distance intervals, and the trace-interval should be at least 5 cm (2 cm for the 500 MHz central frequency and 5 cm for the 250 MHz central frequency). To measure the profile lengths and to control the trace distance interval, an odometer wheel is usually attached to the GPR system. However, some stone masonry arch bridges have a surface pavement composed of large irregular blocks, which make measurements difficult, as the continuous movement of the survey wheel is often interrupted. In such circumstances, the survey wheel should be previously calibrated on-site to avoid inaccuracies in spatial trace positioning. Moreover, for these ancient structures that have suffered several restorations or reconstructions throughout history, causing heterogeneity in filling, it is recommended to survey more than one profile line in order to prevent inappropriate data interpretation. Note that GPR profiles should be collected at a prudent distance from the parapets of the bridge that may lead to airwave events (noise) coupled to the reflected signal.

Testing standards for masonry arch bridge inspection were provided by the International Union of Laboratories and Experts in Construction Materials, Systems and Structures (RILEM) Technical Committee (TC) 127-MS "Non-Destructive Tests for Masonry Materials and Structures – test MS.D.3: Radar Investigation of Masonry" [314].

There are a large variety of processing filters to reduce clutter and any unwanted noise in the GPR data collected over stone masonry bridges, in which complex reflection patterns occur due to internal heterogeneity. The processing objectives are: (i) to correct the downshifting of the radar section due to the air-ground interface (e.g., time-zero correction or max. phase correction), (ii) to amplify the received signal in order to mitigate possible losses or attenuations (e.g., gain function, automatic gain control or geometrical divergence compensation), (iii) to suppress the continuous component (e.g., DC removal or dewow), (iv) to remove horizontal continuous low-frequency reflectors (e.g., subtracting average or background removal), (v) to remove both low- and high-frequency noise (e.g., band pass), and (vi) to suppress strong clutter (e.g., deconvolution or migration). Additionally, ancient bridges usually present a double-slope or steep profile, and topographic corrections are therefore required to improve the accuracy in imaging subsurface features.

Some limitations were reported when measuring the thickness of voussoirs (or ring stones) composing an arch from GPR data collected over the bridge's pathway [144,145]. The conjunction of both the internal heterogeneity in filling and the more irregular internal shape of the voussoirs yields heavy signal scattering that often masks the reflection produced at the filling-stone interface. To overcome this drawback, some authors [143] have proposed to collect the GPR profiles directly through the inferior face of the vault. However, this methodology is only operational if the vault intrados surface is accessible. Another crucial piece of information for more accurate analytical calculations and structural assessment of bridge structures is the thickness of the spandrel walls [143,149]. In this case, the GPR profile lines should be conducted through the vertical walls of the bridge. It is important to also mention the technical difficulties experienced during data acquisition, firstly due to the difficult accessibility to the upper stonework of the bridge and, secondly, due to the excessive weight of the antennas, especially with the 500 MHz antenna. Hence, the use of a ladder is recommended to access the higher parts of the structure, as well as the support of at least two operators in order to maintain both the GPR antenna and the survey wheel in contact with the surface [149]. Further developments should include the design of a measurement device and innovative methods for the acquisition of GPR data through the less accessible parts of a bridge structure (e.g., vaults' intrados, piers, and spandrel walls).

The analysis and interpretation of GPR data is often difficult due to the complexity of stone masonry bridge structures and the heterogeneity of the building materials. In order to understand the propagation of the GPR signal through media and its response (scattering), several electromagnetic simulators have been developed, in which particular scenarios and soil properties can be modelled. One of the most employed is the finite-difference time-domain (FDTD) algorithm, which is based on a discretization of both space- and time-domain. Previous studies have demonstrated that realistic and large-scale models can thus be simulated, which allows for a better approximation and comparison to real data in order to assist interpretation. Solla et al. [130,133,137] present an interesting approach to elaborate FDTD models using real geometries based on orthoimages of the bridge structure provided by geomatics techniques (photogrammetric or laser scanning methods). However, this simulation in fine detail requires excessive computational resources and time. In this regard, Diamanti et al. [315,316] develop a numerical approach in which the subgridding method was implemented into the standard FDTD method. Thus, micro regions can be created, with different spatial-steps (or space discretization). Very fine spatial-steps would be exclusively used to simulate smaller geometries such as cracks or delamination; hence, the calculation time required to simulate the entire model is reduced.

The combination of GPR with complementary geophysical techniques (e.g., sonic/ultrasonic testing, electric resistivity, infrared thermography, etc.) is highly recommended. Each particular method provides different information owing to the physical properties of the construction materials, which allows for a more detailed investigation in the diagnosis of bridge stability. Orbán and Gutermann [124] combine GPR, sonic testing, and infrared tomography (IRT) to investigate hidden geometry, voids and cracks and moisture distribution in masonry and fill, as well as to define wet areas. Trela et al. [134] use geoelectrical measurements of spectral induced polarization (SIP), combined with GPR, to evaluate the moisture distribution inside the masonry. Fauchard et al. [135] present a combination of GPR and electrical resistivity tomography (ERT). The GPR revealed internal damage such as voids and delamination, while the ERT allowed one to characterize the transition zone between the bridge and the riverbanks. Solla et al. [317] combine GPR and IRT, together with FDTD modelling, to analyze the moisture in a medieval arch bridge. More recently, Biscarini et al. [318] use GPR and IRT, together with UAV photogrammetry (RGB imaging), to analyze material degradation and its causes in a Roman masonry bridge. Additional information was found regarding the construction phases and restoring interventions of the bridge. Furthermore, Alani et al. [319] present a holistic approach using GPR and InSAR technologies to investigate both the subsurface geometry and the structural displacements of masonry arch bridges certainly caused by water floods and the riverbed soil expansions. Apart from non-invasive geophysics, there are a large variety of conventional NDT methods that complements the GPR surveys. In this frame, Arêde et al. [145] conduct an integral campaign using GPR with other non-destructive and slightly-destructive techniques oriented to estimate physical and mechanical parameters such as on-site stress and masonry deformability (dynamic probing super heavy (DPSH), Ménard pressuremeter (PMT) and flat-jack tests). Finally, it is possible to find several studies focused on the structural assessment of stone masonry arch bridges that combine the use of GPR with complementary geomatics techniques. Generally, these works use geomatics and GPR data to create the model of the bridge structure to be simulated using the finite element method (FEM) analysis [143,146,149,150]. Hence, the complex external geometry (metric information) of the structure is provided through geomatics—terrestrial laser scanning (TLS) or photogrammetry—while the GPR gives information of its internal composition [320]. More exhaustive and robust modelling integrates GPR and complementary NDT, such as vibrant ambient noise [136,140,141,148,151], sonic testing [141,148,151], impact–echo tests [141,151] and thermography [136].

8.4.2. Concrete Bridges

GPR surveying of concrete bridges is mainly focused on rebar detection and mapping, the estimation of deck thickness, damage detection such as cracking or delamination, moisture and corrosion; as well as other deeper aspects such as foundations and pier assessments. The GPR systems, testing methodologies and analysis to be applied for all these aspects are herein discussed.

Depending on the extension of the surveyed area, the depth required and the traffic flow conditions, both ground-coupled and air-coupled antenna systems can be employed in inspecting this type of bridge. Herein lies the difficulty of providing good quality data sets while being completed rapidly and safely for both operators and users. Firstly, in order to avoid traffic disruption, the air-coupled systems allow measuring at a higher speed (with lower resolution measurements) as they are generally mounted on a van or a vehicle. Next, measurements can be conducted for the whole length of the bridge or only for selected areas of interest in both longitudinal and transversal directions. For small survey areas and vertical surfaces, when a more detailed analysis is required (e.g., abutments, girders and piers), single ground-coupled antennas are most commonly used. Conversely, for large survey areas, and particularly to map utilities and reinforcement, the use of multi-channel systems is highly recommended. Array multi-channel systems are preferred because they cover wider width measurements, which shorten the surveying time, increasing the productivity, and benefiting 3D visualization and the reconstruction of underground network. Moreover, these systems may use antennas arranged with different polarizations. For instance, the optimization of antenna orientation to take advantage of signal polarization is an important feature for successfully locating reinforcement; hence, a hyperbolic reflection is obtained if the GPR profile line is conducted perpendicular to the object direction.

Most commonly, the GPR was used to specifically locate the presence of any cracking and delamination, moisture and corrosion, and to locate the rebar and ducts within the concrete slab. For this purpose, the GPR systems generally used have frequencies higher than 1 GHz (up to 2.6 GHZ). The 1 GHz frequency is capable of investigating up to 100–150 cm depth from the concrete surface, while the higher radar frequency has lower depth of investigation, up to 15–20 cm, but means more resolution and accurate data acquisition (it supports a higher sampling rate). For example, regarding resolution, the 2.0 GHz frequency can guarantee a theoretical resolution of 1.25 cm (quarter of a wavelength criteria) in concrete.

Based on 20 years of operational experiences, Rhee et al. [175] propose a sampling rate for bridge deck inspection using 1.0 GHz air-coupled GPR antenna sets between 12 and 14 scans/m depending on the highway speed. A sampling rate of 12 scans/m means one scan per every 80 mm (with the vehicle where the GPR system is mounted driven at the speed of 80–100 km/h). This sampling rate is much lower than the common sampling rate for ground-coupled GPR operated by normal walking speed, about 50–200 scans/m. In this regard, the recommendation in the Mara Nord's project guidelines [152] states that with air-coupled antenna systems, the sampling rate should be at least 10 scans/m, while with ground-coupled antenna systems, a sampling rate of 100 scans/m is recommended.

The ground-coupled GPR system is usually mounted on a survey cart with an integrated encoder wheel that serves to measure the profile lengths and to control the trace-interval. However, to obtain a good reconstruction of the surveyed area, the system requires a good positioning, using a high precision global positioning system (GPS), a total station, or even a local system of coordinates. On the other hand, the air-coupled GPR systems are most commonly mounted on vehicles or other automatic measuring equipment. The latter employs a differential RTK GPS and an encoder-based distance measuring instrument (DMI) for real-time positioning. This global positioning eliminates the need for predefined profile lines and provides a safer data acquisition, because the driver of the mobile system can pay attention to the traffic flow instead of a predefined reference, without reducing the accuracy of relative positioning. Another benefit of a good global positioning is the fact that georeferencing allows for the integration of results obtained from multiple technologies in a geographic information systems (GIS) environment, which contributes to identify the location and extent of defects [169].

The presentation of GPR data is usually carried out in the form of 3D maps, or time-slices, at different depths. According to the recommendations provided by Mara Nord's project guidelines [152], for an accurate 3D data acquisition, the distance between parallel profile lines should be a maximum of 50 cm with a 2D system and a maximum of 12 cm with a 3D system. Then, the most generalized results on surveying bridge decks are 3D maps of amplitude, which can be produced showing two types of information at the same time [172]. Firstly, the rebar will appear as the most prominent (the steel/metal is a complete reflector of the radar energy). Secondly, the corroded areas will be detected as having low amplitudes (due to the lower dielectric permittivity than healthy rebar areas). Additionally, potential areas of deterioration appear as zones of signal attenuation, and the rebar will appear deeper (lower signal velocity) due to higher moisture and chloride content. Color contour maps of deterioration can be therefore produced, mainly defining moisture, delamination and corrosion. Deterioration maps should be created at different depths in the concrete deck slab in order to estimate the extent of deterioration.

There are different regulations and standard test methods describing processing methodologies to quantitatively analyze the GPR data that generally employ reflection amplitudes. Stryk et al. [321] present an overview of existent recommendations for bridge inspection, namely: the American standard ASTM D6087-08 "Standard Test Method for Evaluating Asphalt-Covered Concrete Bridge Deck Using Ground Penetrating Radar" [53], the American SHRP 2- report S2-R06A-RR-1 "Nondestructive Testing to Identify Concrete Bridge Deck Deterioration" [322], the Mara Nord Project "Recommendations for guidelines for the use of GPR in bridge deck surveys" [152], the British technical specifications DMRB 3.1.7 "Design Manual for Roads and Bridges, Advice notes on the non-destructive testing of highway structures—advice note 3.5 BA 86/2006: Ground Penetrating Radar (GPR)" [5], the German BASt-report B55 "Examination of GPR in combination with magnetic techniques for the determination of moisture and salinity of concrete bridge decks with asphalt cover" [323], and the German document B10 "Recommendation for nondestructive testing of civil engineering structures by GPR" [324]. Additional standards including the assessment of bridge decks are: The American AASHTO R 37-04 "Standard Practice for Application of Ground Penetrating Radar (GPR) to Highways" [325], the American ACI 228.2R-98 "Nondestructive Test Methods for Evaluation of Concrete in Structures" [326], and the American NCHRP RR 848 "Inspection Guidelines for Bridge Post-Tensioning and Stay Cable Systems Using NDE Methods" [327].

The typical post-processing includes time-zero correction (set ground surface), background removal (remove horizontal banding), band pass filtering (remove high- and low-frequency noise), and gain (amplify the received signal across depth range). Migration is also recommended when identifying the rebar in a concrete slab in order to reduce or eliminate hyperbolic diffraction patterns in the data [180]. Thus, each pixel in the original GPR images will be migrated focusing energy on the true rebar locations; hence, the intensity values of the pixels at a true object location will end up having the sum intensity value of all the pixels on the hyperbolic signature (the energy will be focused therefore on those pixels). Moreover, a procedure known as depth correction is commonly applied [178], where linear regression is used to normalize amplitudes in order to obtain a correct deterioration map. The signal strength (amplitude) decreases with rebar corrosion, but also with variation in the rebar depth. Reflections from deeper rebars are more attenuated than those from shallower rebars due to a larger two-way travel time and, therefore, it is important to apply this correction in order to minimize the effect of the rebar depth such that the overall amplitude variation in the entire bridge deck will be only indicative of corrosion.

Although GPR data collection can be fast and efficient when surveying concrete bridges, the interpretation of GPR data is a very time-consuming task and depends on the interpretational abilities of the operator to provide accurate information and reliable results. For example, for rebar detection application, and for a bridge deck with thousands of rebars, it takes an extensive amount of time to manually identify and measure the amplitude value of each rebar. Extensive literature exists, therefore, regarding the automatic detection of rebars in concrete bridge decks, which is based on detecting reflection hyperbola patterns. There is a large variety of computational methods that can be applied, such as interpolation functions, a histogram of oriented gradients (HOG), a feature descriptors-based multi-layer perceptron (MLP) classifier, a support vector machine (SVM), a convolutional neural network (CNN), the limited and simplified hyperbolic summation (LSHS) technique, partial differential equations (PDEs), the hidden Markov model (HMM), a genetic algorithm, and the fuzzy logic approach. These computational methods are described and applied in some relevant and recent studies [177–181] and, in most cases, with accuracies greater than 95.0%. On the other hand, algorithms are also implemented to extract the GPR amplitudes automatically from both the whole GPR signals and the rebar peaks detected [165], [171], [174], generally after applying a migration procedure.

Some limitations can occur during concrete bridge inspections. Firstly, the steel/metal is a quasi-perfect reflector of the radar-waves, which facilitates the rebar detection, but deeper targets can be masked if it is a tight mesh. In this regard, collecting data in both polarizations (with the dipoles perpendicular and parallel to data collection direction) has benefits because the reflections produced by metallic targets perpendicular to the data collection direction are weakly seen in data collected with dipoles parallel to the scanning direction, so that other potential targets below them can be more easily detected. Secondly, overlapping reflections are usually observed between consecutive bars that is dependent on the spatial resolution of the antenna (higher frequencies provide a higher resolution), which may lead to the misinterpretation of closest or smaller-diameter bars. Resolution problems can also occur in sections with a too small concrete cover when measuring pavement thickness. The selection of the most appropriate central frequency of the antenna is therefore crucial herein. Moreover, another important parameter affecting the location accuracy is the horizontal sampling. Dense horizontal sampling (scan spacing), gives more accuracy on positioning and good quality data for further amplitude analysis, although its main consequence is a decrease in the survey speed. Conversely, using a low sampling rate may limit data visibility in the field and cause inaccurate adjustments of hyperbolas peaks when processing. Thirdly, the 3D data acquisition (especially with single GPR antennas) may incur an incorrect distance encoder calibration, position of the antenna on starting/ending grid lines, improperly configured survey grid, georeferencing, etc. The use of antenna arrays or automatic scanner systems makes the acquisition of 3D data easier and thus encourages a wider use of 3D techniques although, in practice, these systems usually have an excessive cost. For example, a higher sampling rate (higher number of traces recorded) decreases a survey speed, which could result in higher survey costs and, more importantly, could interrupt the traffic flow. Fourthly, the amplitude value is highly dependent on various factors such as different depth to the top mat of the reinforcing steel, weather conditions (e.g., humidity) during data acquisition, and concrete properties (e.g., density, porosity, etc.), which makes the detection of corrosion and delamination only difficult from the analysis of amplitude maps. Confirmation with the additional analysis of other signal attributes (e.g., signal attenuation, signal-to-noise ratio and velocity of propagation) or complementary NDT should be part of the process whenever possible.

Regarding the combined used of GPR with other complementary NDT, Sun et al. [174] compare delamination maps from acoustic surveys with deterioration maps (signal attenuation maps) from GPR. The results demonstrate that combining both techniques will provide a complementary and comprehensive evaluation of concrete bridge decks, although GPR detected more defects than acoustic methods. The acoustic method can only detect near surface delamination, while GPR signals can penetrate concrete. Rathod et al. [162] combine GPR with a rebar detector (profoscope), which uses electromagnetic pulse induction technology to measure rebar diameter, spacing and cover depth which are required to assess the load capacity of bridge decks. The results suggest that GPR gives more consistent and accurate data with regards to spacing and cover. Dabous et al. [169] successfully integrate infrared thermography (IRT) and GPR, mainly in indicating the subsurface delamination of concrete bridge decks. In addition, the study demonstrates the feasibility of integrating the collected data into GIS for enhanced visualization of the inspection results, which allows integrating data and gives a full-scale representation of the condition of the inspected area. Other authors, such as Janků et al. [161] combine IRT and GPR, but also ultrasonic pulse echo (UPE), for the detection of delamination in concrete bridges. The results revealed the limited capability of GPR to detect cavities located just beneath the surface, while UPE failed to detect such defects completely. On the other hand, IRT was very successful at locating cavities near the surface, but only under convenient weather conditions. GPR and UPE are better suited for locating deeper defects inside the material. Dinh et al. [160] employ electrical resistivity (ER) and half-cell potential (HCP) in order to validate the corrosion detected by GPR automatically from amplitude maps. Barnes et al. [164] apply HCP and chain drag data as thresholds for an accurate damage estimate from GPR amplitude maps. Abouhamad et al. [168] propose GPR in conjunction with HCP and hammer sounding resistivity to find the appropriate GPR threshold and use this information to correctly assess bridge deterioration. Alani et al. [183] use GPR and IBIS-S (deflection and vibration detection sensor system with interferometric capability) techniques for bridge assessment, which provide valuable information such as rebar position, moisture ingress and deflection. To determinate the bridge condition, a FEM (finite element model) of the bridge was then created using information obtained from the NDT results and compared with the dynamic behavior of the bridge without defects. Varela-Ortiz et al. [159] combine non-destructive load testing with GPR to obtain the safe load-carrying capacity of concrete bridges. Thus, field load testing allowed one to calibrate a FEM analysis and to accurately characterize the structure's live-load response, while GPR provided reliable information regarding the internal reinforcement and the condition of the concrete. Varnavina et al. [172] use LiDAR technology and ultrasonic surface wave (USW) testing to corroborate the spatial distribution of corrosion estimated by GPR.

8.5. Tunneling Inspection

GPR measurements in tunnels are generally conducted using 250 MHz, 500 MHz and 1 GHz center frequency antennas according to the requirements of resolution and depth. When detecting the backfill grouting (grout mortar is commonly injected between the inner lining and bedrock), low frequencies (200–100 MHz) are recommended to better penetrate reinforcement, although results show a poor resolution. However, high frequencies (from 1 to 2 GHz) have shallower detecting depth, leading to a near-surface survey (e.g., detecting rebar and defects or damage in lining). Both air-coupled and ground-coupled systems can be used in tunnel diagnostics.

Generally, tunnels are inspected manually, by maneuvering the GPR antenna over the surface of the tunnel, with single ground-coupled antennas using the mode of continuous acquisition, which is very slow and inefficient. In the process of tunnel inspection, longitudinal survey lines are most commonly arranged, with a range of three to six serial lines located at the vault, the left hance, the right hance, the left sidewall, the right sidewall and the inflected arch. It should be mentioned here that the connection with external GPS devices for data referencing is obviously limited when surveying the interior of a tunnel structure. Therefore, wheel encoders are used during data acquisition to ensure the accuracy of ranging (trace interval) and location. Typical trace interval distances are: 5.0 cm (< 200 MHz), 2.5 cm (250 MHz), 2.0 cm (500 MHz) and 1.0 cm (> 1 GHz). So far, the following practical problems that typically occur are: (1) loss of contact between the antenna and the surface (the operator should ensure both the antenna and the survey wheel are in contact with the surface, while keeping a constant (uniform speed) and continuous rotation); (2) deviation of the antenna with respect to the radar line (the operator should ensure the antenna position is consistent with the location of the acquisition line); and (3) the presence of cables and conduits on the walls of the tunnel that makes it impossible to collect data in those areas.

Due to the shortcomings of existing techniques using manual (hand-held) antennas or even hydraulic supporting systems (mainly a vehicle bracket with an arm or a lifter to reach the top [188,191–193,195,328]), new GPR devices have been developed for higher speed data collection and complete cross section scanning. In this regard, Balaguer et al. [329] describe the current trends towards fully automated tunnel inspection, and introduce new technologies such as robotic platforms, control algorithms and decision-making strategies. Zan et al. [197] develop a train-mounted GPR system with multi-channel air-launched antennas, and absolute positioning technology using GPS and GIS systems, in order to not interrupt normal railway operation. In this system, recording data is controlled by a distance sensor pulse, allowing for data collection to be fully automated. Using six sets of 300 MHz air-launched antennas forming a full-section detecting system, with a trace-interval of 5 cm, the maximum test speed can reach up to 175 km/h. A set of machinery was designed by Xie et al. [330] consisting of arc-tracks connected to a shield tunnel machine in order to detect grouting in real time during the construction phase. Both 500 MHz bowtie and Vivaldi antennas were used, which automatically move along the tracks.

As previously mentioned, the detection efficiency is quite low when using ground-coupled antennas, as they must be put close to the tunnel wall or roofs, which usually requires traffic disruption due to slower survey speeds. Conversely, air-launched antennas can be mounted on a vehicle with a certain distance to the wall. However, due to the diffusion attenuation, some energy is lost in the air and the probing depth is shallower than ground-coupled antennas. The American SHRP-2 report S2-R06G-RR-1 entitled "Mapping voids, debonding, delamination, moisture, and other defects behind or within tunnel linings" [331] recommends the use of air-coupled GPR antennas for detecting areas of high moisture or low density, whereas ground-coupled antennas are recommended for detecting defects at different cover depths within or behind the tunnel linings.

The extraction of information from the GPR data is often not a simple process. In order to obtain a more distinct image, the radargrams are therefore enhanced using post-processing filters. Most commonly, minimal filters are applied to ensure the minimal loss of data. Time-zero is firstly corrected as the actual detection surface and the DC component is then removed to normalize voltage. The dewow filter, band pass filter and deconvolution are vertical (temporal) filters generally applied to remove DC. Gain is applied afterwards to equalize the signal amplitude (e.g., linear and exponential gain or automatic gain control) and, lastly, horizontal (spatial) filtering (e.g., subtracting average, background removal or running average) can eliminate the noise or uncertain fluctuations in a given window.

Lyu et al. [332] apply the reverse-time migration (RTM) algorithm to improve the detection accuracy of tunnel lining cavities. Compared with the Kirchhoff migration, the RTM can better focus the energy of the diffraction wave on the position of the reflector and reduce clutter (random scattering and multiple reflections from reinforcement), which greatly improves data interpretation.

However, there are some negative factors adversely affecting the GPR detection. Firstly, there are so many noisy signals from utilities (e.g., power cables) or metal (reinforcement) in the tunnel. Shotcrete-containing steel fibers cannot be inspected because fibers generate random electromagnetic scattering. Secondly, detecting grouting thickness requires enough dielectric permittivity contrast between the grout and the soil to easily detect the boundary between layers. In this regard, Zhang et al. [189] report that contrast in the dielectric constant between the grout and the soil appeared to be sufficiently large fourteen days after the grouting treatment.

In complex structures such as tunnels, the comparison of measured field data with synthetic data has been proved to be a useful assisting tool for data interpretation. Numerical simulations are set up to predict the performance of the radar-wave and to recognize typical reflection patterns, which allows one to obtain an accurate interpretation more easily. There are some noteworthy studies showing that additional forward modelling has increased the reliability of GPR results when inspecting internal rebar location and possible damage, or even thickness of segments and grouting [189,194–196].

The GPR method was also applied in combination with other NDT methods in tunnel diagnostics. Karlovšek et al. [333] include an overview of the non-destructive methods used for possible testing bored tunnels' integrity (such as GPR, impact-echo, ultrasounds, etc.), in which their pros and cons are briefly discussed. Cardarelli et al. [202] present an integrated study using GPR, seismic refraction and seismic transmission tomography to investigate tunnel stability. The interpretation of integrated data from GPR and seismic tomography allowed for the detection of discontinuities in the rock (most likely associated to open fractures), while the integration of GPR and seismic refraction is appropriate to identify the loosened zone around the tunnel. Abraham and Dérobert [186] successfully use GPR and seismic refraction to detect damaged zones and specific material properties in fired tunnel walls. Xie and Zeng [193] combine GPR and 3D laser scanning for the evaluation of a shielded tunnel condition, which demonstrates that GPR leads to the detection of cavities behind lining, while laser scanning allows testing on surface situations of tunnel liners (e.g., water seepage).

9. Final Remarks and Future Perspectives

The use of GPR for transport infrastructure assessment is increasingly trustworthy for several purposes as already addressed in this paper. The reliability of the results is enabling a more confident application of GPR together with complementary non-destructive tests for the definition of key performance indicators (KPIs) in order to support the maintenance decisions of transport infrastructure administrations [259,260,263].

Some researchers are focused on the design of innovative devices that facilitates surveying in complex or inaccessible structures such as columns, walls and roofs. One example is the device designed and tested by Hugenschmidt et al. [117] that consists of a support for the antenna in rails moved with an electric motor. The assessment of tall structures, in some cases with difficult accessibility, requires the development of gadgets with the objective of rapid surveying while assuring accurate data acquisition and precise antenna position. In this way, in some cases, the use of drones has been also proposed and tested. The common application consists of the combination of a usual GPR survey with a photogrammetric study by means of an UAV and other techniques [266,318]. However, some authors propose prototypes to mount the GPR on an UAV [334]. This approach requires the design of novel portable antennas [335]. Some proposed devices provide promising results, although the antennas could be improved in order to enhance the signal-to-noise ratio, and additional tests are therefore needed to validate the results and to calibrate the equipment [336]. Further progress could be focused on the development of antennas and UAV systems for surveying structures or parts of the structures with poor accessibility (roofs, tunnels, columns, etc.) and large structures (airports and roads). Data acquisition using an UAV could be used for the surveying of punctual zones of the structures, and combined with more traditional surveys (common offsets, CMP or WARR). Moreover, additional tests could be needed to validate and compare the results with those obtained by using common ground-coupled or air-coupled antennas.

Other interesting field of research is the combination of GPR, and other NDT, with robotic platforms for autonomous surveys. Valuable proposals include a climbing robot supporting a GPR and a camera for the inspection of damage in walls [337,338] or robots specifically designed for disaster management that could integrated different sensors (GPR, LiDAR, cameras, etc.) on the platform [339]. The proposal of a big robotic platform with a large number of sensors for autonomous pavement assessment [340] opens an interesting line of research in the field of structures assessment. The development of those types of devices requires studies for the integration of different sensors in the robotic system and the progress on data fusion to obtain a complete and accurate inspection. An important challenge in this field is the design of low weight equipment and fast and robust protocols of communication, allowing for the visualization of data and interpreted models in real time. The design of antennas to be mounted on robotic platforms for the assessment of zones with difficult accessibility or landmine detection [341] is crucial in the advance of combining cybernetics with NDT surveys. Moreover, most of the ongoing works are tested in simple cases such as detecting rebars or metallic sheets. Most likely, robust and higher resolution antenna, as well as precise sensors for positioning might be developed and tested in different environments.

As has already been referred to, the GPR data processing is time-consuming and still a challenge. There is a trend in using artificial intelligence (AI) in automatic processing and detection instead of the common detailed trace-to-trace processing and subjective data interpretation. During the last decade, artificial neural networks associated with machine learning techniques have been successfully applied to improve the GPR processing [342], namely, the inversion approach to locate subsurface targets [343,344], reconstructing high-quality relative permittivity maps of tunnel lining [345], rebar detection [346], and railway ballast diagnosis through unsupervised processing [347,348].

Innovative technologies and modelling, such as LiDAR and BIM (building information modelling), when combined with GPR, thermography and electromagnetic location, are critical for transport infrastructures subsurface diagnosis, even more significant in dense urban areas [349]. Highly reliable 2D and 3D mapping became possible in this way. Thus, providing an "accurate" view of the occupied subsurface space, disruptions can be prevented in the case of new constructions, consequently improving safety and accuracy of excavation sites [350,351].

Research financed by H2020 also emphasize the importance of this matter, with the following ongoing projects:

- GEOFIT [352] aims to develop a compact GPR for mapping the underground with an accurate positioning measurement device (e.g., GNSS). Regarding the processing, a retrofitting approach is studied, namely to search for meaningful features in the GPR data. For this purpose, based on the library of patterns of interest developed in this project, the pattern that best fits in the GPR image is found. In this way, a higher level of automation will be achieved through automatic recognition of objects and patterns. An additional objective of GEOFIT is to integrate models of retrofitted buildings and construction sites (as-built BIM) with models of the underground situation.
- Asset4Rail [353] aims "to developing a set of cost efficient and cutting-edge asset-specific measuring and monitoring devices ... The information collected by such devices will then be processed to generate relevant maintenance infrastructure-related information to support asset management decision." One of the main outputs expected from the project is to develop a new product for the NDT inspection of tunnel lining based on GPR, LiDAR and thermal cameras, as well as to develop and validate the drone inspection of tunnels and bridges. Additionally, a future trend is to integrate the monitoring data into BIM models, mainly for tunnels and bridges, enabling maintenance decisions function. "The resulting integrated BIM environment will contain all current capabilities - a 3D BIM model carrying geometry, design information, quantities etc., together with relevant linked documents, 5D costs, 6D maintenance plans and similar – as well as a new capability for sensory readings, both real-time and historical, enabling this data to be displayed side-by-side with all other relevant asset and maintenance information, 3D model data and properties."
- IM-SAFE [354] aims to support the European Commission (EC) and the European Committee for Standardization (CEN) to prepare a new standard in monitoring for optimal maintenance and safety of transport infrastructure based on consolidated and accepted knowledge and experience in the EU and worldwide. Aligned with the topic of this article, IM-SAFE includes the review of surveying technologies used in the condition evaluation and diagnosis of bridges and tunnels (satellite imaging, LiDAR, NDT active and passive testing technologies (such as GPR among others)). Moreover, procedures will be adopted for determining damage detection indicators and actions on structures in risk and safety analysis based on condition survey data, as well as procedures for data quality assurance and digitalization (use of BIM, predictive twin and other digital innovations).

To sum up, all the contributions herein presented have proven the benefits of the wide application of GPR to on-site diagnosis, integration with complementary tests and geospatial visualization and management modelling. Nevertheless, there are still open issues mainly focused on the development of algorithms for more efficient data processing and interpretation, in addition to new systems and devices improving data acquisition and quality. Furthermore, digital innovations are expected in a near future, including the integration of GPR data into a GIS/BIM environment, augmented reality and predictive modelling in order to support industry 4.0.

Supplementary Materials: The following are available online at https://www.mdpi.com/2072-429 2/13/4/672/s1, Table S1: Roads: relevant on-site GPR surveys, Table S2: Airports: relevant on-site GPR surveys, Table S3: Railways: relevant on-site GPR surveys, Table S4: Retaining walls: relevant on-site GPR surveys, Table S5: Bridges: relevant on-site GPR surveys, Table S6: Tunneling: relevant on-site GPR surveys.

Author Contributions: M.S., V.P.-G. and S.F. conceived, designed and wrote this review article. All authors have read and agreed to the published version of the manuscript.

Funding: This project has received funding from the European Union's Horizon 2020 research and innovation program under grant agreement No 958171. M.S. acknowledges financial support from the Spanish Ministry of Science and Innovation for a "Ramón y Cajal" contract (RYC2019–026604–I / AEI / 10.13039/501100011033).

Institutional Review Board Statement: Not applicable.

Informed Consent Statement: Not applicable.

Data Availability Statement: Not applicable.

Acknowledgments: The authors would like to thank the Applied Geotechnologies (Geotech) research group of the Universidade de Vigo (Spain), the Geophysics and Earthquake Engineering (GIES) research group of the Universitat Politècnica de Catalunya (Spain), and the Transport Department of the National Laboratory for Civil Engineering (LNEC) in Lisbon (Portugal) to provide the graphic material and results included in this article.

Conflicts of Interest: The authors declare no conflict of interest. The funders had no role in the design of the study; in the collection, analyses, or interpretation of data; in the writing of the manuscript, or in the decision to publish the results.

References

1. Annan, P. *GPR Principles, Procedures & Applications*; Sensors and Software Inc.: Mississauga, ON, Canada, 2003; p. 278.
2. Daniels, D.J. *Ground Penetrating Radar*; The institution of Electrical Engineers: London, UK, 2004; p. 726.
3. Jol, H.M. *Ground Penetrating Radar: Theory and Applications*; Elsevier Science: Amsterdam, The Netherlands, 2009; p. 544.
4. ASTM International. *ASTM D6432-19, Standard Guide for Using the Surface Ground Penetrating Radar Method for Subsurface Investigation*; ASTM International: West Conshohocken, PA, USA, 2019; Available online: www.astm.org (accessed on 11 February 2021). [CrossRef]
5. Department for Transport, Highway Agency. *DMRB 3.1.7.: Design Manual for Roads and Bridges, Advice Notes on the Non-Destructive Testing of Highway Structures—Advice Note 3.5 BA 86/2006: Ground Penetrating Radar (GPR), UK*; Department for Transport, Highway Agency: Birmingham, UK, 2006; ISBN 0115527788.
6. Highway Agency. *DMRB 7.3.2.: Design Manual for Roads and Bridges, Data for Pavement Assessment—Annex 6 HD 29/2008: Ground-Penetrating Radar (GPR), UK*; Highway Agency: Birmingham, UK, 2008.
7. European GPR Association. *GS1601: The European GPR Association Guidelines for Pavement Structural Surveys*; European GPR Association, 2016.
8. Comitato Elettrotecnico Italiano. *CEI 306-8: Impiego del Radar per Introspezione del Suolo per Prospezioni ad Opera di Posa di Servizi ed Infrastrutture Sotterranee*; Comitato Elettrotecnico Italiano: Milano, MI, USA, 2004.
9. Morey, R.M. Continuous subsurface profiling by impulse radar. In *Proceedings of the Conference on Subsurface Exploration for Underground Excavation and Heavy Construction, Henniken, NH, United States, 11–16 August 1974*; American Society of Civil Engineers: Reston, VA, USA, 1974; pp. 213–232.
10. Morey, R.M. Detection of subsurface cavities by ground penetrating radar. *Highw. Geol. Symp.* **1976**, *27*, 28–30.
11. Caldecott, R.; Poirier, M.; Scofea, D.; Svoboda, D.E.; Terzuoli, A.J. Underground mapping of utility lines using impulse radar. *Inst. Electr. Eng. Proc. F Commun. Radar Signal Process.* **1988**, *135*, 343–361. [CrossRef]
12. Osumi, N.; Ueno, K. Detection of buried plant. *Inst. Electr. Eng. Proc. F Commun. Radar Signal Process.* **1988**, *135*, 330–342. [CrossRef]

13. Hironaka, M.C.; Hitchcock, R.D.; Forrest, J.B. *Detection of Voids Underground and Under Pavements*; Report No. CEL-TN-1449; Naval Civil Engineering Laboratory: Port Hueneme, CA, USA, 1976.

14. Kovacs, A.; Morey, R.M. *Detection of Cavities under Concrete Pavement*; Report No. CRREL 83-18; Cold Regions Research and Engineering Laboratory, Department of the Army: Hanover, Germany, 1983.

15. Benedetto, A.; Pajewski, L. *Civil Engineering Applications of Ground Penetrating Radar*; Springer Transactions in Civil and Environmental Engineering; Springer International: New York, NY, USA, 2015. [CrossRef]

16. Wai-Lok Lai, W.; Dérobert, X.; Annan, P. A Review of Ground Penetrating Radar Application in Civil Engineering: A 30-Year Journey from Locating and Testing to Imaging and Diagnosis. *NDT E Int.* **2018**, *96*, 58–78. [CrossRef]

17. Pajewski, L.; Fontul, S.; Solla, M. Ground-penetrating radar for the evaluation and monitoring of transport infrastructures. In *Innovation in Near-Surface Geophysics. Instrumentation, Application, and Data Processing Methods*; Elsevier: Amsterdam, The Netherlands, 2019. [CrossRef]

18. FORMAT poject "Fully Optimised Road Maintenance". In *Assessment of High Speed Monitoring Prototype Equipment*; Deliverable Report D12; 2004.

19. Hopman, V.; Beuving, E. Repeatability, reproducibility and accuracy of GPR measurements. In Proceedings of the 6th International Conference on BCRRA, Lisbon, Portugal, 24–26 June 2002; Branco, C., Ed.; Balkema: Lisbon, Portugal, 2002.

20. Highways Agency. *Pavement Design and Maintenance. Pavement Maintenance Assessment. Structural Assessment Methods*; DMRB Volume 7 Section 3 Part 2 (HD 29/94); The Stationary Office: London, UK, 2001.

21. Lorenzo, H.; Rial, F.I.; Pereira, M.; Solla, M. A full non-metallic trailer for GPR road surveys. *J. Appl. Geophys.* **2011**, *75*, 490–497. [CrossRef]

22. Plati, C.; Loizos, A. Using ground-penetrating radar for assessing the structural needs of asphalt pavements. *Nondestruct. Test. Eval.* **2012**, *27*, 273–284. [CrossRef]

23. AASHTO. Part 2. Chapter 5. Evaluation of Existing Pavements for Rehabilitation. In *National Cooperative Highway Research Program 1-37A: Guide for Mechanistic-Empirical Design of New and Rehabilitated Pavement Structures*; AASHTO: Washington, DC, USA, 2004.

24. Maser, K.; Scullion, T.; Briggs, R.C. *Use of Radar Technology for Pavement Layer Evaluation*; Research Report 0930-SF; Research Study Number 2-18-88-930; Texas Transportation Institute, Texas University: Austin, TX, USA, 1991.

25. Al-Qadi, I.L.; Hazim, O.A.; Su, W.; Riad, S.M. Dielectric properties of Portland cement concrete at low radio frequencies. *J. Mater. Civ. Eng.* **1995**, *7*, 192–198. [CrossRef]

26. Tannous, B.S. Investigation of Electrical Properties of Earth Materials by Ground Penetrating Radar. Retrospective Theses and Dissertations. Master's Thesis, University of Central Florida, Oviedo, FL, USA, 1987; p. 5082. Available online: https://stars.library.ucf.edu/rtd/5082 (accessed on 11 February 2021).

27. Robert, A. Dielectric permittivity of concrete between 50 MHz and 1 GHz and GPR measurements for building materials evaluation. *J. Appl. Geophys.* **1998**, *40*, 89–94. [CrossRef]

28. Saarenketo, T. Using ground-penetrating radar and dielectric probe measurements in pavement density quality control. *Transp. Res. Rec.* **1997**, *1575*, 34–41. [CrossRef]

29. Maser, K.R. *Ground Penetrating Radar Surveys to Characterize Pavement Layer Thickness Variations at GPS Sites (No. SHRP-P-397)*; Strategic Highway Research Program, National Research Council: Washington, DC, USA, 1994.

30. Gordon, M.O.; Broughton, K.; Hardy, M.S.A. The assessment of the value of GPR imaging of flexible pavements. *NDT E Int.* **1998**, *31*, 429–438. [CrossRef]

31. Maser, K.R.; Scullion, T. *Automated Pavement Subsurface Profiling Using Radar: Case Studies of Four Experimental Field Sites*; Transportation Research Record 1344; Texas Transportation Institute, Texas A&M University: Bryan, TX, USA, 1992.

32. Maser, K.R. Condition assessment of transportation infrastructure using ground-penetrating radar. *J. Infrastruct. Syst.* **1996**, *2*, 94–101. [CrossRef]

33. Hugenschmidt, J.; Partl, M.N.; De Witte, H. GPR inspection of a mountain motorway in Switzerland. *J. Appl. Geophys.* **1998**, *40*, 95–104. [CrossRef]

34. Grote, K.; Hubbard, S.; Harvey, J.; Rubin, Y. Evaluation of infiltration in layered pavements using surface GPR reflection techniques. *J. Appl. Geophys.* **2005**, *57*, 129–153. [CrossRef]

35. Benedetto, A. Water content evaluation in unsaturated soil using GPR signal analysis in the frequency domain. *J. Appl. Geophys.* **2010**, *71*, 26–35. [CrossRef]

36. Loizos, A.; Plati, C. Accuracy of pavement thicknesses estimation using different ground penetrating radar analysis approaches. *NDT E Int.* **2007**, *40*, 147–157. [CrossRef]

37. Varela-González, M.; Solla, M.; Martínez-Sánchez, J.; Arias, P. A semi-automatic processing and visualisation tool for ground-penetrating radar pavement thickness data. *Autom. Constr.* **2014**, *45*, 42–49. [CrossRef]

38. Lahouar, S.; Al-Qadi, I.L. Automatic detection of multiple pavement layers from GPR data. *NDT E Int.* **2008**, *41*, 69–81. [CrossRef]

39. Wang, S.; Zhao, S.; Al-Qadi, I.L. Continuous real-time monitoring of flexible pavement layer density and thickness using ground penetrating radar. *NDT E Int.* **2018**, *100*, 48–54. [CrossRef]

40. Solla, M.; Lagüela, S.; González-Jorge, H.; Arias, P. Approach to identify cracking in asphalt pavement using GPR and infrared thermographic methods: Preliminary findings. *NDT E Int.* **2014**, *62*, 55–65. [CrossRef]

41. Rasol, M.A.; Pérez-Gracia, V.; Fernandes, F.M.; Pais, J.C.; Santos-Assunçao, S.; Santos, C.; Sossa, V. GPR laboratory tests and numerical models to characterize cracks in cement concrete specimens, exemplifying damage in rigid pavement. *Measurement* **2020**, *158*, 107662. [CrossRef]

42. Rasol, M.A.; Pérez-Gracia, V.; Solla, M.; Pais, J.C.; Fernandes, F.M.; Santos, C. An experimental and numerical approach to combine Ground Penetrating Radar and computational modelling for the identification of early cracking in cement concrete pavements. *NDT E Int.* **2020**, *115*, 102293. [CrossRef]

43. Benedetto, A.; Tosti, F.; Ortuani, B.; Giudici, M.; Mele, M. Mapping the spatial variation of soil moisture at the large scale using GPR for pavement applications. *Near Surf. Geophys.* **2015**, *13*, 269–278. [CrossRef]

44. Fernandes, F.M.; Fernandes, A.; Pais, J. Assessment of the density and moisture content of asphalt mixtures of road pavements. *Constr. Build. Mater.* **2017**, *154*, 1216–1225. [CrossRef]

45. Zhang, J.; Yang, X.; Li, W.; Zhang, S.; Jia, Y. Automatic detection of moisture damages in asphalt pavements from GPR data with deep CNN and IRS method. *Autom. Constr.* **2020**, *113*, 103119. [CrossRef]

46. Zhao, S.; Al-Qadi, I. Pavement drainage pipe condition assessment by GPR image reconstruction using FDTD modeling. *Constr. Build. Mater.* **2017**, *154*, 1283–1293. [CrossRef]

47. Pedret-Rodes, J.; Martínez-Reguero, A.; Pérez-Gracia, V. GPR Spectra for Monitoring Asphalt Pavements. *Remote Sens.* **2020**, *12*, 1749. [CrossRef]

48. Rodés, J.P.; Perez-Gracia, V.; Martínez-Reguero, A. Evaluation of the GPR frequency spectra in asphalt pavement assessment. *Constr. Build. Mater.* **2015**, *96*, 181–188. [CrossRef]

49. Liu, H.; Deng, Z.; Han, F.; Xia, Y.; Liu, Q.H.; Sato, M. Time-frequency analysis of air-coupled GPR data for identification of delamination between pavement layers. *Constr. Build. Mater.* **2017**, *154*, 1207–1215. [CrossRef]

50. ASTM. *Standard Test Method for Determining the Thickness of Bound Pavement Layers Using Short-Pulse Radar*; Non-destructive testing of pavement structures; ASTM D4748; ASTM: West Conshohocken, PA, USA, 2004.

51. Saarenketo, T. *Recommendations for guidelines for the use of GPR in asphalt air voids content measurement. Mara Nord Project*; Europeiska Unionen: Brussels, Belgium, 2012.

52. Pajewski, L.; Vrtunski, M.; Bugarinovic, Ž.; Ristic, A.; Govedarica, M.; Van der Wielen, A.; Grégoire, C.; Van Geem, C.; Dérobert, X.; Borecky, V.; et al. GPR system performance compliance according to COST Action TU1208 guidelines. *Ground Penetrating Radar* **2018**, *1*, 104–122. [CrossRef]

53. ASTM. ASTM D6087-08(2015)e1. In *Standard Test Method for Evaluating Asphalt-Covered Concrete Bridge Decks Using Ground Penetrating Radar*; ASTM International: West Conshohocken, PA, USA, 2015; Available online: www.astm.org (accessed on 11 February 2021). [CrossRef]

54. Silva, L.A.; Borges, W.R.; Cunha, L.S.; Branco, M.G.C.; Farias, M.M. Use of GPR to identify metal bars and layer thickness in a rigid pavement. *Geotech. Geophys. Site Charact.* **2013**, *4*, 1341–1346.

55. Al-Qadi, I.L.; Lahouar, S. Measuring rebar cover depth in rigid pavements with ground-penetrating radar. *Transp. Res. Rec.* **2005**, *1907*, 80–85. [CrossRef]

56. Rasol, M.A.; Pérez-Gracia, V.; Fernandes, F.M.; Pais, J.C.; Solla, M.; Santos, C. NDT assessment of rigid pavement damages with ground penetrating radar: Laboratory and field tests. *Int. J. Pavement Eng.* **2020**, 1–16. [CrossRef]

57. Alani, A.M.; Lantini, L. Recent Advances in Tree Root Mapping and Assessment Using Non-destructive Testing Methods: A Focus on Ground Penetrating Radar. *Surv. Geophys.* **2020**, *41*, 605–646. [CrossRef]

58. Vafidis, A.; Economou, N.; Dimitriadis, K. Time varying zero-phase filtering of GPR data for imaging pavement layers. In Proceedings of the 73rd EAGE Conference and Exhibition Incorporating SPE EUROPEC 2011 (pp. cp-238), Vienna, Austria, 23–27 May 2011; European Association of Geoscientists & Engineers: Houten, The Netherlands, 2011. [CrossRef]

59. Grégoire, C.; Van Geem, C. Use of radar in road investigation BRRC experience. In Proceedings of the 7th International Workshop on Advanced Ground Penetrating Radar, Nantes, France, 2–5 July 2013; IEEE: New York, NY, USA, 2013. [CrossRef]

60. Al-Qadi, I.L.; Lahouar, S. Measuring layer thicknesses with GPR–Theory to practice. *Constr. Build. Mater.* **2005**, *19*, 763–772. [CrossRef]

61. Le Bastard, C.; Baltazart, V.; Wang, Y.; Saillard, J. Thin-pavement thickness estimation using GPR with high-resolution and superresolution methods. *IEEE Trans. Geosci. Remote Sens.* **2007**, *45*, 2511–2519. [CrossRef]

62. Loulizi, A.; Al-Qadi, I.L.; Lahouar, S. Optimization of ground-penetrating radar data to predict layer thicknesses in flexible pavements. *J. Transp. Eng.* **2003**, *129*, 93–99. [CrossRef]

63. Liu, H.; Sato, M. In situ measurement of pavement thickness and dielectric permittivity by GPR using an antenna array. *NDT E Int.* **2014**, *64*, 65–71. [CrossRef]

64. Stryk, J. Road diagnostics-ground penetrating radar possibilities. *Intersectii/Intersect.* **2008**, *5*, 48–57.

65. Diamanti, N.; Redman, D. Field observations and numerical models of GPR response from vertical pavement cracks. *J. Appl. Geophys.* **2012**, *81*, 106–116. [CrossRef]

66. Krysiński, L.; Sudyka, J. GPR abilities in investigation of the pavement transversal cracks. *J. Appl. Geophys.* **2013**, *97*, 27–36. [CrossRef]

67. Fernandes, F.M.; Pais, J.C. Laboratory observation of cracks in road pavements with GPR. *Constr. Build. Mater.* **2017**, *154*, 1130–1138. [CrossRef]

68. Torbaghan, M.E.; Li, W.; Metje, N.; Burrow, M.; Chapman, D.N.; Rogers, C.D. Automated detection of cracks in roads using ground penetrating radar. *J. Appl. Geophys.* **2020**, *179*, 104118. [CrossRef]

69. Saarenketo, T.; Scullion, T. Road evaluation with ground penetrating radar. *J. Appl. Geophys.* **2000**, *43*, 119–138. [CrossRef]

70. Al-Qadi, I.L.; Lahouar, S.; Loulizi, A. Successful application of ground-penetrating radar for quality assurance-quality control of new pavements. *Transp. Res. Rec.* **2003**, *1861*, 86–97. [CrossRef]

71. Saarenketo, T. *Measuring Electromagnetic Properties of Asphalt for Pavement Quality Control and Defect Mapping*; Roadscanners: Ravaniemi, Finland, 2009.

72. Poikajärvi, J.; Peisa, K.; Herronen, T.; Aursand, P.O.; Maijala, P.; Narbro, A. GPR in road investigations–equipment tests and quality assurance of new asphalt pavement. *Nondestruct. Test. Eval.* **2012**, *27*, 293–303. [CrossRef]

73. Grote, K.; Hubbard, S.; Rubin, Y. GPR monitoring of volumetric water content in soils applied to highway construction and maintenance. *Lead. Edge* **2002**, *21*, 482–504. [CrossRef]

74. Plati, C.; Loizos, A. Estimation of in-situ density and moisture content in HMA pavements based on GPR trace reflection amplitude using different frequencies. *J. Appl. Geophys.* **2013**, *97*, 3–10. [CrossRef]

75. Fontul, S. Structural Evaluation of Flexible Pavements Using Non-Destructive Tests. Ph.D. Thesis, University of Coimbra, Coimbra, Portugal, 2004.

76. Song, X.; Wu, R.; Liu, J. Underground diseases identification of airport runway using GPR. In Proceedings of the Progress in Electromagnetics Research, Xi'an, China, 22–26 March 2010; Volume 93.

77. Graczyk, M.; Krysinski, L.; Topczewski, Ł.; Sudyka, J. The use of three-dimensional analysis of GPR data in evaluation of operational safety of airfield pavements. *Transp. Res. Procedia* **2016**, *14*, 3704–3712. [CrossRef]

78. Ni, J.C.; Cheng, W.C. Trial grouting under rigid pavement: A case history in Magong Airport, Penghu. *J. Test. Eval.* **2012**, *40*, 107–118. [CrossRef]

79. Doler, D.; Kovačič, B. Improved decision-making geo-information system for continuous monitoring of deformations on airport infrastructure. *Isprs Int. J. Geo-Inf.* **2019**, *8*, 1. [CrossRef]

80. Marecos, V.; Fontul, S.; Antunes, M.L.; Solla, M. Assessment of a concrete pre-stressed runway pavement with ground penetrating radar. In Proceedings of the 8th International Workshop on Advanced Ground Penetrating Radar (IWAGPR), Florence, Italy, 7–10 July 2015; pp. 1–4. [CrossRef]

81. Yi, L.; Zou, L.; Sato, M. Practical approach for high-resolution airport pavement inspection with the Yakumo multistatic array ground-penetrating radar system. *Sensors* **2018**, *18*, 2684. [CrossRef] [PubMed]

82. Borges, W.R.; da Silva, L.A.; da Cunha, L.S.; Branco, R.M.G.C.; de Farias, M.M. GPR Applied to Rigid Pavement from Santos Dumont Airport, RJ. *Braz. J. Geophys.* **2014**, *32*, 225–234. [CrossRef]

83. Mocnik, A.; Dossi, M.; Forte, E.; Zambrini, R.; Zamariolo, A.; Pipan, M. Ground Penetrating Radar applications for roads and airport pavements investigations. *Geophysics* **2015**, *56*, 951–960.

84. Fontul, S.; Antunes, M.L.; Fortunato, E.; Oliveira, M. Practical application of GPR in transport infrastructure survey. In Proceedings of the International Conference on Advanced Characterisation of Pavement and Soil Engineering Materials, Athens, Greece, 20–22 June 2007.

85. Zou, L.; Yi, L.; Sato, M. On the use of lateral wave for the interlayer debonding detecting in an asphalt airport pavement using a multistatic GPR system. *IEEE Trans. Geosci. Remote Sens.* **2020**, *58*, 4215–4224. [CrossRef]

86. Kadioglu, S. Research of the sea-filled airport OGU in the Black Sea, Turkey, using ground penetrating radar method. *Constr. Build. Mater.* **2018**, *158*, 1123–1133. [CrossRef]

87. Jørgensen, A.S.; Andreasen, F. Mapping of permafrost surface using ground-penetrating radar at Kangerlussuaq Airport, western Greenland. *Cold Reg. Sci. Technol.* **2007**, *48*, 64–72. [CrossRef]

88. Yannan, L.; Zhaoran, X. Study on the Application of GPR in the Void Detection of Airport Pavement. *J. Henan Sci. Technol.* **2015**, *13*, 38.

89. Sussman, T. Application of Ground-Penetrating Radar to Railway Track Substructure Maintenance Management. Ph.D. Thesis, University of Massachusetts Amherst, Amherst, MA, USA, 1999.

90. Basye, C.; Wilk, S.; Gao, Y. *Ground Penetrating Radar (GPR) Technology Evaluation and Implementation*; Federal Railroad Administration, Office of Research, Development and Technology: Washington, DC, USA, 2020; p. 20590.

91. Kathage, A.; Niessen, J.; White, G.; Bell, N. Fast Inspection of Railway Ballast By Means of Impulse GPR Equipped with Horn Antennas. Railway Engineering-2005, The Eighth International Conference. Available online: https://www.ndt.net/article/v10n09/kathage/kathage.htm (accessed on 11 February 2021).

92. Manacorda, G.; Morandi, D.; Sarri, A.; Staccone, G. A customized GPR system for railroad tracks verification. In Proceedings of the SPIE—The International Society for Optical Engineering, Santa Barbara, CA, USA, 12 April 2002. [CrossRef]

93. Al-Qadi, I.L.; Xie, W.; Jones, D.L.; Roberts, R. Development of a time–frequency approach to quantify railroad ballast fouling condition using ultrawide band ground-penetrating radar data. *Int. J. Pavement Eng.* **2010**, *11*, 269–279. [CrossRef]

94. Khakiev, Z.; Shapovalov, V.L.; Kruglikov, A.; Yavna, V. GPR determination of physical parameters of railway structural layers. *J. Appl. Geophys.* **2014**, *106*, 139–145. [CrossRef]

95. Fontul, S.; Fortunato, E.; De Chiara, F.; Burrinha, R.; Marco Baldeiras, M. Railways Track Characterization Using Ground Penetrating Radar. Advances in Transportation Geotechnics 3. The 3rd International Conference on Transportation Geotechnics (ICTG 2016). *Procedia Eng.* **2016**, *143*, 1193–1200. [CrossRef]

96. Plati, C.; Loizos, A.; Papavasiliou, V. Inspection of railroad ballast using geophysical method. *Int. J. Pavement Eng.* **2010**, *11*, 309–317. [CrossRef]
97. Hyslip, J.P.; Smith, S.S.; Olhoeft, G.R.; Selig, E.T. Assessment of Railway Track Substructure Condition Using Ground Penetrating Radar. In Proceedings of the Annual Conference of AREMA, Chicago, IL, USA, 5–7 October 2003.
98. Xiao, J.; Liu, L. Permafrost Subgrade Condition Assessment Using Extrapolation by Deterministic Deconvolution on Multifrequency GPR Data Acquired Along the Qinghai-Tibet Railway. *IEEE J. Sel. Top. Appl. Earth Obs. Remote Sens.* **2016**, *9*, 83–90. [CrossRef]
99. Saarenketo, T.; Silvast, M.; Noukka, J. Using GPR on Railways to Identify Frost Susceptible Areas. Proceedings of 6th International Conference on Railway Engineering, London, UK, 3 April–1 May 2003; p. 11.
100. Vorster, D.J.; Gräbe, P.J. The use of ground-penetrating radar to develop a track substructure characterisation model. *J. S. Afr. Inst. Civ. Eng.* **2013**, *55*, 69–78.
101. Leng, Z.; Al-Qadi, I.L. Railroad Ballast Evaluation Using Ground-Penetrating Radar: Laboratory Investigation and Field Validation. *Transp. Res. Rec. J. Transp. Res. Board* **2010**, *2159*, 110–117. [CrossRef]
102. De Chiara, F. Improving of Railway Track Diagnosis Using Ground Penetrating Radar. Ph.D. Thesis, Sapienza University of Rome, Rome, Italy, 2014.
103. De Bold, R.; O'connor, G.; Morrissey, J.P.; Forde, M. Benchmarking large scale GPR experiments on railway ballast. *Constr. Build. Mater.* **2015**, *92*, 31–42. [CrossRef]
104. Roberts, R.; Schutz, A.; Al-Qadi, I.L.; Tutumluer, E.; Boyle, J. Characterizing Railroad Ballast Using GPR: Recent Experiences in the United States. In Proceedings of the 2007 4th International Workshop on Advanced Ground Penetrating Radar, IWAGPR, Naples, Italy, 27–29 June 2007. [CrossRef]
105. Fernandes, F.; Pereira, M.; Gomes Correia, A.; Lourenço, P.; Caldeira, L. Assessment of layer thickness and uniformity in railway embankments with ground penetrating radar. In *Adv. Transp. Geotech.*; Ellis, E., Thom, N., Yu, H.-S., Dawson, A., McDowell, G., Eds.; CRC Press: Boca Raton, FL, USA, 2008; pp. 571–575.
106. Bianchini Ciampoli, L.; Calvi, A.; D'Amico, F. Railway Ballast Monitoring by GPR: A Test Site Investigation. *Remote Sens.* **2019**, *11*, 2381. [CrossRef]
107. Benedetto, F.; Tosti, F.; Alani, A.M. An Entropy-Based Analysis of GPR Data for the Assessment of Railway Ballast Conditions. *IEEE Trans. Geosci. Remote Sens.* **2017**, *55*, 3900–3908. [CrossRef]
108. Solla, M.; Fontul, S. Non-destructive tests for railway evaluation: Detection of fouling and joint interpretation of GPR data and track geometric parameters. *Ground Penetrating Radar* **2018**, *1*, 75–103. [CrossRef]
109. Tosti, F.; Bianchini Ciampoli, L.; Calvi, A.; Alani, A.M.; Benedetto, A. An investigation into the railway ballast dielectric properties using different GPR antennas and frequency systems. *NDT E Int.* **2018**, *93*, 131–140. [CrossRef]
110. Su, L.J.; Indraratna, B.; Rujikiatkamjorn, C. Non-destructive assessment of rail track condition using ground penetrating radar. In *Proceedings of the 13th International Conference of the International Association for Computer Methods and Advances in Geomechanics, Melbourne, Australia, 9–11 May 2011*; Khalili, N., Oeser, M., Eds.; Centre for Infrastructure Engineering and Safety: Sydney, Australia, 2011; pp. 478–482.
111. Ciampoli, L.B.; Artagan, S.S.; Tosti, F.; Gagliardi, V.; Alani, A.M.; Benedetto, A. A comparative investigation of the effects of concrete sleepers on the GPR signal for the assessment of railway ballast. In Proceedings of the 17th International Conference on Ground Penetrating Radar (GPR), Rapperswil, Switzerland, 18–21 June 2018; pp. 1–4. [CrossRef]
112. Colla, C.; Krause, M.; Maierhofer, C.; Hohberger, J.J.; Simmer, H. Combination of NDT techniques for site investigation of non-ballasted railway tracks. *NDT E Int.* **2002**, *35*, 95–105. [CrossRef]
113. Huang, J.; Su, Q.; Liu, T.; Wang, W. Behavior and Control of the Ballastless Track-Subgrade Vibration Induced by High-Speed Trains Moving on the Subgrade Bed with Mud Pumping. *Hindawishock Vib.* **2019**, *2019*, 1–14. [CrossRef]
114. Brutus, O.; Tauber, G. *Guide to Asset Management of Earth Retaining Structures*; US Department of Transportation, Federal Highway Administration, Office of Asset Management: Washington, DC, USA, 2009; pp. 1–120.
115. Huston, D.R.; Pelczarski, N.V.; Esser, B. Inspection of bridge columns and retaining walls with electromagnetic waves. In Proceedings of the Smart Structures and Materials, Smart Systems for Bridges, Structures, and Highways, Newport Beach, CA, USA, 30 July 2001; International Society for Optics and Photonics: Bellingham, WA, USA, 2001; Volume 4330, pp. 66–76.
116. Hugenschmidt, J.; Kalogeropoulos, A. The inspection of retaining walls using GPR. *J. Appl. Geophys.* **2009**, *67*, 335–344. [CrossRef]
117. Hugenschmidt, J.; Kalogeropoulos, A.; Soldovieri, F.; Prisco, G. Processing strategies for high-resolution GPR concrete inspections. *NDT E Int.* **2010**, *43*, 334–342. [CrossRef]
118. Beben, D.; Anigacz, W.; Ukleja, J. Diagnosis of bedrock course and retaining wall using GPR. *NDT E Int.* **2013**, *59*, 77–85. [CrossRef]
119. Ukleja, J.; Bęben, D.; Anigacz, W. Determination of the railway retaining wall dimensions and its foundation in difficult terrain and utility. *Agh J. Min. Geoengin.* **2012**, *36*, 299–308.
120. Solla, M.; González-Jorge, H.; Álvarez, M.X.; Arias, P. Application of non-destructive geomatic techniques and FDTD modeling to metrical analysis of stone blocks in a masonry wall. *Constr. Build. Mater.* **2012**, *36*, 14–19. [CrossRef]
121. Santos-Assunçao, S.; Dimitriadis, K.; Konstantakis, Y.; Perez-Gracia, V.; Anagnostopoulou, E.; Gonzalez-Drigo, R. Ground-penetrating radar evaluation of the ancient Mycenaean monument Tholos Acharnon tomb. *Near Surf. Geophys.* **2016**, *14*, 197–205. [CrossRef]

122. McCann, D.M.; Forde, M.C. Review of NDT methods in the assessment of concrete and masonry structures. *NDT E Int.* **2001**, *34*, 71–84. [CrossRef]
123. Bhandari, N.M.; Kumar, P. Structural health monitoring and assessment of masonry arch bridges. Proceedings of National Conference on Advances in Bridge Engineering (ABE), Roorkee, India, 24–25 March 2006; pp. 115–132.
124. Orbán, Z.; Gutermann, M. Assessment of masonry arch railway bridges using nondestructive in-situ testing methods. *Eng. Struct.* **2009**, *31*, 2287–2298. [CrossRef]
125. Rehman, S.K.U.; Ibrahim, Z.; Memon, S.A.; Jameel, M. Nondestructive test methods for concrete bridges: A review. *Constr. Build. Mater.* **2016**, *107*, 58–86. [CrossRef]
126. Dabous, S.A.; Feroz, S. Condition Monitoring of Bridge Infrastructure Using Non-Contact Testing Technologies: A Comprehensive Review. *Autom. Constr.* **2020**, *116*, 103224. [CrossRef]
127. Ural, A.; Oruç, S.; Dogangün, A.; Tulik, O.I. Turkish historical arch bridges and their deteriorations and failures. *Eng. Fail. Anal.* **2008**, *15*, 43–53. [CrossRef]
128. Solla, M.; Lorenzo, H.; Novo, A.; Rial, F.I. Ground-penetrating Radar Assessment of the Medieval Arch Bridge of San Antón, Galicia, Spain. *Archaeol. Prospect.* **2010**, *17*, 223–232. [CrossRef]
129. Solla, M.; Lorenzo, H.; Rial, F.I.; Novo, A. GPR evaluation of the Roman masonry arch bridge of Lugo (Spain). *NDT E Int.* **2011**, *44*, 8–12. [CrossRef]
130. Solla, M.; Riveiro, B.; Lorenzo, H.; Armesto, J. Ancient Stone bridge surveying by ground-penetrating radar and numerical modeling methods. *J. Bridge Eng.* **2014**, *19*, 110–119. [CrossRef]
131. Solla, M.; Lorenzo, H.; Riveiro, B.; Rial, F.I. Non-destructive methodologies in the assessment of the masonry arch bridge of Traba, Spain. *Eng. Fail. Anal.* **2011**, *18*, 828–835. [CrossRef]
132. Alani, A.M.; Tosti, F.; Banks, K.; Biancini-Ciampoli, L.; Benedetto, A. Non-destructive assessment of a historic masonry arch bridge using ground penetrating radar and 3D laser scanner. In Proceedings of the IMEKO International Conference on Metrology for Archaeology and Cultural Heritage, Lecce, Italy, 23–25 October 2017.
133. Solla, M.; Asorey-Cacheda, R.; Núñez-Nieto, X.; Conde-Carnero, B. Evaluation of historical bridges through recreation of GPR models with the FDTD algorithm. *NDT E Int.* **2016**, *77*, 19–27. [CrossRef]
134. Trela, C.; Wöstmann, J.; Kruschwitz, S. Contribution of radar measurements to the inspection and condition assessment of railway bridges—Case study at a historic masonry arch bridge in Oleśnica/Poland. *Wit Trans. Built Environ.* **2008**, *97*, 535–544. [CrossRef]
135. Fauchard, C.; Antoine, R.; Bretar, F.; Lacogne, J.; Fargier, Y.; Maisonnave, C.; Guilbert, V.; Marjerie, P.; Thérain, P.F.; Dupont, J.P.; et al. Assessment of an ancient bridge combining geophysical and advanced photogrammetric methods: Application to the Pont De Coq, France. *J. Appl. Geophys.* **2013**, *98*, 100–112. [CrossRef]
136. Bergamo, O.; Campione, G.; Donadello, S.; Russo, G. In-situ NDT testing procedure as an integral part of failure analysis of historical masonry arch bridges. *Eng. Fail. Anal.* **2015**, *57*, 31–55. [CrossRef]
137. Solla, M.; Lorenzo, H.; Rial, F.I.; Novo, A. Ground-penetrating radar for the structural evaluation of masonry bridges: Results and interpretational tools. *Constr. Build. Mater.* **2012**, *29*, 458–465. [CrossRef]
138. Kalogeropoulos, A.; Brühwiler, E. High-resolution 3D condition survey of a masonry arch bridge using Ground Penetrating Radar. In Proceedings of the First Middle East Conference on Smart Monitoring, Assessment and Rehabilitation of Civil Structures, Dubai, United Arab Emirates, 8–10 February 2011.
139. Arias, P.; Armesto, J.; Di-Capua, D.; González–Drigo, R.; Lorenzo, H.; Pérez-Gracia, V. Digital photogrammetry, GPR and computational analysis of structural damages in a mediaeval bridge. *Eng. Fail. Anal.* **2007**, *14*, 1444–1457. [CrossRef]
140. Pérez-Gracia, V.; Di Capua, D.; Caselles, O.; Rial, F.; Lorenzo, H.; González-Drigo, R.; Armesto, J. Characterization of a Romanesque bridge in Galicia (Spain). *Int. J. Archit. Herit.* **2011**, *5*, 251–263. [CrossRef]
141. Sánchez-Aparicio, L.J.; Bautista-De Castro, A.; Conde, B.; Carrasco, P.; Ramos, L.F. Non-destructive means and methods for structural diagnosis of masonry arch bridges. *Autom. Constr.* **2019**, *104*, 360–382. [CrossRef]
142. Lubowiecka, I.; Arias, P.; Riveiro, B.; Solla, M. Multidisciplinary approach to the assessment of historic structures based on the case of a masonry bridge in Galicia (Spain). *Comput. Struct.* **2011**, *89*, 1615–1627. [CrossRef]
143. Stavroulaki, M.E.; Riveiro, B.; Drosopoulos, G.A.; Solla, M.; Koutsianitis, P.; Stavroulakis, G.E. Modelling and strength evaluation of masonry bridges using terrestrial photogrammetry and finite elements. *Adv. Eng. Softw.* **2016**, *101*, 136–148. [CrossRef]
144. Solla, M.; Lorenzo, H.; Novo, A.; Caamaño, J.C. Structural analysis of the Roman Bibei Bridge (Spain) based on GPR data and numerical modelling. *Autom. Constr.* **2012**, *22*, 334–339. [CrossRef]
145. Arêde, A.; Costa, C.; Gomes, A.T.; Menezes, J.E.; Silva, R.; Morais, M.; Gonçalves, R. Experimental characterization of the mechanical behaviour of components and materials of stone masonry railway bridges. *Constr. Build. Mater.* **2017**, *153*, 663–681. [CrossRef]
146. Lubowiecka, I.; Armesto, J.; Arias, P.; Lorenzo, H. Historic bridge modelling using laser scanning, ground penetrating radar and finite element methods in the context of structural dynamics. *Eng. Struct.* **2009**, *31*, 2667–2676. [CrossRef]
147. Costa, C.; Ribeiro, D.; Jorge, P.; Silva, R.; Arêde, A.; Calçada, R. Calibration of the numerical model of a stone masonry railway bridge based on experimentally identified modal parameters. *Eng. Struct.* **2016**, *123*, 354–371. [CrossRef]
148. Russo, S. Integrated assessment of monumental structures through ambient vibrations and ND tests: The case of rialto bridge. *J. Cult. Herit.* **2016**, *19*, 402–414. [CrossRef]

149. Conde, B.; Ramos, L.F.; Oliveira, D.V.; Riveiro, B.; Solla, M. Structural assessment of masonry arch bridges by combination of non-destructive testing techniques and three-dimensional numerical modelling: Application to Vilanova bridge. *Eng. Struct.* **2017**, *148*, 621–638. [CrossRef]
150. Cortés-Pérez, J.P.C.; de Sanjosé Blasco, J.J.; Atkinson, A.D.; del Río Pérez, L.M. Assessment of the structural integrity of the Roman bridge of Alcántara (Spain) using TLS and GPR. *Remote Sens.* **2018**, *10*, 387. [CrossRef]
151. Bautista-De Castro, A.; Sánchez-Aparicio, L.J.; Carrasco-García, P.; Ramos, L.F.; González-Aguilera, D. A multidisciplinary approach to calibrating advanced numerical simulations of masonry arch bridges. *Mech. Syst. Signal Process.* **2019**, *129*, 337–365. [CrossRef]
152. Saarenketo, T.; Maijala, P.; Leppäl¨, A. Recommendations for Guidelines for the Use of GPR in Bridge Deck Surveys; Publications of Mara Nord Project. 18p. 2011. Available online: http://maranord.ramk.fi (accessed on 11 February 2021).
153. Hugenschmidt, J. Concrete bridge inspection with a mobile GPR system. *Constr. Build. Mater.* **2002**, *16*, 147–154. [CrossRef]
154. Hugenschmidt, J.; Mastrangelo, R. GPR inspection of concrete bridges. *Cem. Concr. Compos.* **2006**, *28*, 384–392. [CrossRef]
155. Hasan, M.I.; Yazdani, N. Ground penetrating radar utilization in exploring inadequate concrete covers in a new bridge deck. *Case Stud. Constr. Mater.* **2014**, *1*, 104–114. [CrossRef]
156. Beben, D.; Mordak, A.; Anigacz, W. Identification of viaduct beam paremeters using the Ground Penetrating Radar (GPR) technique. *NDT E Int.* **2012**, *49*, 18–26. [CrossRef]
157. Simi, A.; Manacorda, G.; Benedetto, A. Bridge deck survey with high resolution Ground Penetrating Radar. In Proceedings of the 14th International Conference on Ground Penetrating Radar (GPR), Shanghai, China, 4–8 June 2012. [CrossRef]
158. Alani, A.M.; Aboutalebi, M.; Kilic, G. Applications of ground penetrating radar (GPR) in bridge deck monitoring and assessment. *J. Appl. Geophys.* **2013**, *97*, 45–54. [CrossRef]
159. Varela-Ortiz, W.; Lugo-Cintrón, C.Y.; Velázquez, G.I.; Stanton, T.R. Load testing and GPR assessment for concrete bridges on military installations. *Constr. Build. Mater.* **2013**, *38*, 1255–1269. [CrossRef]
160. Dinh, K.; Gucunski, N.; Zayed, T. Automated visualization of concrete bridge deck condition from GPR data. *NDT E Int.* **2019**, *102*, 120–128. [CrossRef]
161. Janků, M.; Cikrle, P.; Grošek, J.; Anton, O.; Stryk, J. Comparison of infrared thermography, ground-penetrating radar and ultrasonic pulse echo for detecting delaminations in concrete bridges. *Constr. Build. Mater.* **2019**, *225*, 1098–1111. [CrossRef]
162. Rathod, H.; Debeck, S.; Gupta, R.; Chow, B. Applicability of GPR and a rebar detector to obtain rebar information of existing concrete structures. *Case Stud. Constr. Mater.* **2019**, *11*, e00240. [CrossRef]
163. Sławski, Ł.; Kosno, Ł.; Świt, G. Evaluation of Precast Pre-post-tensioned Concrete Bridge Beams with the Use of GPR Method. *Procedia Eng.* **2016**, *156*, 443–450. [CrossRef]
164. Barnes, C.L.; Trottier, J.F.; Forgeron, D. Improved concrete bridge deck evaluation using GPR by accounting for signal depth-amplitude effects. *NDT E Int.* **2008**, *41*, 427–433. [CrossRef]
165. Benedetto, A. A three dimensional approach for tracking cracks in bridges using GPR. *J. Appl. Geophys.* **2013**, *97*, 37–44. [CrossRef]
166. Varnavina, A.V.; Khamzin, A.K.; Sneed, L.H.; Torgashov, E.V.; Anderson, N.L.; Maerz, N.H.; Boyko, K.J. Concrete bridge deck assessment: Relationship between GPR data and concrete removal depth measurements collected after hydrodemolition. *Constr. Build. Mater.* **2015**, *99*, 26–38. [CrossRef]
167. Varnavina, A.V.; Khamzin, A.K.; Torgashov, E.V.; Sneed, L.H.; Goodwin, B.T.; Anderson, N.L. Data acquisition and processing parameters for concrete bridge deck condition assessment using ground-coupled ground penetrating radar: Some considerations. *J. Appl. Geophys.* **2015**, *114*, 123–133. [CrossRef]
168. Abouhamad, M.; Dawood, T.; Jabri, A.; Alsharqawi, M.; Zayed, T. Corrosiveness mapping of bridge decks using image-based analysis of GPR data. *Autom. Constr.* **2017**, *80*, 104–117. [CrossRef]
169. Dabous, S.A.; Yaghi, S.; Alkass, S.; Moselhi, O. Concrete bridge deck condition assessment using IR Thermography and Ground Penetrating Radar technologies. *Autom. Constr.* **2017**, *81*, 340–354. [CrossRef]
170. Diamanti, N.; Annan, P.; Redman, J.D. Concrete bridge deck deterioration assessment using ground penetrating radar (GPR). *J. Environ. Eng. Geophys.* **2017**, *22*, 121–132. [CrossRef]
171. Dinh, K.; Gucunski, N.; Kim, J.; Duong, T. Method for attenuation assessment of GPR data from concrete bridge decks. *NDT E Int.* **2017**, *92*, 50–58. [CrossRef]
172. Varnavina, A.V.; Sneed, L.H.; Khamzin, A.K.; Torgashov, E.V.; Anderson, N.L. An attempt to describe a relationship between concrete deterioration quantities and bridge deck condition assessment techniques. *J. Appl. Geophys.* **2017**, *142*, 38–48. [CrossRef]
173. Ghodoosi, F.; Bagchi, A.; Zayed, T.; Hosseini, M.R. Method for developing and updating deterioration models for concrete bridge decks using GPR data. *Autom. Constr.* **2018**, *91*, 133–141. [CrossRef]
174. Sun, H.; Pashoutani, S.; Zhu, J. Nondestructive Evaluation of Concrete Bridge Decks with Automated Acoustic Scanning System and Ground Penetrating Radar. *Sens. (Basel)* **2018**, *18*, 1955. [CrossRef]
175. Rhee, J.-Y.; Choi, J.-J.; Kee, S.-H. Evaluation of the Depth of Deteriorations in Concrete Bridge Decks with Asphalt Overlays Using Air-Coupled GPR: A Case Study from a Pilot Bridge on Korean Expressway. *Int. J. Concr. Struct. Mater.* **2019**, *13*, 23. [CrossRef]
176. Agred, K.; Klysz, G.; Balayssac, J.P. Location of reinforcement and moisture assessment in reinforced concrete with a double receiver GPR antenna. *Constr. Build. Mater.* **2018**, *188*, 1119–1127. [CrossRef]
177. Wang, Z.W.; Zhou, M.; Slabaugh, G.G.; Zhai, J.; Fang, T. Automatic Detection of Bridge Deck Condition From Ground Penetrating Radar Images. *IEEE Trans. Autom. Sci. Eng.* **2011**, *8*, 633–640. [CrossRef]

178. Kaur, P.; Dana, K.J.; Romero, F.A.; Gucunski, N. Automated GPR rebar analysis for robotic bridge deck evaluation. *IEEE Trans. Cybern.* **2016**, *46*, 2265–2276. [CrossRef]

179. Dinh, K.; Gucunski, N.; Duong, T.H. An algorithm for automatic localization and detection of rebars from GPR data of concrete bridge decks. *Autom. Constr.* **2018**, *89*, 292–298. [CrossRef]

180. Dinh, K.; Gucunski, N.; Duong, T.H. Migration-based automated rebar picking for condition assessment of concrete bridge decks with ground penetrating radar. *NDT E Int.* **2018**, *98*, 45–54. [CrossRef]

181. Asadi, P.; Gindy, M.; Álvarez, M.; Asadi, A. A computer vision based rebar detection chain for automatic processing of concrete bridge deck GPR data. *Autom. Constr.* **2020**, *112*, 103106. [CrossRef]

182. Ma, X.; Liu, H.; Wang, M.L.; Birken, R. Automatic detection of steel rebar in bridge decks from ground penetrating radar data. *J. Appl. Geophys.* **2018**, *158*, 93–102. [CrossRef]

183. Alani, A.M.; Aboutalebi, M.; Kilic, G. Integrated health assessment strategy using NDT for reinforced concrete bridges. *NDT E Int.* **2014**, *61*, 80–94. [CrossRef]

184. ITA/AITES. Settlements induced by tunneling in Soft Ground. *Tunn. Undergr. Space Technol.* **2007**, *22*, 119–149. [CrossRef]

185. ASTM. ASTM D6429-99(2011)e1. In *Standard Guide for Selecting Surface Geophysical Methods (Withdrawn 2020)*; ASTM International: West Conshohocken, PA, USA, 2011. [CrossRef]

186. Abraham, O.; Dérobert, X. Non-destructive testing of fired tunnel walls: The Mont-Blanc Tunnel case study. *NDT E Int.* **2003**, *36*, 411–418. Available online: www.astm.org (accessed on 11 February 2021). [CrossRef]

187. Xie, X.; Liu, Y.; Huang, H.; Du, J.; Zhang, F.; Liu, L. Evaluation of grout behind the lining of shield tunnels using ground-penetrating radar in the Shanghai Metro Line, China. *J. Geophys. Eng.* **2007**, *4*, 253–261. [CrossRef]

188. Parkinson, G.; Ékes, C. Ground Penetrating Radar Evaluation of Concrete Tunnel Linings. In Proceedings of the 12th International Conference on Ground Penetrating Radar (GPR), Birmingham, UK, 16–19 June 2008.

189. Zhang, F.; Xie, X.; Huang, H. Application of ground penetrating radar in grouting evaluation for shield tunnel construction. *Tunn. Undergr. Space Technol.* **2010**, *25*, 99–107. [CrossRef]

190. Li, C.; Li, M.-J.; Zhao, Y.-G.; Liu, H.; Wan, Z.; Xu, J.-C.; Xu, X.-P.; Chen, Y.; Wang, B. Layer recognition and thickness evaluation of tunnel lining based on ground penetrating radar measurements. *J. Appl. Geophys.* **2011**, *73*, 45–48. [CrossRef]

191. Liu, X.; Sun, Q.; Liu, P. The application of Ground Penetrating Radar in detection of the water delivery tunnel lining. In Proceedings of the International Conference on Remote Sensing, Environment and Transportation Engineering, Nanjing, China, 24–26 June 2011. [CrossRef]

192. Yu, H.; Ouyang, Y.; Chen, H. Application of Ground Penetrating Radar to Inspect the Metro Tunnel. In Proceedings of the 14th International Conference on Ground Penetrating Radar (GPR), Shanghai, China, 4–8 June 2012. [CrossRef]

193. Xie, X.; Zeng, C. Non-destructive evaluation of shield tunnel condition using GPR and 3D laser scanning. In Proceedings of the 14th International Conference on Ground Penetrating Radar (GPR), Shanghai, China, 4–8 June 2012. [CrossRef]

194. Xiong, X.; Zhou, Q.; Zhou, J.; Ma, Y.; Wang, K. Application of GPR Technique and Research on high-speed railway tunnel. In Proceedings of the 14th International Conference on Ground Penetrating Radar (GPR), Shanghai, China, 4–8 June 2012. [CrossRef]

195. Lei, X.; Hui-lin, Z.; Zhen, S.; Si-hao, T.; Guo-qing, L.; Jian, Z. GPR evaluation of the Damaoshan highway tunnel: A case study. *NDT E Int.* **2013**, *59*, 68–76. [CrossRef]

196. Yu, Q.-M.; Zhou, H.-L.; Wang, Y.-H.; Duan, R.-X. Quality monitoring of metro grouting behind segment using ground penetrating radar. *Constr. Build. Mater.* **2016**, *110*, 189–200. [CrossRef]

197. Zan, Y.W.; Su, G.F.; Li, Z.L. A Train-mounted GPR System for Fast and Efficient Monitoring of Tunnel Health Conditions. In Proceedings of the 16th International Conference on Ground Penetrating Radar (GPR), Hong Kong, China, 13–16 June 2016. [CrossRef]

198. Xiongyao, X.; Li, Z.; Biao, Z. Real-time detection technology of synchronous grouting for shield tunnel and analysis of grouting effect. In Proceedings of the 17th International Conference on Ground Penetrating Radar (GPR), Rapperswill, Switzerland, 18–21 June 2018. [CrossRef]

199. Cao, Y.; Liu, Q.; Tao, L. Application of Ground Penetrating Radar for Detecting Grouting Quality in Highway Tunnel. In Proceedings of the 8th Joint International Information Technology and Artificial Intelligence Conference (ITAIC), Chongqing, China, 24–26 May 2019. [CrossRef]

200. Tang, S.; Bai, X. Study on Application of GPR to Quality Detection of Tunnel Lining. *Appl. Mech. Mater.* **2013**, *333–335*, 1611–1616. [CrossRef]

201. Kilic, G.; Eren, L. Neural network based inspection of voids and karst conduits in hydro–electric power station tunnels using GPR. *J. Appl. Geophys.* **2018**, *151*, 194–204. [CrossRef]

202. Cardarelli, E.; Marrone, C.; Orlando, L. Evaluation of tunnel stability using integrated geophysical methods. *J. Appl. Geophys.* **2003**, *52*, 93–102. [CrossRef]

203. Arosio, D.; Munda, S.; Zanzi, L.; Longoni, L.; Papini, M. GPR investigations to assess the state of damage of a concrete water tunnel. *J. Environ. Eng. Geophys.* **2012**, *17*, 159–169. [CrossRef]

204. Prego, F.J.; Solla, M.; Núñez-Nieto, X.; Arias, P. Assessing the Applicability of Ground-Penetrating Radar to Quality Control in Tunneling Construction. *J. Constr. Eng. Manag.* **2016**, *142*, 06015006. [CrossRef]

205. Ismail, M.A.; Abas, A.A.; Arifin, M.H.; Ismail, M.N.; Othman, N.A.; Setu, A.; Ahmad, M.R.; Shah, M.K.; Amin, S.; Sarah, T. Integrity inspection of main access tunnel using ground penetrating radar. In Proceedings of the IOP Conference Series: Materials Science and Engineering 271, Global Congress on Construction, Material and Structural Engineering (GCoMSE), Johor Bahru, Malaysia, 28–29 August 2017. [CrossRef]
206. Zan, Y.; Li, Z.; Su, G.; Zhang, X. An innovative vehicle-mounted GPR technique for fast and efficient monitoring of tunnel lining structural conditions. *Case Stud. Nondestruct. Test. Eval.* **2016**, *6*, 63–69. [CrossRef]
207. Alani, A.M.; Banks, K. Applications of Ground Penetrating Radar in the Medway Tunnel-Inspection of Structural Joints. In Proceedings of the 15th International Conference on Ground Penetrating Radar (GPR), Brussels, Belgium, 30 June–4 July 2014. [CrossRef]
208. Alani, A.M.; Tosti, F. GPR applications in structural detailing of a major tunnel using different frequency antenna systems. *Constr. Build. Mater.* **2018**, *158*, 1111–1122. [CrossRef]
209. Stryk, J.; Matula, R.; Pospíšil, K.; Dérobert, X.; Simonin, J.M.; Alani, A.M. Comparative measurements of ground penetrating radars used for road and bridge diagnostics in the Czech Republic and France. *Constr. Build. Mater.* **2017**, *154*, 1199–1206. [CrossRef]
210. *ME91/16: Methodologies for the Use of Ground-Penetrating Radar in Pavement Condition Surveys*; Belgian Road 445 Research Centre: Brussels, Belgium, 2016.
211. Solla, M.; González-Jorge, H.; Lorenzo, H.; Arias, P. Uncertainty evaluation of the 1 GHz GPR antenna for the estimation of concrete asphalt thickness. *Measurement* **2013**, *46*, 3032–3040. [CrossRef]
212. Li, X.T.; Wang, D.Y.; Zhang, X.N. A high-accuracy calibration method for thickness measurements of asphalt pavement using ground penetrating radar. *SATC* **2006**, *2006*, 305–312.
213. Fauchard, C.; Dérobert, X.; Cariou, J.; Côte, P. GPR performances for thickness calibration on road test sites. *NDT E Int.* **2003**, *36*, 67–75. [CrossRef]
214. Hu, J.; Vennapusa, P.K.; White, D.J.; Beresnev, I. Pavement thickness and stabilised foundation layer assessment using ground-coupled GPR. *Nondestruct. Test. Eval.* **2016**, *31*, 267–287. [CrossRef]
215. Zhao, S.; Al-Qadi, I.L. Development of an analytic approach utilizing the extended common midpoint method to estimate asphalt pavement thickness with 3-D ground-penetrating radar. *NDT E Int.* **2016**, *78*, 29–36. [CrossRef]
216. Lahouar, S.; Al-Qadi, I.L.; Loulizi, A.; Clark, T.M.; Lee, D.T. Approach to determining in situ dielectric constant of pavements: Development and implementation at interstate 81 in Virginia. *Transp. Res. Rec.* **2002**, *1806*, 81–87. [CrossRef]
217. Al-Qadi, I.L.; Lahouar, S. Use of GPR for thickness measurement and quality control of flexible pavements. *J. Assoc. Asph. Paving Technol.* **2004**, *73*, 501–528.
218. Al-Qadi, I.L.; Lahouar, S.; Loulizi, A. *Ground-Penetrating Radar Calibration at the Virginia Smart Road and Signal Analysis to Improve Prediction of Flexible Pavement Layer Thicknesses*; Virginia Center for Transportation Innovation and Research: Charlottesville, VA, USA, 2005.
219. Diamanti, N.; Annan, A.P. Air-launched and ground-coupled GPR data. In *Proceedings of the 2017 11th European Conference on Antennas and Propagation (EUCAP), Paris, France, 19–24 March 2017*; IEEE: New York, NY, USA, 2017; pp. 1694–1698. [CrossRef]
220. Dong, Z.; Ye, S.; Gao, Y.; Fang, G.; Zhang, X.; Xue, Z.; Zhang, T. Rapid detection methods for asphalt pavement thicknesses and defects by a vehicle-mounted ground penetrating radar (GPR) system. *Sensors* **2016**, *16*, 2067. [CrossRef]
221. Hoegh, K.; Khazanovich, L.; Dai, S.; Yu, T. Evaluating asphalt concrete air void variation via GPR antenna array data. *Case Stud. Nondestruct. Test. Eval.* **2015**, *3*, 27–33. [CrossRef]
222. Dérobert, X.; Fauchard, C.; Côte, P.; Le Brusq, E.; Guillanton, E.; Dauvignac, J.Y.; Pichot, C. Step-frequency radar applied on thin road layers. *J. Appl. Geophys.* **2001**, *47*, 317–325. [CrossRef]
223. Marecos, V.; Solla, M.; Fontul, S.; Antunes, V. Assessing the pavement subgrade by combining different non-destructive methods. *Constr. Build. Mater.* **2017**, *135*, 76–85. [CrossRef]
224. Diallo, M.C.; Cheng, L.Z.; Rosa, E.; Gunther, C.; Chouteau, M. Integrated GPR and ERT data interpretation for bedrock identification at Cléricy, Québec, Canada. *Eng. Geol.* **2019**, *248*, 230–241. [CrossRef]
225. Colagrande, S.; Ranalli, D.; Tallini, M. Ground penetrating radar assessment of flexible road pavement degradation. *Int. J. Geophys.* **2011**, *2011*, 989135. [CrossRef]
226. Plati, C.; Loizos, A.; Gkyrtis, K. Integration of non-destructive testing methods to assess asphalt pavement thickness. *NDT E Int.* **2020**, *115*, 102292. [CrossRef]
227. Scullion, T.; Saarenketo, T. Integrating ground penetrating radar and falling weight deflectometer technologies in pavement evaluation. In *Nondestructive Testing of Pavements and Backcalculation of Moduli*; ASTM International: West Conshohocken, PA, USA, 2000; Volume 3. [CrossRef]
228. Marecos, V.; Fontul, S.; de Lurdes Antunes, M.; Solla, M. Evaluation of a highway pavement using non-destructive tests: Falling weight deflectometer and ground penetrating radar. *Constr. Build. Mater.* **2017**, *154*, 1164–1172. [CrossRef]
229. Evans, R.D.; Frost, M.; Stonecliffe-Jones, M.; Dixon, N. A review of pavement assessment using ground penetrating radar (GPR). In Proceedings of the 12th International Conference on Ground Penetrating Radar, Birmingham, UK, 16–19 June 2008.
230. Marecos, V.; Fontul, S.; Solla, M.; de Lurdes Antunes, M. Evaluation of the feasibility of Common Mid-Point approach for air-coupled GPR applied to road pavement assessment. *Measurement* **2018**, *128*, 295–305. [CrossRef]

231. Maser, K.R. Measurement of as-built conditions using ground penetrating radar. In Proceedings of the Structural Materials Technology: An NDT Conference, San Diego, CA, USA, 20–23 February 1996; pp. 61–67.

232. Al-Qadi, I.L.; Lahouar, S.; Jiang, K.; MeGhee, K.K.; Mokarem, D. Validation of ground penetration radar accuracy for estimating pavement layer thicknesses. In Proceedings of the Transportation Research Board 84th Annual Meeting, Washington, DC, USA, 9–13 January 2005.

233. Schmidtgen, B.D.; Milne, T.I.; Saarenketo, T. Road management data assimilation and quality control of asphalt and bituminous pavements using GPR. In Proceedings of the 10th Conference on Asphalt Pavements for Southern Africa [CAPSA11], KwaZulu-Natal, South Africa, 11–14 September 2011.

234. Benedetto, A.; Benedetto, F.; Tosti, F. GPR applications for geotechnical stability of transportation infrastructures. *Nondestruct. Test. Eval.* **2012**, *27*, 253–262. [CrossRef]

235. Lagüela, S.; Solla, M.; Puente, I.; Prego, F.J. Joint use of GPR, IRT and TLS techniques for the integral damage detection in paving. *Constr. Build. Mater.* **2018**, *174*, 749–760. [CrossRef]

236. Le Bastard, C.; Pan, J.; Wang, Y.; Sun, M.; Todkar, S.S.; Baltazart, V.; Pinel, N.; Ihamouten, A.; Derobert, X.; Bourlier, C. A Linear Prediction and Support Vector Regression-Based Debonding Detection Method Using Step-Frequency Ground Penetrating Radar. *IEEE Geosci. Remote Sens. Lett.* **2018**, *16*, 367–371. [CrossRef]

237. Todkar, S.S.; Le Bastard, C.; Baltazart, V.; Ihamouten, A.; Dérobert, X. Performance assessment of SVM-based classification techniques for the detection of artificial debondings within pavement structures from stepped-frequency A-scan radar data. *NDT E Int.* **2019**, *107*, 102128. [CrossRef]

238. Ercoli, M.; Di Matteo, L.; Pauselli, C.; Mancinelli, P.; Frapiccini, S.; Talegalli, L.; Cannata, A. Integrated GPR and laboratory water content measures of sandy soils: From laboratory to field scale. *Constr. Build. Mater.* **2018**, *159*, 734–744. [CrossRef]

239. Venmans, A.A.; van de Ven, R.; Kollen, J. Rapid and non-intrusive measurements of moisture in road constructions using passive microwave radiometry and GPR–full scale test. *Procedia Eng.* **2016**, *143*, 1244–1251. [CrossRef]

240. Bianchini Ciampoli, L.; Tosti, F.; Economou, N.; Benedetto, F. Signal processing of GPR data for road surveys. *Geosciences* **2019**, *9*, 96. [CrossRef]

241. Puente, I.; Solla, M.; González-Jorge, H.; Arias, P. Validation of mobile LiDAR surveying for measuring pavement layer thicknesses and volumes. *NDT E Int.* **2013**, *60*, 70–76. [CrossRef]

242. Khamzin, A.K.; Varnavina, A.V.; Torgashov, E.V.; Anderson, N.L.; Sneed, L.H. Utilization of air-launched ground penetrating radar (GPR) for pavement condition assessment. *Constr. Build. Mater.* **2017**, *141*, 130–139. [CrossRef]

243. Shangguan, P.; Al-Qadi, I.L. Calibration of FDTD simulation of GPR signal for asphalt pavement compaction monitoring. *IEEE Trans. Geosci. Remote Sens.* **2014**, *53*, 1538–1548. [CrossRef]

244. Bala, D.C.; Garg, R.D.; Jain, S.S. Rebar detection using GPR: An emerging non-destructive QC approach. *International Journal of Engineering Research and Applications (IJERA)* **2011**, *1*, 2111–2117.

245. Zhou, F.; Chen, Z.; Liu, H.; Cui, J.; Spencer, B.F.; Fang, G. Simultaneous estimation of rebar diameter and cover thickness by a GPR-EMI dual sensor. *Sensors* **2018**, *18*, 2969. [CrossRef] [PubMed]

246. Kim, N.; Kim, K.; An, Y.K.; Lee, H.J.; Lee, J.J. Deep learning-based underground object detection for urban road pavement. *Int. J. Pavement Eng.* **2018**, 1–13. [CrossRef]

247. Solla, M.; Lagüela, S.; Fernández, N.; Garrido, I. Assessing Rebar Corrosion through the Combination of Nondestructive GPR and IRT Methodologies. *Remote Sens.* **2019**, *11*, 1705. [CrossRef]

248. Sossa, V.; Pérez-Gracia, V.; González-Drigo, R.; A Rasol, M. Lab Non Destructive Test to Analyze the Effect of Corrosion on Ground Penetrating Radar Scans. *Remote Sens.* **2019**, *11*, 2814. [CrossRef]

249. Hong, S.; Lai, W.W.L.; Wilsch, G.; Helmerich, R.; Helmerich, R.; Günther, T.; Wiggenhauser, H. Periodic mapping of reinforcement corrosion in intrusive chloride contaminated concrete with GPR. *Constr. Build. Mater.* **2014**, *66*, 671–684. [CrossRef]

250. Baltrušaitis, A.; Vaitkus, A.; Smirnovs, J. Asphalt Layer Density and Air Voids Content: GPR and Laboratory Testing Data Reliance. *Balt. J. Road Bridge Eng.* **2020**, *15*, 93–110. [CrossRef]

251. Zhong, Y.; Wang, Y.; Zhang, B.; Li, X.; Li, S.; Zhong, Y.; Hao, M.; Gao, Y. Prediction Model of Asphalt Content of Asphalt Mixture Based on Dielectric Properties. *Adv. Civ. Eng.* **2020**, *2020*, 1–10. [CrossRef]

252. Gao, J.; Yuan, D.; Tong, Z.; Yang, J.; Yu, D. Autonomous pavement distress detection using ground penetrating radar and region-based deep learning. *Measurement* **2020**, *164*, 108077. [CrossRef]

253. Tong, Z.; Gao, J.; Yuan, D. Advances of deep learning applications in ground-penetrating radar: A survey. *Constr. Build. Mater.* **2020**, *258*, 120371. [CrossRef]

254. Giannakis, I.; Giannopoulos, A.; Warren, C. A machine learning-based fast-forward solver for ground penetrating radar with application to full-waveform inversion. *IEEE Trans. Geosci. Remote Sens.* **2019**, *57*, 4417–4426. [CrossRef]

255. Tong, Z.; Yuan, D.; Gao, J.; Wei, Y.; Dou, H. Pavement-distress detection using ground-penetrating radar and network in networks. *Constr. Build. Mater.* **2020**, *233*, 117352. [CrossRef]

256. Ozkaya, U.; Seyfi, L. Deep dictionary learning application in GPR B-scan images. *Signalimage Video Process.* **2018**, *12*, 1567–1575. [CrossRef]

257. Tong, Z.; Gao, J.; Zhang, H. Recognition, location, measurement, and 3D reconstruction of concealed cracks using convolutional neural networks. *Constr. Build. Mater.* **2017**, *146*, 775–787. [CrossRef]

258. Kim, N.; Kim, S.; An, Y.K.; Lee, J.J. A novel 3D GPR image arrangement for deep learning-based underground object classification. *Int. J. Pavement Eng.* **2019**, 1–12. [CrossRef]
259. Muller, W. A comparison of TSD, FWD and GPR field measurements. In Proceedings of the International Symposium Non-Destructive Testing in Civil Engineering (NDT-CE), Berlin, Germany, 15–17 September 2015.
260. Maser, K.; Schmalzer, P.; Gerber, A.; Poorbaugh, J. *Implementation of the Traffic Speed Deflectometer (TSD) for Network Level Pavement Management*; Infrasense. Pavement Evaluation: Woburn, MA, USA, 2019.
261. Maser, K.; Schmalzer, P.; Shaw, W.; Carmichael, A. Integration of Traffic Speed Deflectometer and Ground-Penetrating Radar for Network-Level Roadway Structure Evaluation. *Transp. Res. Rec.* **2017**, *2639*, 55–63. [CrossRef]
262. Katicha, S.; Shrestha, S.; Flintsch, G.; Diefenderfer, B. *Network Level Pavement Structural Testing with the Traffic Speed Deflectometer*; Final Report VTRC 21–R4; Virginia Transportation Research Council: Charlottesville, VA, USA, 2020.
263. Lenngren, C.; Bergström, J.; Ersson, B. Using Ground Penetrating Radar for Assessing Highway Pavement Thickness. In Proceedings of the SPIE—The International Society for Optical Engineering, San Diego, CA, USA, 6 July 2020. [CrossRef]
264. Yuan, C.; Cai, H. Spatial reasoning mechanism to enable automated adaptive trajectory planning in ground penetrating radar survey. *Autom. Constr.* **2020**, *114*, 103157. [CrossRef]
265. Miccinesi, L.; Pieraccini, M. A GPR Able to Detect Its Own Position Using Fixed Corner Reflectors on Surface. *IEEE Trans. Geosci. Remote Sens.* **2020**, 1–8. [CrossRef]
266. Šarlah, N.; Podobnikar, T.; Mongus, D.; Ambrožič, T.; Mušič, B. Kinematic GPR-TPS Model for Infrastructure Asset Identification with High 3D Georeference Accuracy Developed in a Real Urban Test Field. *Remote Sens.* **2019**, *11*, 1457. [CrossRef]
267. Gabryś, M.; Ortyl, Ł. Georeferencing of Multi-Channel GPR—Accuracy and Efficiency of Mapping of Underground Utility Networks. *Remote Sens.* **2020**, *12*, 2945. [CrossRef]
268. Wu, Y.; Shen, F.; Xu, D. Ground-penetrating radar-based underground environmental perception radar for robotic system. *Int. J. Adv. Robot. Syst.* **2020**, *17*, 1729881420921642. [CrossRef]
269. Eskandari Torbaghan, M.; Kaddouh, B.; Abdellatif, M.; Metje, N.; Liu, J.; Jackson, R.; Rogeres, C.D.; Chapman, D.N.; Fuentes, R.; Miodownik, M.; et al. Robotic and autonomous systems for road asset management: A position paper. *Proc. Inst. Civ. Eng.-Smart Infrastruct. Constr.* **2020**, *172*, 83–93. [CrossRef]
270. Rodriguez-Vaqueiro, Y.; Vázquez-Cabo, J.; Gonzalez-Valdes, B.; Pino, A.; Álvarez, Y.; Garcia-Fernandez, M.; Las-Heras, F.; Arboleya, A. Array of Antennas for a GPR system onboard a UAV. In *Proceedings of the 2019 IEEE International Symposium on Antennas and Propagation and USNC-URSI Radio Science Meeting, Atlanta, GA, USA, 7–12 July 2019*; IEEE: New York, NY, USA, 2019; pp. 821–822. [CrossRef]
271. Pan, Y.; Zhang, X.; Sun, M.; Zhao, Q. Object-based and supervised detection of potholes and cracks from the pavement images acquired by UAV. *Int. Arch. Photogramm. Remote Sens. Spat. Inf. Sci.* **2017**, *42*, 209–217. [CrossRef]
272. Garcia-Fernandez, M.; Alvarez-Lopez, Y.; Las Heras, F.; Gonzalez-Valdes, B.; Rodriguez-Vaqueiro, Y.; Pino, A.; Arboleya-Arboleya, A. GPR system onboard a uav for non-invasive detection of buried objects. In *Proceedings of the 2018 IEEE International Symposium on Antennas and Propagation & USNC/URSI National Radio Science Meeting, Boston, MA, USA, 8–13 July 2018*; IEEE: New York, NY, USA, 2018; pp. 1967–1968.
273. EN 13848-5: 2017. In *Railway Applications-Track-Track Geometry Quality—Part 5: Geometric Quality Levels—Plain Line, Switches and Crossings*; BSI: London, UK, 2017.
274. Esveld, C. *Modern Railway Track*; MRT-Productions: Zaltbommel, The Netherlands, 2001.
275. Fortunato, E. Renovação de Plataformas Ferroviárias. Estudos Relativos à Capacidade de Carga. Ph.D. Thesis, University of Porto, Porto, Portugal, 2005. (In Portuguese).
276. Saarenketo, T. Electrical Properties of Road Materials and Subgrade Soils and the Use of Ground Penetrating Radar in Traffic Infrastructure Surveys. Ph.D. Thesis, Faculty of Science, University of Oulu, Oulu, Finland, 2006.
277. Loizos, A.; Silvast, M.; Dimitrellou, S. Railway trackbed assessment using the GPR technique. *Adv. Charact. Pavement Soil Eng. Mater.* **2007**, *1*, 1817–1826.
278. De Chiara, F.; Fontul, S.; Fortunato, E. GPR Laboratory Tests for Railways Materials Dielectric Properties Assessment. *Remote Sens.* **2014**, *6*, 9712–9728. [CrossRef]
279. Liu, S.; Lu, Q.; Li, H.; Wang, Y. Estimation of Moisture Content in Railway Subgrade by Ground Penetrating Radar. *Remote Sens.* **2020**, *12*, 2912. [CrossRef]
280. Shao, W.; Bouzerdoum, A.; Phung, S.L.; Su, L.; Indraratna, B. Automatic Classification of Ground-Penetrating-Radar Signals for Railway-Ballast Assessment. *IEEE Trans. Geosci. Remote Sens.* **2011**, *49*, 3961–3972. [CrossRef]
281. Al-Qadi, I.L.; Xie, W.; Roberts, R. Scattering Analysis of Railroad Ballast Using Ground Penetrating Radar. *J. Nondestruct. Test. Eval.* **2008**, *41*, 441–447. [CrossRef]
282. Xie, W.; Al-Qadi, I.L.; Jones, D.L.; Roberts, R. Development of a Time-Frequency Approach to Quantify Railroad Ballast Fouling Condition Using UWB GPR Data. In Proceedings of the 87th TRB Annual Meetings, Paper No. 08-1846, Washington, DC, USA, 13–17 January 2008.
283. Fontul, S.; Paixão, A.; Solla, M.; Pajewski, L. Railway Track Condition Assessment at Network Level by Frequency Domain Analysis of GPR Data. *Remote Sens.* **2018**, *10*, 559. [CrossRef]
284. Al-Qadi, I.; Xie, W.; Roberts, R. Optimization of antenna configuration in multiple-frequency ground penetrating radar system for railroad substructure assessment. *NDT E Int.* **2010**, *43*, 20–28. [CrossRef]

285. Anbazhagan, P.; Su, L.; Indraratna, B.; Rujikiatkamjorn, C. Model track studies on fouled ballast using ground penetrating radar and multichannel analysis of surface wave. *J. Appl. Geophys.* **2011**, *74*, 175–184. [CrossRef]

286. Fortunato, E.; Fontul, S.; Paixão, A.; Cruz, N.; Cruz, J.; Asseiceiro, F. Reabilitação de uma Linha Férrea de Mercadorias—Aspetos Geotécnicos da via (Rehabilitation of a Freight Railway Line—Geotechnical Aspects of the Track). Jornal Geotecnia Nº 131, July 2014; Sociedade Portuguesa de Geotecnia; pp. 95–112. Available online: https://spgeotecnia.pt/uploads/jgeot201413106.pdf (accessed on 11 February 2021). (In Portuguese).

287. Roadscanners. *GPR Survey on Railway Sections between Montevideo—Paso de los Toros in Uruguay*; Project Report; Roadscanners: Rovaniemi, Finland, 2017; (consulted on October 2020).

288. Riveiro, B.; Solla, M. (Eds.) *Non-Destructive Techniques for the Evaluation of Structures and Infrastructure*; CRC Press: London, UK, 2016; ISBN 9781138028104.

289. Zarembski, A.; Yurlov, D.; Palese, J.; Attoh-Okine, N. *Relationship between Track Geometry Defects and Measured Track Subsurface Condition*; Federal Railroad Administration Report DOT/FRA/ORD-20/07; Federal Railroad Administration: Washington, DC, USA, 2020.

290. Smekal, A.; Berggren, E.G.; Silvast, M. Monitoring and substructure condition assessment of existing railway lines for upgrading to higher axle loads and speeds. In Proceedings of the 7th World Congress on Railway Research, Montreal, QC, Canada, 5–7 June 2006.

291. Brough, M.; Stirling, A.; Ghataora, G.; Madelin, K. Evaluation of railway trackbed and formation: A case study. *NDT E Int.* **2003**, *36*, 145–156. [CrossRef]

292. Lamas-Lopez, F.; Cui, Y.J.; Calon, N.; Costa d'Aguiar, S. Geotechnical auscultation of a French conventional railway track-bed for maintenance purposes. *Soils Found* **2016**, *56*, 225–240. [CrossRef]

293. Zarembski, A.M.; Grissom, G.T.; Euston, T.L. On the Use of Ballast Inspection Technology for the Management of Track Substructure. *Transp. Infrastruct. Geotech.* **2014**, *1*, 83–109. [CrossRef]

294. Hu, F.; Leijen, F.J.v.; Chang, L.; Wu, J.; Hanssen, R.F. Monitoring Deformation along Railway Systems Combining Multi-Temporal InSAR and LiDAR Data. *Remote Sens.* **2019**, *11*, 2298. [CrossRef]

295. Jung, J.; Kim, D.-J.; Palanisamy Vadivel, S.K.; Yun, S.-H. Long-term deflection monitoring for bridges using X and C band time-series SAR interferometry. *Remote Sens.* **2019**, *11*, 1258. [CrossRef]

296. Bianchini Ciampoli, L.; Gagliardi, V.; Ferrante, C.; Calvi, A.; D'Amico, F.; Tosti, F. Displacement monitoring in airport runways by persistent scatterers SAR Interferometry. *Remote Sens.* **2020**, *12*, 3564. [CrossRef]

297. Bianchini Ciampoli, L.; Gagliardi, V.; Clementini, C.; Latini, D.; Del Frate, F.; Benedetto, A. Transport infrastructure monitoring by InSAR and GPR data fusion. *Surv. Geophys.* **2020**, *41*, 371–394. [CrossRef]

298. D'Amico, F.; Gagliardi, V.; Bianchini Ciampoli, L.; Tosti, F. Integration of InSAR and GPR techniques for monitoring transition areas in railway bridges. *NDT E Int.* **2020**, *115*, 102291. [CrossRef]

299. TRACKSCAN—A Practical, Portable and Robust Scanning System Using Infrastructure Inspection Radar (IIR) for the Investigation of Rail Track Substructure, Ballast and Tunnel Infrastructure. H20202 Project Grant agreement ID: 700138. Cordis. 2016. Available online: https://cordis.europa.eu/project/id/700138/reporting (accessed on 11 February 2021).

300. Railview. Ground Penetrating Radar for the Rail Industry. 2020. Available online: https://www.railview.co.uk/sevices (accessed on 11 February 2021).

301. DESTination RAIL Decision Support Tool for Rail Infrastructure Managers. Project Reference: 636285. Available online: http://www.destinationrail.eu/ (accessed on 11 February 2021).

302. Hugenschmidt, J.; Mastrangelo, R. The inspection of large retaining walls using GPR. In *Proceedings of the 2007 4th International Workshop on Advanced Ground Penetrating Radar, Napoli, Italy, 27–29 June 2007*; IEEE: New York, NY, USA, 2007; pp. 267–271.

303. Lualdi, M.; Zanzi, L.; Binda, L. Acquisition and processing requirements for high quality 3D reconstructions from GPR investigations. In Proceedings of the Proceedings, International Symposium Non-Destructive Testing in Civil Engineering (NDT-CE), Berlin, Germany, 16–19 September 2003.

304. Soldovieri, F.; Persico, R.; Utsi, E.; Utsi, V. The application of inverse scattering techniques with ground penetrating radar to the problem of rebar location in concrete. *NDT E Int.* **2006**, *39*, 602–607. [CrossRef]

305. Soldovieri, F.; Hugenschmidt, J.; Persico, R.; Leone, G. A linear inverse scattering algorithm for realistic GPR applications. *Near Surf. Geophys.* **2007**, *5*, 29–42. [CrossRef]

306. Li, S.; Feng, X.T.; Li, Z.; Zhang, C.; Chen, B. Evolution of fractures in the excavation damaged zone of a deeply buried tunnel during TBM construction. *Int. J. Rock Mech. Min. Sci.* **2012**, *55*, 125–138. [CrossRef]

307. Liu, L.; Shi, Z.; Peng, M.; Tsoflias, G.P. Investigation of Geological Anomalies at Pile Foundation Location in Urban Karst Areas Using Single Borehole Radar. *Geosciences* **2020**, *10*, 232. [CrossRef]

308. Kavanagh, J.; Bowers, K.H.; Brady, K.C. *The Use of Ground Penetrating Radar in the Investigation of Masonry Retaining Walls*; Traffic and Transport Resource Centre, Transport Research Laboratory: Hong Kong, China, 1999.

309. Athanasopoulos Zekkos, A.; Lynch, J.; Zekkos, D.; Grizi, A.; Admassu, K.; Benhamida, B.; Spino, R.J.; Mikolajczyk, M. *Asset Management for Retaining Walls*; Technical Report; University of Michigan: Ann Arbor, MI, USA; Volume 1001, pp. 42125–48109.

310. Wiggenhauser, H.; Behrens, M.; Moser, D.; Moryson, R.M.; Pudovikov, S.; Herrmann, H.G. Non-destructive assessment of retaining wall of former coal mine plant: Zerstörungsfreie Prüfung der Stützmauer einer ehemaligen Kohlegrube. *Mauerwerk* **2018**, *22*, 175–186. [CrossRef]

311. Santos-Assunçao, S.; Perez-Gracia, V.; Salinas, V.; Caselles, O.; Gonzalez-Drigo, R.; Pujades, L.G.; Lantada, N. GPR backscattering intensity analysis applied to detect paleochannels and infilled streams for seismic nanozonation in urban environments. *IEEE J. Sel. Top. Appl. Earth Obs. Remote Sens.* **2015**, *9*, 167–177. [CrossRef]

312. Salinas Naval, V.; Santos-Assunçao, S.; Pérez-Gracia, V. GPR clutter amplitude processing to detect shallow geological targets. *Remote Sens.* **2018**, *10*, 88. [CrossRef]

313. Bao, Y.; Ghasr, M.T.; Ying, K.; Chen, G.; Zoughi, R. Microwave Synthetic Aperture Radar Imaging for Nondestructive Evaluation of Mechanically Stabilized Earth Walls. *Mater. Eval.* **2017**, *75*, 177–184.

314. RILEM TC 127-MS: Non Destructive Tests for Masonry Materials and Structures. Materials and Structures/Matériaux et Constructions, Volume 34, pp. 134–143. April 2001. Available online: www.rilem.org (accessed on 11 February 2021).

315. Diamanti, N.; Giannopoulos, A.; Forde, M. Numerical modelling and experimental verification of GPR to investigate ring separation in brick masonry arch bridges. *NDT E Int.* **2008**, *41*, 354–363. [CrossRef]

316. Diamanti, N.; Giannopoulos, A. Employing ADI-FDTD subgrids for GPR numerical modelling and their application to study ring separation in brick masonry arch bridges. *Near Surf. Geophys.* **2011**, *9*, 245–256. [CrossRef]

317. Solla, M.; Lagüela, S.; Riveiro, B.; Lorenzo, H. Non-destructive testing for the analysis of moisture in the masonry arch bridge of Lubians (Spain). *Struct. Control Health Monit.* **2013**, *20*, 1366–1376. [CrossRef]

318. Biscarini, C.; Catapano, I.; Cavalagli, N.; Ludeno, G.; Pepe, F.A.; Ubertini, F. UAV photogrammetry, infrared thermography and GPR for enhancing structural and material degradation evaluation of the Roman masonry bridge of Ponte Lucano in Italy. *NDT E Int.* **2020**, *115*, 102287. [CrossRef]

319. Alani, A.M.; Tosti, F.; Bianchini Ciampoli, L.; Gagliardi, V.; Benedetto, A. An integrated investigative approach in health monitoring of masonry arch bridges using GPR and InSAR technologies. *NDT E Int.* **2020**, *115*, 102288. [CrossRef]

320. Puente, I.; Solla, M.; González-Jorge, H.; Arias, P. NDT documentation and evaluation of the Roman bridge of Lugo using GPR and mobile and static LiDAR. *J. Perform. Constr. Facil.* **2015**, *29*, 06014004. [CrossRef]

321. Stryk, J.; Alani, A.M.; Matula, R.; Pospisil, K. Innovative Inspection Procedures for Effective GPR Surveying of Critical Transport Infrastructures (Pavements, Bridges and Tunnels). In *Civil Engineering Applications of Ground Penetrating Radar*; Springer Transactions in Civil and Environmental Engineering: Berlin/Heidelberg, Germany, 2015. [CrossRef]

322. SHRP 2- report S2-R06A-RR-1. In *Nondestructive Testing to Identify Concrete Bridge Deck Deterioration*; Transportation Research Board: Washington, DC, USA, 2013; Available online: www.TRB.org/SHRP2 (accessed on 11 February 2021).

323. BASt-Report B55. Examination of GPR in Combination with Magnetic Techniques for the Determination of Moisture and Salinity of Concrete Bridge Decks with Asphalt Cover; Federal Highway Research Institute; Bundesanstalt für Straßenwesen. 2007. Available online: www.bast.de (accessed on 11 February 2021).

324. *Document B10—Recommendation for Nondestructive Testing of Civil Engineering Structures by GPR*; German Society for Non-Destructive Testing (DGZfP): Berlin, Germany, 2008.

325. *AASHTO R 37-04. Standard Practice for Application of Ground Penetrating Radar (GPR) to Highways*; American Association of State and Highway Transportation Officials: Washington, DC, USA, 2004.

326. *ACI 228.2R-98 (Reapproved 2004). Nondestructive Test Methods for Evaluation of Concrete in Structures*; American Concrete Institute: Farmington Hills, MI, USA, 1998.

327. NCHRP Research Report 848. In *Inspection Guidelines for Bridge Post-Tensioning and Stay Cable Systems Using NDE Methods*; TRB's National Cooperative Highway Research Program: Washington, DC, USA, 2017. [CrossRef]

328. Lalagüe, A.; Lebens, M.A.; Hoff, I.; Grøv, E. Detection of rockfall on a tunnel concrete lining with ground-penetrating radar (GPR). *Rock Mech. Rock Eng.* **2016**, *49*, 2811–2823. [CrossRef]

329. Balaguer, C.; Montero, R.; Victores, J.G.; Martínez, S.; Jardón, A. Towards fully automated tunnel inspection: A survey and future trends. In Proceedings of the 31st International Symposium on Automation and Robotics in Construction and Mining (ISARC 2014), Sydney, Australia, 9–11 July 2014. [CrossRef]

330. Xie, X.Y.; Chen, Y.F.; Zhou, B. Data processing of backfill grouting detected by GPR in shield tunnel and research on equipment of GPR antenna. In Proceedings of the 16th International Conference on Ground Penetrating Radar (GPR), Hong Kong, China, 13–16 June 2016. [CrossRef]

331. SHRP-2 Report S2-R06G-RR-1. In *Mapping Voids, Debonding, Delaminations, Moisture, and Other Defects behind or within Tunnel Linings*; Transportation Research Board: Washington, DC, USA, 2014; Available online: www.TRB.org/SHRP2 (accessed on 11 February 2021).

332. Lyu, Y.; Wang, H.; Gong, J. GPR detection of tunnel lining cavities and reverse-time migration imaging. *Appl. Geophys.* **2020**, *17*, 1–7. [CrossRef]

333. Karlovšek, J.; Scheuermann, A.; Willimas, D.J. Investigation of voids and cavities in Bored Tunnels using GPR. In Proceedings of the 14th International Conference on Ground Penetrating Radar (GPR), Shanghai, China, 4–8 June 2012. [CrossRef]

334. Garcia-Fernandez, M.; Alvarez-Lopez, Y.; Las Heras, F. Autonomous airborne 3D SAR imaging system for subsurface sensing: UWB-GPR on board a UAV for landmine and IED detection. *Remote Sens.* **2019**, *11*, 2357. [CrossRef]

335. Wu, K.; Rodriguez, G.A.; Zajc, M.; Jacquemin, E.; Clément, M.; De Coster, A.; Lambot, S. A new drone-borne GPR for soil moisture mapping. *Remote Sens. Environ.* **2019**, *235*, 111456. [CrossRef]

336. Šipoš, D.; Gleich, D. A Lightweight and Low-Power UAV-Borne Ground Penetrating Radar Design for Landmine Detection. *Sensors* **2020**, *20*, 2234. [CrossRef]

337. Howlader, M.O.F.; Sattar, T.P.; Dudley, S. Development of a Wall Climbing Robotic Ground Penetrating Radar System for Inspection of Vertical Concrete Structures. *Int. J. Mech. Mechatron. Eng.* **2016**, *10*, 1382–1388.
338. Yangí, L.; Yang, G.; Liu, Z.; Chang, Y.; Jiang, B.; Awad, Y.; Xiao, J. Wall-climbing robot for visual and GPR inspection. In *Proceedings of the 2018 13th IEEE Conference on Industrial Electronics and Applications (ICIEA), Wuhan, China, 31 May–2 June 2018*; IEEE: New York, NY, USA, 2018; pp. 1004–1009. [CrossRef]
339. Bertolino, M.; Tanzi, T.J. Towards 3D Simulation for Disaster Intervention Robot Behaviour Assessment. *Adv. Radio Sci.* **2020**, *18*, 23–32. [CrossRef]
340. Gibb, S.; La, H.M.; Le, T.; Nguyen, L.; Schmid, R.; Pham, H. Nondestructive evaluation sensor fusion with autonomous robotic system for civil infrastructure inspection. *J. Field Robot.* **2018**, *35*, 988–1004. [CrossRef]
341. Pochanin, G.; Varianytsia-Roshchupkina, L.; Ruban, V.; Pochanina, I.; Falorni, P.; Borgioli, G.; Capineri, L.; Bechtel, T. Design and simulation of a "single transmitter-four receiver" impulse GPR for detection of buried landmines. In *Proceedings of the 2017 9th International Workshop on Advanced Ground Penetrating Radar (IWAGPR), Edinburgh, UK, 28–30 June 2017*; IEEE: New York, NY, USA, 2017; pp. 1–5. [CrossRef]
342. Travassos, X.L.; Avila, S.L.; Ida, N. Artificial Neural Networks and Machine Learning techniques applied to Ground Penetrating Radar: A review. *Appl. Comput. Inform.* **2020**. [CrossRef]
343. Liu, T.; Su, Y.; Huang, C. Inversion of Ground Penetrating Radar Data Based on Neural Networks. *Remote Sens.* **2018**, *10*, 730. [CrossRef]
344. Kang, M.-S.; An, Y.-K. Frequency–Wavenumber Analysis of Deep Learning-based Super Resolution 3D GPR Images. *Remote Sens.* **2020**, *12*, 3056. [CrossRef]
345. Liu, B.; Ren, Y.; Liu, H.; Xu, H.; Wang, Z.; Cohn, A.; Jiang, P. *GPRInvNet: Deep Learning-Based Ground Penetrating Radar Data Inversion for Tunnel Lining*; Cornell University: Ithaca, NY, USA, 2020.
346. Asadi, P.; Gindy, M.; Alvarez, M. A Machine Learning Based Approach for Automatic Rebar Detection and Quantification of Deterioration in Concrete Bridge Deck Ground Penetrating Radar B-scan Images. *Ksce J. Civ. Eng.* **2019**, *23*, 2618–2627. [CrossRef]
347. Zhang, Y.; Venkatachalam, A.; Xia, T. Ground-penetrating radar railroad ballast inspection with an unsupervised algorithm to boost the region of interest detection efficiency. *J. Appl. Remote Sens.* **2015**, *9*, 1–19. [CrossRef]
348. Xu, X.; Lei, Y.; Yang, F. Railway Subgrade Defect Automatic Recognition Method Based on Improved Faster R-CNN. *Sci. Program.* **2018**, *2018*, 4832972. [CrossRef]
349. Deshpande, J. *Geographical Information System & Building Information Modelling*; ADCC Infocad IT Services: Dubai, United Arab Emirates, 2020.
350. Solla, M.; Gonçalves, L.M.S.; Gonçalves, G.; Francisco, C.; Puente, I.; Providência, P.; Gaspar, F.; Rodrigues, H. A Building Information Modeling Approach to Integrate Geomatic Data for the Documentation and Preservation of Cultural Heritage. *Remote Sens.* **2020**, *12*, 4028. [CrossRef]
351. Šarlah, N.; Podobnikar, T.; Ambrožič, T.; Mušič, B. Application of Kinematic GPR-TPS Model with High 3D Georeference Accuracy for Underground Utility Infrastructure Mapping: A Case Study from Urban Sites in Celje, Slovenia. *Remote Sens.* **2020**, *12*, 1228. [CrossRef]
352. GEOFIT "Deployment of Novel GEOthermal Systems, Technologies and Tools for Energy Efficient Building retroFITting" Project (Grant Agreement ID 792210) (May 2018–April 2022). Available online: https://cordis.europa.eu/project/id/792210 (accessed on 24 December 2020).
353. Assets4Rail "Measuring, Monitoring and Data Handling for Railway Assets; Bridges, Tunnels, Tracks and Safety Systems" Project (Grant Agreement ID 826250) (December 2018–May 2021). Available online: https://cordis.europa.eu/project/id/826250 (accessed on 24 December 2020).
354. IM-SAFE "Harmonised Transport Infrastructure Monitoring in Europe for Optimal Maintenance and Safety" Project (Grant Agreement ID 958171). (November 2020–April 2023). Available online: https://cordis.europa.eu/project/id/958171/pl (accessed on 30 November 2020).

remote sensing

Article

GPR Monitoring of Artificial Debonded Pavement Structures throughout Its Life Cycle during Accelerated Pavement Testing

Xavier Dérobert [1,*], Vincent Baltazart [1], Jean-Michel Simonin [1], Shreedhar Savant Todkar [1], Christophe Norgeot [2] and Ho-Yan Hui [3]

1   GERS-GeoEND, Université Gustave Eiffel, Campus Nantes, IFSTTAR, F-44344 Bouguenais, France; vincent.baltazart@univ-eiffel.fr (V.B.); jean-michel.simonin@univ-eiffel.fr (J.-M.S.); shreedhar.todkar@univ-eiffel.fr (S.S.T.)
2   MDS—Le Matériel de Sondage, F-92400 Courbevoie, France; christophe@mds-paris.com
3   Department of Land Surveying and Geo-Informatics, The Hong Kong Polytechnic University, Hung Hom, KLN, Hong Kong, China; hoyan.hui@connect.polyu.hk
*   Correspondence: xavier.derobert@univ-eiffel.fr; Tel.: +33-240845911

Abstract: The paper gives an overview of a ground penetrating radar (GPR) experiment to survey debonding areas within pavement structure during accelerated pavement tests (APT) conducted on the university Gustave Eiffel's fatigue carrousel. Thirteen artificial defect sections composed of three types of defects (Tack-free, Geotextile, and Sand-based) were embedded during the construction phase between the top and the base layers. The data were collected in two stages covering the entire life cycle of the pavement structure using four GPR systems: An air-coupled ultra-wideband GPR (SF-GPR), two wideband 2D ground coupled GPRs (a SIR-4000 with a 1.5 GHz antenna and a 2.6 GHz-StructureScan from GSSI manufacturer), and a wideband 3D GPR (from 3D-radar manufacturer). The first stage of the experiments took place in 2012–2013 and lasted up to 300 K loadings. During this stage, the pavement structure presented no clear degradation. The second stage of experiments was conducted in 2019 and continued until the pavement surface demonstrated a strong degradation, which was observed at 800 K loadings. At the end of the GPR experiments, several trenches were cut at various sections to get the ground truth of the pavement structure. Finally, the GPR data are processed using the conventional amplitude ratio test to study the evolution of the echoes coming from the debonded areas.

Keywords: GPR; pavement; debonding; accelerated pavement testing (APT); database

Citation: Dérobert, X.; Baltazart, V.; Simonin, J.-M.; Todkar, S.S.; Norgeot, C.; Hui, H.-Y. GPR Monitoring of Artificial Debonded Pavement Structures throughout Its Life Cycle during Accelerated Pavement Testing. *Remote Sens.* **2021**, *13*, 1474. https://doi.org/10.3390/rs13081474

Academic Editors: Mercedes Solla, Vega Perez-Gracia and Simona Fontul

Received: 4 March 2021
Accepted: 1 April 2021
Published: 11 April 2021

Publisher's Note: MDPI stays neutral with regard to jurisdictional claims in published maps and institutional affiliations.

1. Introduction

Evaluation of road structures is of major importance to maintain their durability and extend their lifetime [1]. Damages due to heavy traffic may result from a weak or defective bonding between asphalt layers [2,3]. So, early detection of delamination in asphalt pavement is a challenge for appropriate maintenance or rehabilitation strategy. In this context, ground-penetrating radar (GPR) is part of efficient non-destructive testing (NDT) for the evaluation of road structures, for thicknesses estimation, and in particular for crack and debonding damages [4–8].

One project of the second Strategic Highway Research Program (SHRP-2) from the US has focused on asphalt pavement delamination. The results of research and experiments are gathered in several reports, dedicated to modeling and experimental tests performed on test and real sections by different NDT including radar systems from three manufacturers [9,10]. Among their conclusions, confirming those from [4–8], they state that GPR can detect moderate to severe delamination as interpretation of coherent anomalies at specific depths. Moreover, this detection is facilitated when water is present and induces damages (stripping) even if not particularly sensitive to severity.

In asphalt pavements, if two layers of HMA are well bonded, the only detectable effect in the GPR signal would be caused by the difference in dielectric properties between the two layers. Amplitude of such an echo is normally low as asphalt materials constituting the bond layers are very similar (coming from the same asphalt plant that uses the same bitumen and local aggregate type). When delamination occurs, the damage and water infiltration at the debonded area can produce stronger anomalous reflections, which can potentially be detected with GPR. Nevertheless, a limitation of debonding and crack detection can occur while radar wavelengths remain much larger than the thin layer of degradation at the interface of two bituminous layers.

The RILEM committee (TC 241-MCD) has conducted a state-of-the-art review for a fundamental understanding on the mechanism of cracking and debonding in asphalt and composite pavements [11]. It highlights in particular that water infiltration through cracks and other defects have significant influence on the road structure, while reducing bond strength drastically, and that temperature effects on bond behavior have to be studied more. These two points are of particular interest for accelerated pavement testing (APT). Such full-scale APT facilities enable to monitor the state of a section road all along its lifetime and be comparable to real structure under heavy traffic.

As these damages are due to traffic, a full-scale experiment, done with the pavement fatigue carrousel of the university Gustave Eiffel, has focused on the detectability of different kinds of artificial debonded areas in a pavement structure test by few GPR under a controlled traffic. The interest of such study is to monitor the test section along its service life, to create a large GPR database while studying the evolution of the test structure in terms of damage level and lateral extension, of the defects at different loading cycles.

The data collection was organized in a two-stage experiments and covers the full life-cycle of the pavement structure. During the first stage, which took place in 2012 [12], 300 K loading cycles were performed, and degradation, either in extension or evolution, was expected. Unfortunately, no visible surface degradation was observed at the end of this first stage of experiments. As a result, the ongoing experiment has prompted 6 years of several different actions, including internal and international research projects [13–15]. Recently, a final series of loading has occurred, with the support of three national contributions, to complete this experiment, allowing to reach an advanced level of degradation requiring repair from the point of view of the engineers managing road networks.

The article is organized as follows: Section 2 presents the accelerated pavement testing (APT), the GPR systems used during these campaigns, and the recognition of the pavement damage at the end of the experiment. Section 3 presents all the GPR results detailed and commented on per central frequency and per defect. Lastly, Section 4 describes the autopsy of the pavement structure and discusses the link between this ground truth information and GPR evaluation.

2. Design of the Accelerated Pavement Testing

2.1. The Fatigue Carrousel

The univ. Eiffel's pavement fatigue carrousel is a full-scale road traffic simulator, with accelerated pavement testing, composed of four arms carrying the moving heavy loads (65 kN on twin wheel for this experiment) at a maximum speed of 90 km/h [12]. The test track associated with this experiment is a 25 m structure of road composed of two bituminous layers (6 cm thick wearing course and 8 cm thick base layer) over a granular sub-base.

Thirteen rectangular patches of materials (sand, geotextile, and tack-coat free interface), simulating debonded areas, were inserted at the interface between the two asphalt layers [13]. The 3 types of defect were chosen to represent at least 2 levels of interface damage that can be observed by coring during our road expertise's or during experiments on accelerated pavement testing. The first level, simulated by I3 and I13, corresponds to a simple debonding of the layers, which separate during coring. The second level, simulated by the other insertions, corresponds to a worse defect where disintegrated materials are observed at the interface in a thin layer. Figure 1 presents the positioning of the 13 defects on the test track and Table 1 provides the detailed description of their characteristics. Designs of the accelerated pavement testing facility and the road structure test are detailed in [16,17].

The dynamic traffic, simulated by a half-axle load for twin wheels, has been performed centered on the majority of the defects on a radius of 16 m (Figure 2a). Three hundred thousand (or 300 K) loadings were realized in 2012 [12]. As no obvious defect was detected on the structure, from a road engineer point-of-view, 500 K more loads were carried out in 2019–2020, resulting in a degradation level officially requiring repairs.

To avoid unrealistic rutting, the twin wheel was laterally displaced on 11 positions, spaced 10.5 cm apart for a 1.65 m footprint width. Figure 2b shows the location and density of lateral traffic due to lateral wandering, done in the first- and second-stage experiments, considering the 62 cm width of the twin wheel.

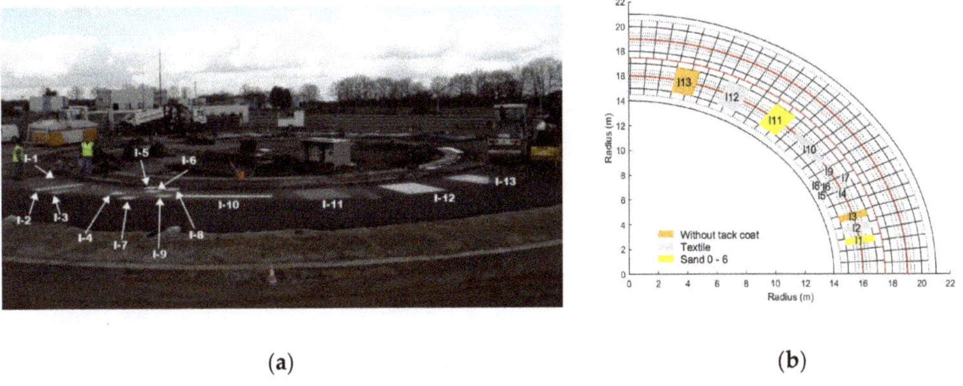

(a)   (b)

Figure 1. Test section, (**a**) with the defects made during construction, (**b**) in the form of a site plan.

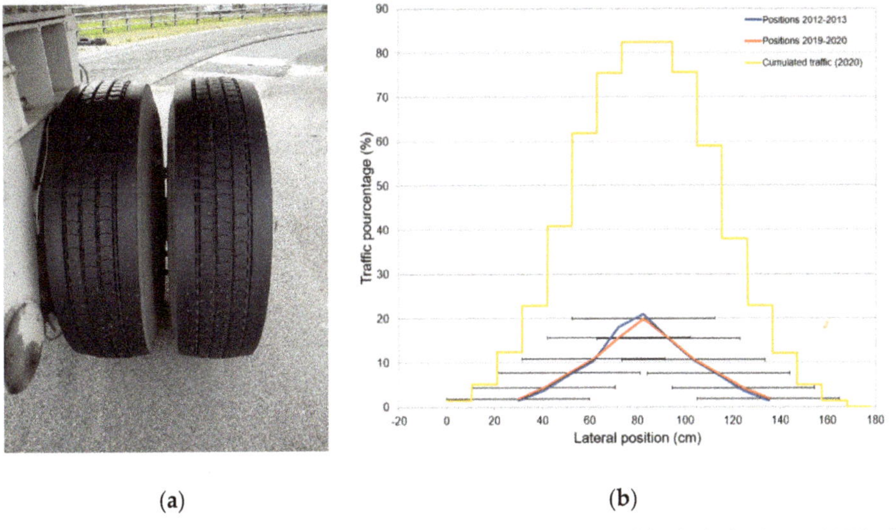

(a)   (b)

Figure 2. (**a**) Dual wheels of the carrousel, (**b**) Traffic density vs. lateral positions of the dual wheels per 100 K loadings.

Table 1. Characteristics of the artificial defects along the test track.

Name	Type	Length (m)	Width (m)
I-1	Sand	0.5	2
I-2	Geotextile	0.5	2
I-3	Tack-coat free	0.5	2
I-4 to I-9	Geotextile	0.5	0.5
I10	Geotextile	3	0.5
I-11	Sand	1.5	2
I-12	Geotextile	1.5	2
I-13	Tack-coat free	1.5	2

2.2. GPR Systems

In this section, the different GPR systems used during the two experiment stages are presented. A first group is characterized as impulse radar systems. As commercial common systems, they correspond to standards for major classical applications. Thus, three ground-coupled impulse radar were operated during this experiment, using GSSI systems; a SIR3000 device associated to a 2.6 GHz antenna during the first series in 2012, and for the 2019 series, a SIR4000 system combined with a 1.5 GHz antenna and a 2.6 GHz StructureScan (Figure 3).

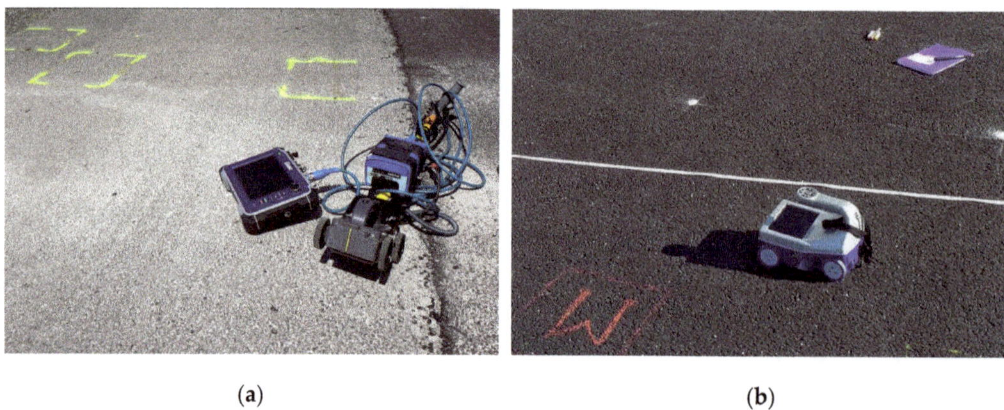

(a)   (b)

Figure 3. Impulse radar systems: (**a**) SIR4000 model with a 1.5 GHz antenna, (**b**) 2.6 GHz StructureScan.

In the second group, stepped-frequency systems were used—an experimental one developed at university Gustave Eiffel working with an equivalent 5 GHz central frequency, and a second one, which is a commercial array system, working with an equivalent 1.5 GHz central frequency (Figure 4). In this paper, the "central frequency" of the SFR pulse is defined from the timely radar pulse signature as the inverse of the wavelength period. The latter is roughly determined from the time shift between successive zero crossing (or from the time difference between successive amplitude extrema). For the SFR parameters at hand, this practical definition provides intermediate value between the peak energy of the pulse in the Fourier domain (~3.4 GHz) and the center of the GPR bandwidth (~5.8 GHz).

The univ. Eiffel's stepped-frequency radar (SFR) relies on ultra-wide-band (UWB) radar technology. Data were collected in the frequency domain within the bandwidth 0.8 GHz–10.8 GHz using a vector network analyzer (VNA) and two air-coupled UWB Vivaldi antennas [16]. Inverse fourier transform is conventionally used to provide time domain radar data (B-scans), comparable to a 5 GHz impulse system, in the time domain.

The 3D-radar manufacturer-provided stepped-frequency 3D radar array system is composed of 21 ground-coupled antennas separated by 7.5 cm providing a sweep width of ~1.40 m. The frequency band is ranging from 40 to 3000 MHz, working with the highest range of frequency, leading, after an inverse Fourier transform, to radar data equivalent to multiple 1.5 GHz B-scan. The array system is pulled behind a vehicle and localized by an RTK centimetric global positioning system.

(a)

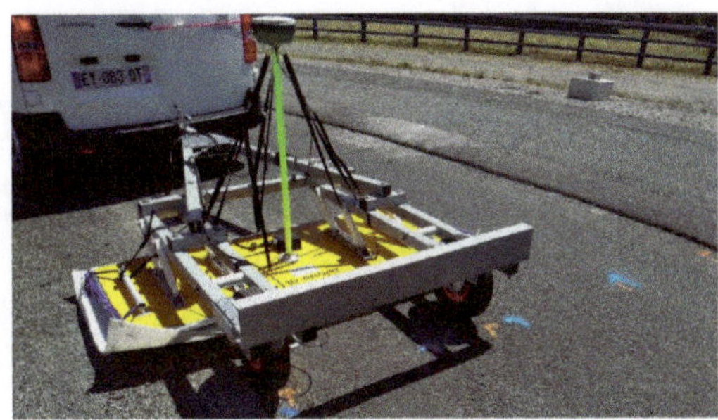

(b)

Figure 4. (a) Univ. Eiffel's robotic antenna-holder system associated to the stepped-frequency radar (SFR) with ultra-wide-band (UWB) antennas (surrounding blue cones are electromagnetic (EM) absorbing foams). (b) 3D-radar system during acquisition.

2.3. Acquisition and Processing Methodologies

During the experiment, radar surveys were performed at periodical loading stages. In 2012, they were focused on the largest defects, and done after the construction of the test zone, at 10 K loadings (the structure being considered as consolidated), and at around 50 K, 100 K, 200 K, 250 K, and 300 K loadings. In 2019–2020, the experiment was completed by measurements done on every defect (except the univ. Eiffel's SFR focused on I-11, I-12, and I-13) at about 310 K, 396 K, 420 K, 500 K, 600 K, 720 K, and 800 K loadings (end of the APT). All the data were realized in the summer period, except for 720 K loading stage done end of January.

For the data collection, raw B-scans from impulse radar systems and univ. Eiffel's SFR were taken at each loading stage, in two major directions: Transverse and longitudinal at the center of the defects. Transverse profiles were performed from the inner to the outer radius of the traffic path, and the longitudinal ones in the direction I-13 to I1. For the 3D-radar system, data are presented as horizontal maps (C-scans) of the surface echoes and the ones from the interface between the two bituminous layers. From these data, longitudinal B-scans have been extracted from the center of the largest defects.

The data pre-processing differs from the radar technology. The data pre-processing steps for impulse radar is specified in Table 2. Because of the technology of acquisition, the stepped-frequency radar data are basically required to perform inverse Fourier transform to provide the temporal B-scan images (with some time gating to limit the time horizon) and some calibration steps beforehand to take into account the antenna response in free space.

Table 2. Pre-processing steps performed on 2.6 GHz GSSI data.

Steps	Setting Parameters
Subtract DC-shift	Start time: 0 ns
	Stop time: 6 ns
Subtract mean (Dewow)	Time window: 3.5 ns
Bandpass filter	Lower cut-off: 520 MHz
	Upper cut-off: 5200 MHz
Remove header gain	Yes

Figure 5a,b compares the pre-processed B-scan data obtained by the 2.6 GHz impulse radar and the 5 GHz SFR radar on I-11 (sand) at 720 K loadings. For the sake of clarity, the vertical time scales have been harmonized. The surface echo at the top shows the strongest amplitude, and it is called direct wave (DW) for both data types. The second strongest echo is the reflected wave (RW) from the sand-based debonded interface. The reflected echo from the healthy area shows a weaker amplitude at roughly the same time-depth.

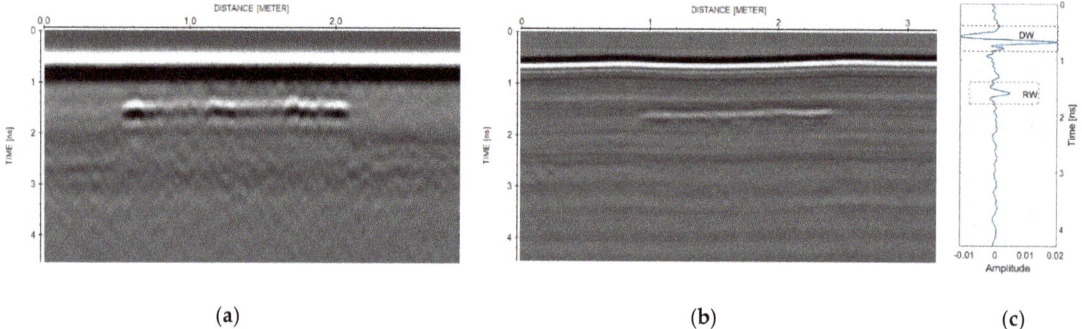

(a)   (b)   (c)

Figure 5. Examples of B-scans performed at 720 K loadings on the defect I11 (sand) by (**a**) the ground-coupled 2.6 GHz impulse system and (**b**) the air-coupled 5 GHz stepped-frequency system. (**c**) A-scan extracted from Figure 5b on the defect and showing the direct wave (DW) and the reflected wave (RW).

The next step in detection debonding requires to compute, for each A-scan, the conventional amplitude ratio test (ART) as the ratio between the reflected wave (RW) and the direct wave (DW), namely, ART = RW/DW [15,18]. Then, the data processing basically consists of automatically picking the maximum (or minimum) amplitude of the two latter echoes with some existing commercial radar software. In terms of pavement monitoring, any debonding provides additional echoes, which mostly interact constructively with each other, resulting in an increased signal amplitude of the reflected signal (RW) compared to the one of the healthy zone. The pavement monitoring then turns on analyzing the amplitude ratio variation w.r.t. traffic loading.

Moreover, as the first layer is subject to strong mechanical stresses, which are not transferred to the second layer in the zones of defects, cracking and micro-cracking appear vs. traffic much more quickly. Then, the analysis of the amplitude of the surface direct wave itself was carried out, as characterizing the first layer is also a subject of interest.

2.4. State of the Road Section at the End of the Experiment

The experiments were stopped in August 2020 after noticeable damage was seen on the pavement surface, which, in a real situation, would require some repairs. Three major types of damages occurred: Cracking, micro-cracking, and rutting. The cracks initially appeared around 500 K loading over major defect zones beginning with I13 followed by I12 and I11. Beyond this loading, the cracking and micro-cracking evolved towards an unacceptable density (see Figure 6), while the healthy zones did not present any surface cracks.

Moreover, several cross-section profiles, done with a laser rugosimeter system, were performed showing the rutting resulting from the heavy traffic. Figure 7 presents, as an example, the cross-section profiles done in the center of I-13 and in two healthy zones (between I-11 and I-13 and between I-3 and I-4), from the inner to the outer radius of the traffic path. Measurements show rutting of above 2 cm in depth at the center of the largest defects, while it remains under 1 cm in the healthy zones.

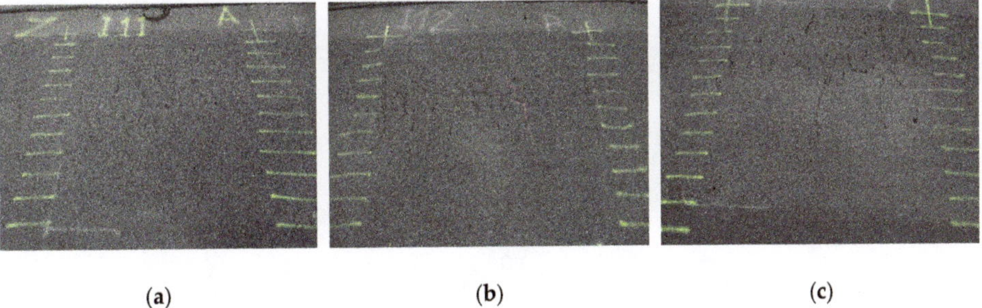

(a)   (b)   (c)

Figure 6. Photographs of defects (**a**) I-11 (sand), (**b**) I-12 (geotextile), and (**c**) I-13 (tack-coat free), at stage 800 K loadings. Paints correspond to the presumed longitudinal limits of the defects.

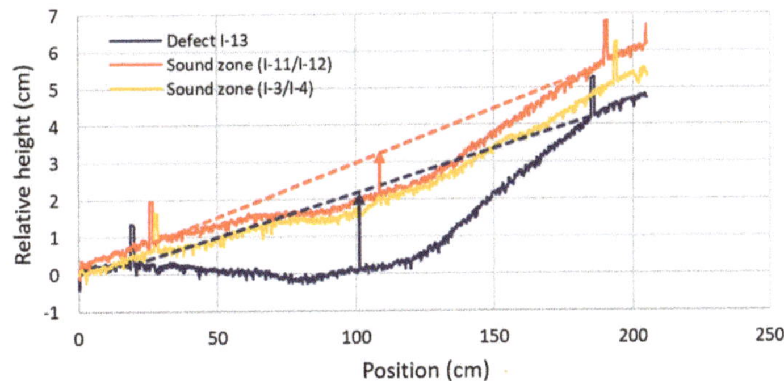

Figure 7. Example of cross section profiles done with a laser rugosimeter after 800 K loadings, with one centered on defect I-13 and two others done on healthy zones (between defects).

3. GPR Results

3.1. Introduction

This global experiment, done in two series over a period of eight years, led to the acquisition of more than 360 B-scans from impulse systems and more than 140 from stepped-frequency ones, with open-access data available in [16,17]. Processing methodology, presented in Section 2.3, has been performed on most of these data. Nevertheless, for reasons of clarity, only the major ones are presented and commented on, with the others being referenced in Appendices A–D.

A general overview is proposed thanks to GPR maps (C-scans) done by the 3D-radar system. Then, results are presented defect by defect, for the largest ones, then together for the narrowest ones.

3.2. GPR Amplitude Maps

Figure 8 presents a longitudinal profile along the center of the test section at 2.6 GHz. We note that every defect is detected (I-1 location being more ambiguous) and that defects composed of geotextile (I-4, I-9, I-10, and I-12, but not I-2) show similar echoes stronger than the others, explained by stronger dielectric contrasts. Several sensors, embedded to monitor the stress under the heavy traffic, are located in the center of I-13 and between I-12 and I-13. This is why results presented for defect I-13 in the next section will show a lack of values in the center of the data.

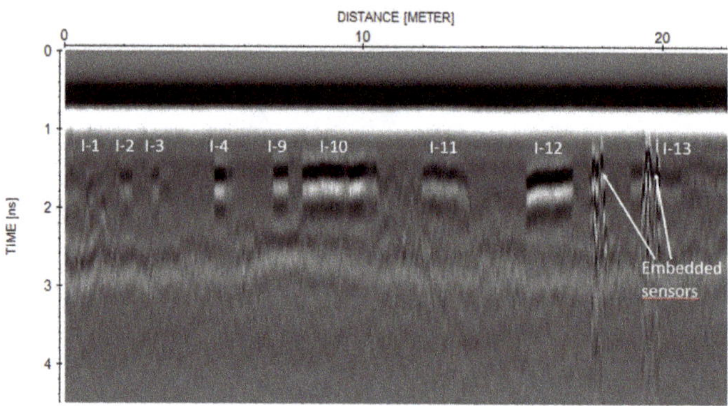

Figure 8. Example of longitudinal B-scan performed at 2.6 GHz along the complete test section.

C-scans were performed at 396 K, 500 K, 600 K, 720 K, and 800 K loading stages, using the commercial software 3D-radar Examiner. Figure 9 shows the horizontal slices of the C-scans taken at the interface (or defect depth). These slices give an overview of the global test section and its general evolution. Most of the defects are clearly visible and their geometry well drawn. Nevertheless, the thickness and the EM state of the top layer varies due to the implementation and evolution of the asphalt layer. Therefore, they can present some bias in estimating the importance of the defects. So, it is interesting to study the amplitude of the surface direct wave (DW) and the echo at the interface of the two asphalt layers (RW), as shown in Figures 10 and 11, picked in the area of I-11 to I-13.

From these maps, some general observations can be made. Firstly, the heavy traffic has not induced strong lateral expansion of the defects although numerous surface cracks that appeared in the two last loading stages. The evolution of the RW amplitudes is visible on Figure 10 for increasing vs. degradation for defects I-11 and I-13 until 720 K loading stage, with defect I-12 presenting a surprisingly high amplitude level at 600 K loading stage. The last measurement stage, done in July 2020 five months after the 720 K loading one (imposed COVID confinement), shows a decreasing trend. Such a decrease of EM contrast suggests an internal moisture evaporation and possible auto-repair of the structure.

Lastly, we note that at the internal radius area the RW amplitudes are stronger. It could be due to the transverse slope towards the center of the carrousel, as shown in Figure 6, which trap the moisture at the border of the defects.

The DW maps from Figure 11 show that stronger amplitudes from the central zones of defects I-13 and I-11 (and a second-order of defect I-12) may be related to traffic. They could be attributed to greater concentration of constraints inducing visible cracking in Figure 6, and so, an increase of surface porosity and a lower permittivity. Another possibility could be the effect of the rutting (several millimeters) on the direct wave due to loss of contact of the antenna array on the ground.

Complementary measurements done by bi-static systems on the centers of the defects make comparisons possible.

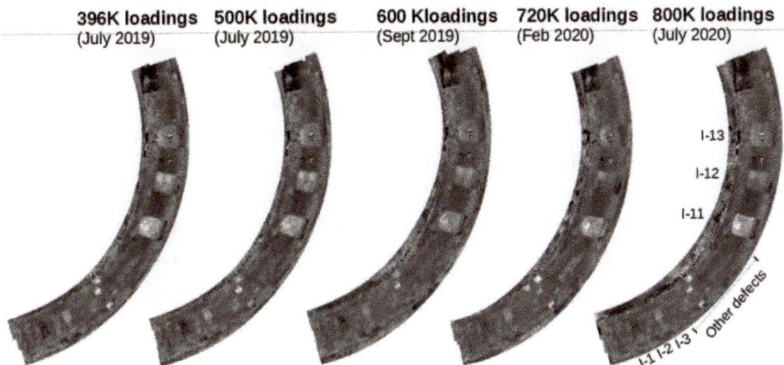

Figure 9. RW amplitude maps of the global test zone at the defect depth, extracted from 1.5 GHz C-scans.

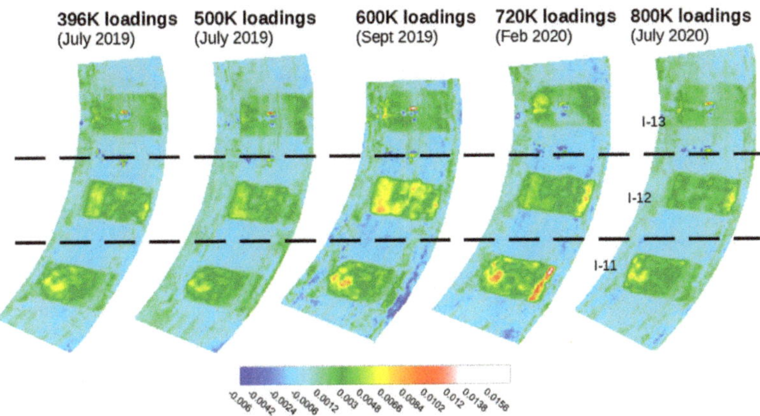

Figure 10. Maps of the reflected echo at the defect interface extracted from C-scans in the area of I-11 to I-13 (values in Legend refer to the amplitude of the received signal).

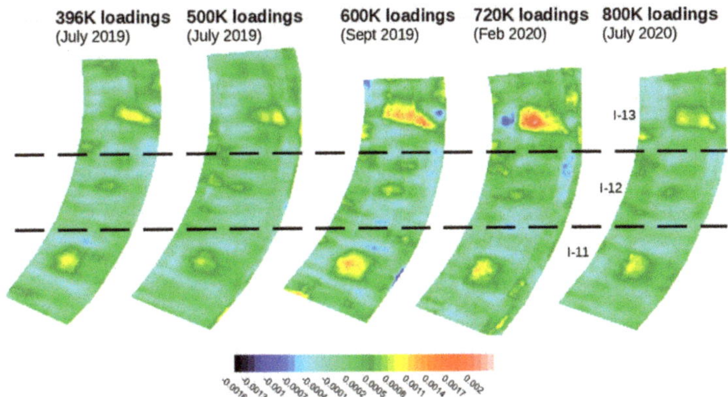

Figure 11. Maps of the surface direct wave extracted from C-scans in the area of I-11 to I-13 (values in Legend refer to the amplitude of the received signal).

3.3. Study of the Largest Defects I-11 to I-13

The processed maps from the C-scans show variations in the DW, and thus, suggest an evolution of the state of the pavement layer. This also reinforces the use of amplitude ratio approach to process all the B-scans. The approach further enables to estimate the evolution, the degradation of RW above defects, and possibly compare different central frequency results.

The following sub-sections detail the results per central frequency, from 2.6 GHz considered as a commonly used central frequency for this application, to 1.5 and 5 GHz. Appendix A gives some B-scans examples, done on the I-12 defect, for every central frequency.

3.3.1. Study of Defects I-11 to I-13 at 2.6 GHz

Figures 12–14 gather normalized RW amplitudes, namely RW/DW as defined in Section 2.3, picked on the defects I-11, I-12 and I-13 from the 2.6 GHz data over all the loading stages.

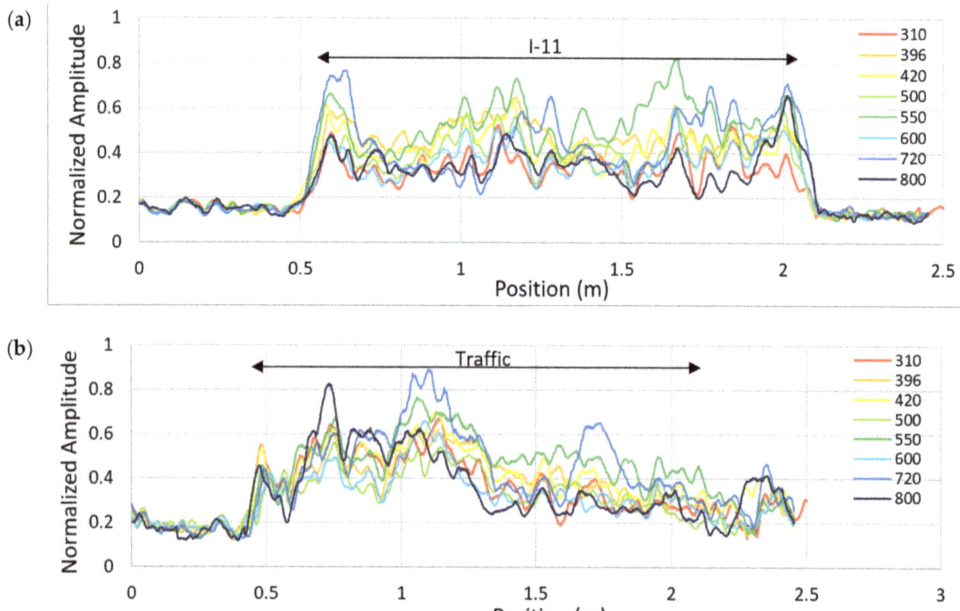

Figure 12. 2.6 GHz Norm. RW amplitudes on sand (I-11) in the (**a**) long. and (**b**) transv. direction vs. K loadings.

3.3.3. Study of Defects I-11 to I-13 at 5 GHz

When measurements are done with air-coupled antennas, the polarity of the DW and RW are inverted and the amplitude ratios present negative values. We observed, in Figures 19–21, similar results to those from 2.6 GHz, with stable values for longitudinal profiles and a general trend for the transverse ones. As the acquisition length of the univ. Eiffel's robotic antenna-holder system is limited to 1.6 m, longitudinal acquisitions were performed twice per defect. For the processing of these B-scans, they were gathered before picking, with the results being shown in Figures 19a, 20a and 21a. Results obtained at 5 GHz show the same trend as the 2.6 and 1.5 GHz ones.

While studying normalized amplitudes, we noted that 2.6 GHz corresponded to the best central frequency for debonding detection of the pavement layer. Indeed, on the defect I-12 for example, we obtained average amplitudes of about 0.5–0.55 for 2.6 GHz, 0.45 for 1.5 GHz, and 0.5–0.55 for 5 GHz. When comparing these with the 1.5 GHz results, higher frequency, and smaller wavelength, shows better sensitivity to the very thin layer as debonding. This observation is no longer valid as, for the wavelength at 5 GHz, asphalt concrete cannot be considered as homogeneous. At such high frequencies, EM waves are scattered by the biggest aggregates (diam = 10 mm) inducing an attenuation that counterbalances their sensitivity to very-thin-layer detection.

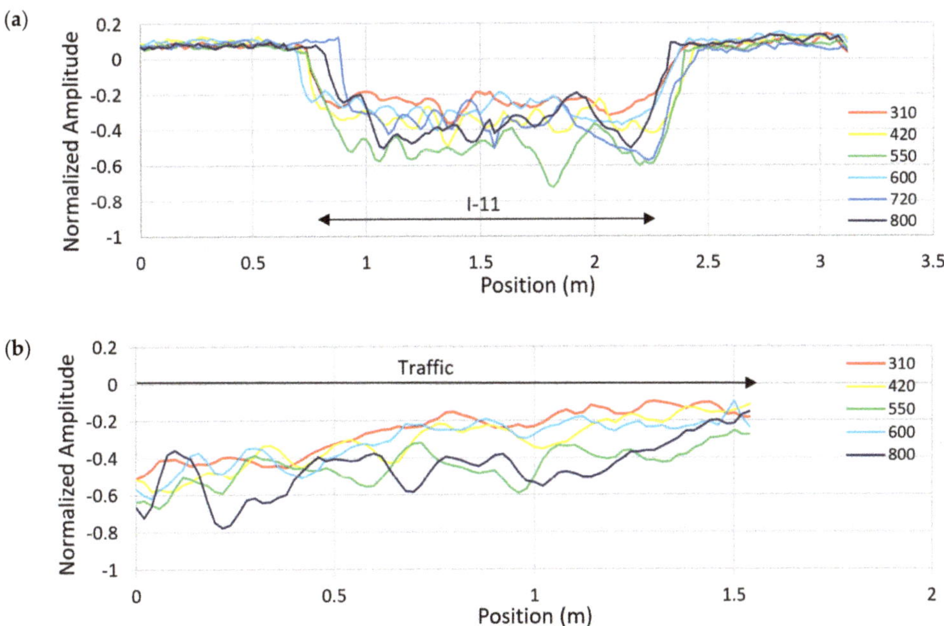

Figure 19. 5 GHz Norm. RW amplitudes on sand (I-11) in the (**a**) long. and (**b**) transv. direction vs. K loadings.

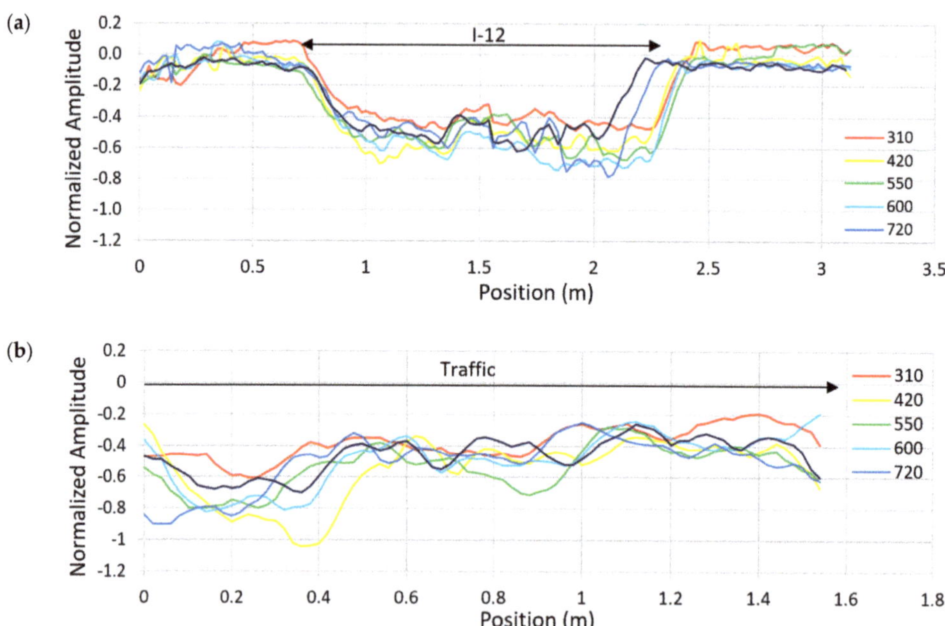

Figure 20. 5 GHz Norm. RW amplitudes on geotextile (I-12) in the (**a**) long. and (**b**) transv. direction vs. K loadings.

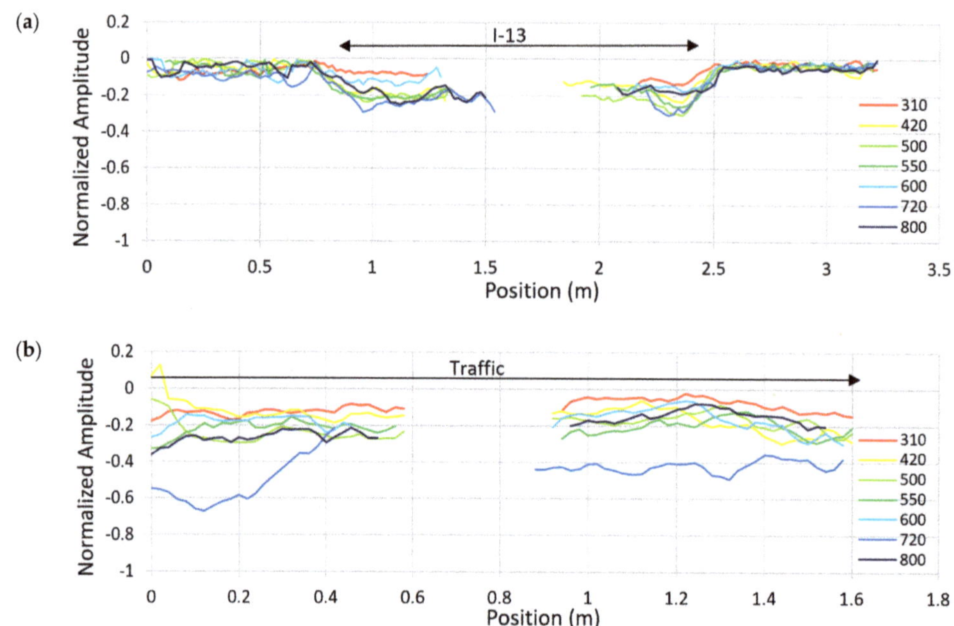

Figure 21. 5 GHz Norm. RW amplitudes on tack-coat free (I-13) in the (**a**) long. and (**b**) transv. direction vs. K loadings.

3.4. Direct Waves in the Zone I-11 to I-13 at 2.6 GHz

Similarly, DW amplitudes are picked on the B-scans for each defect and are then normalized by averaged values of DW coming from the healthy zone, outside the traffic, from the beginning of the corresponding transverse profiles. This parametric study has been performed with 2.6 GHz data (presented in this section) and 1.5 GHz data (presented in Appendix C). Finally, the DW at 5 GHz is not included in this paper as the data are not reliable due to the variations caused by rutting.

Figures 22–24 respectively present the longitudinal and transversal variation in normalized DW amplitude for 2.6 GHz data over I-11, I-12, and I-13 defects. Results show that the first layer has already suffered damage from the first series of loading in 2012, remaining from the start of the second series of loading.

Longitudinal data do not show clear trend between healthy and weak zones for the defect I-11. However, detection is more visible for the two other defects (due to higher EM contrasts).

Concerning transverse data, we can note a general trend proportional to the traffic density, shown in Figures 22b, 23b and 24b, in a V shape with the minimum amplitude corresponding to the maximum of traffic. As the data come from ground-coupled acquisition, the decrease of the normalized amplitudes could be associated with an increase of the relative permittivity of the first asphalt, which narrows the radiation pattern and thus decreases the DW amplitude. This finding, when associated with the increase of cracks in the first layer, can be interpreted as an increase of moisture content trapped in this opened porosity.

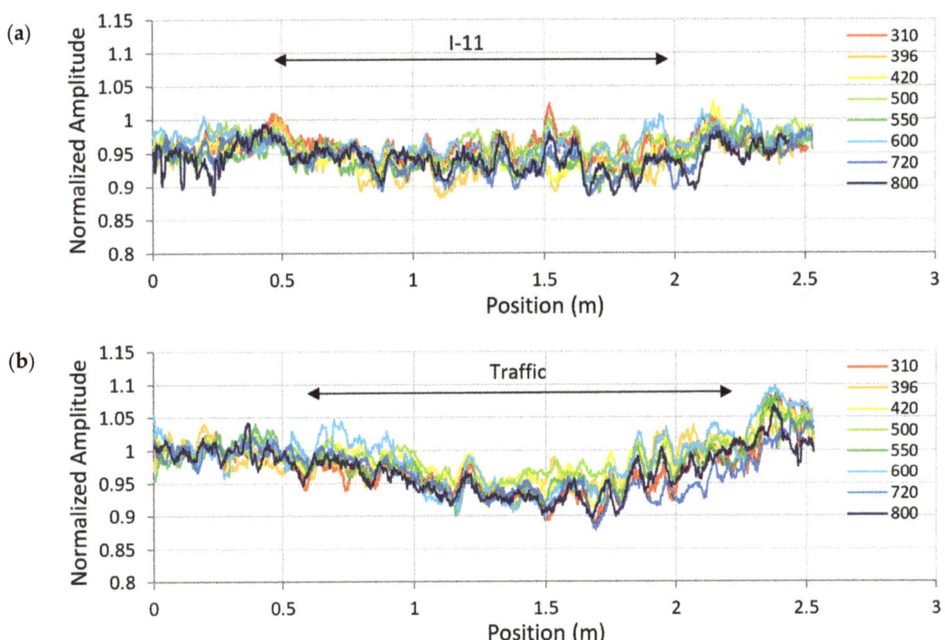

Figure 22. 2.6 GHz Norm. DW amplitudes on sand (I-11) in the (**a**) long. and (**b**) transv. direction vs. K loadings.

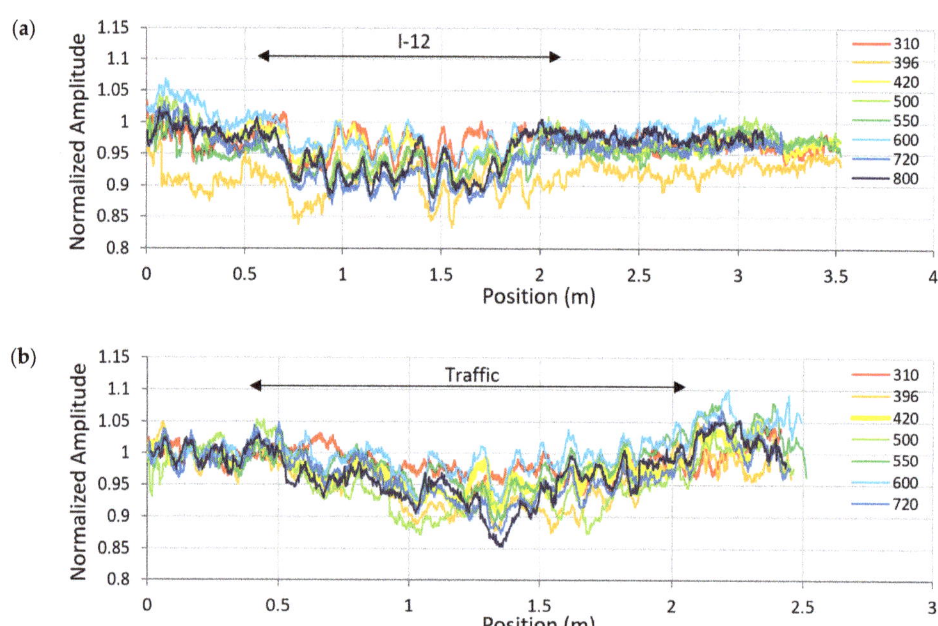

Figure 23. 2.6 GHz Norm. DW amplitudes on geotextile (I-12) in the (**a**) long. and (**b**) transv. direction vs. K loadings.

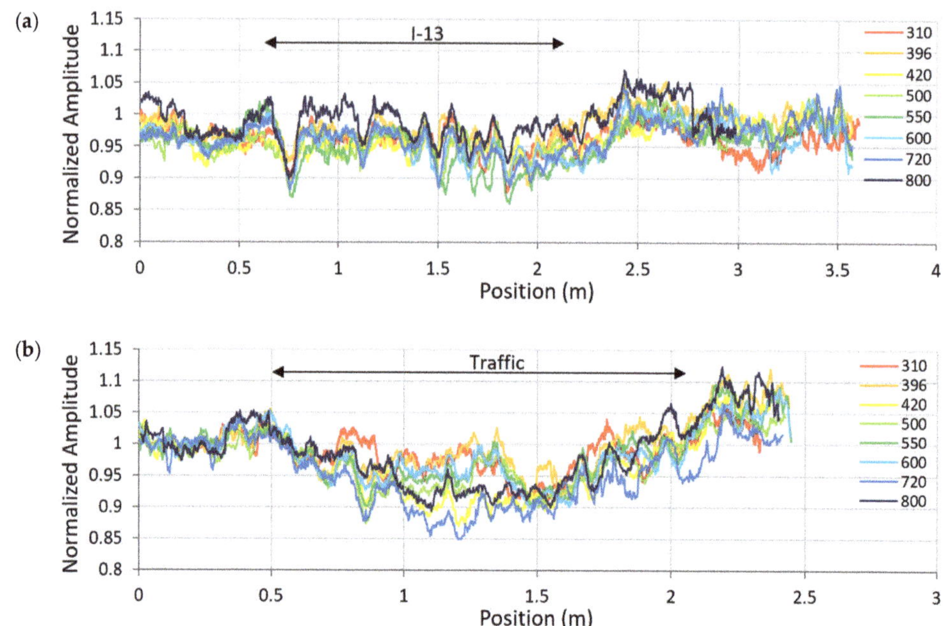

Figure 24. 2.6 GHz Normalized DW amplitudes on tack-coat free (I-13) in the (**a**) long. and (**b**) transv. direction vs. K loadings.

3.5. Study of Defects I-1 to I-3

Defects I-1 to I-3 correspond respectively to defects I-11 to I-13 with a narrow width of 50 cm. Measurements were only done at 1.5 and 2.6 GHz, with GSSI systems during the second stages of experiment. Only 2.6 GHz data are processed and presented in this section as the 1.5 GHz and 2.6 GHz data are very similar. From Figures 25–28, we note that the narrowness of the defects induces different behavior under traffic than I-11 to I-13. The amplitudes do not appear to be stable along the longitudinal traces of I1 to I3. Additionally, no V-shaped variation in trends is seen for the transversal results over the largest defects. This phenomenon could be explained by the loading transfer in the asphalt partially supported by the healthy borders, and then not damaging the surveyed interface.

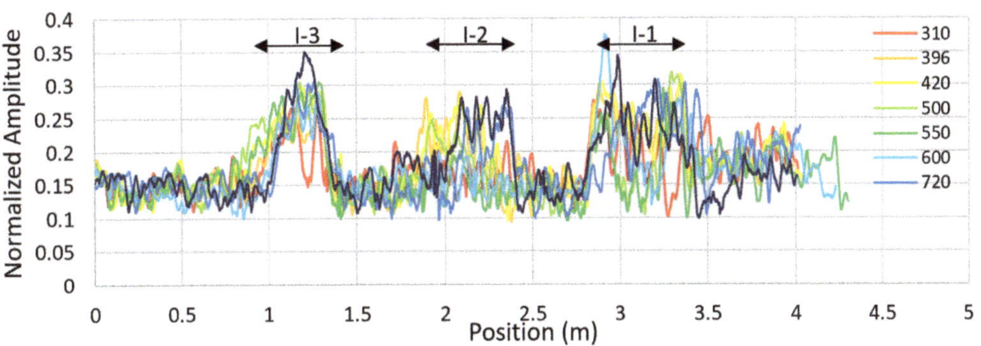

Figure 25. 2.6 GHz Normalized RW amplitudes on defects I-3 to I-1, in the longitudinal direction vs. K loadings.

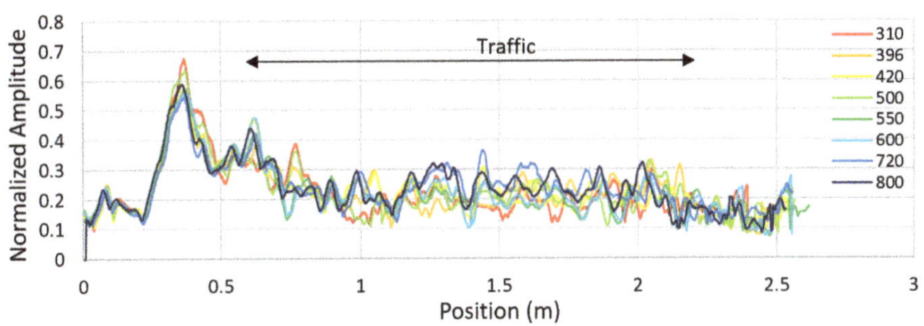

Figure 26. 2.6 GHz Normalized RW amplitudes on sand (I-1), in the transversal direction vs. K loadings.

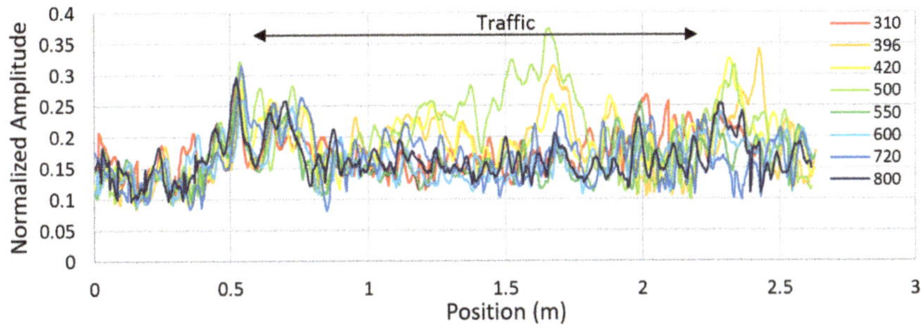

Figure 27. 2.6 GHz Normalized RW amplitudes on geotextile (I-2), in the transversal direction vs. K loadings.

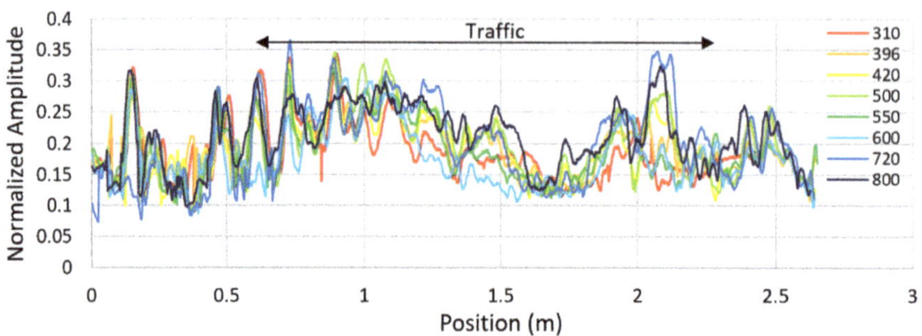

Figure 28. 2.6 GHz Normalized RW amplitudes on tack-coat free (I-3), in the transversal direction vs. K loadings.

3.6. Study of Defects I-4 to I-10

This panel of defects presents a similar interface (geotextile-based), with small or narrow dimensions and some of them being shifted from main traffic. As for defects I-1 to I-3, measurements were performed with 1.5 and 2.6 GHz GSSI systems, and only 2.6 GHz results are presented herein.

Figures 29–38 show strong variations of normalized amplitudes from defect I-4 to I-10, depending on the loading step and the location of the defect. As a reminder, defects I-4, I-9, and I-10 are centered on the axis of the major loading traffic. Defects I-6 and I-7 are off-center, even out of traffic, and at last, defects I-5 and I-8 are out of traffic.

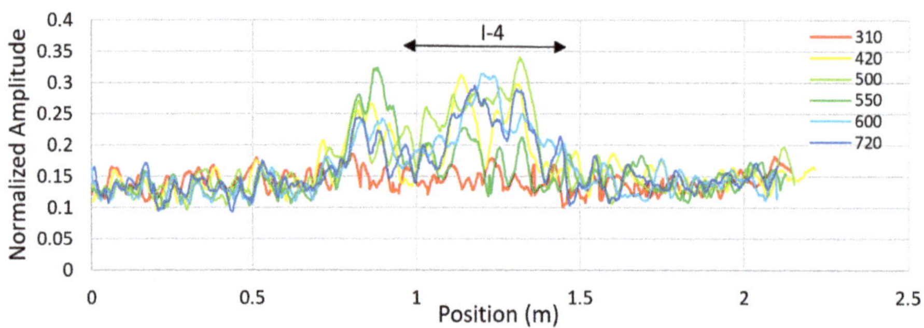

Figure 29. 2.6 GHz Normalized RW amplitudes on defect I-4, in the longitudinal direction vs. K loadings.

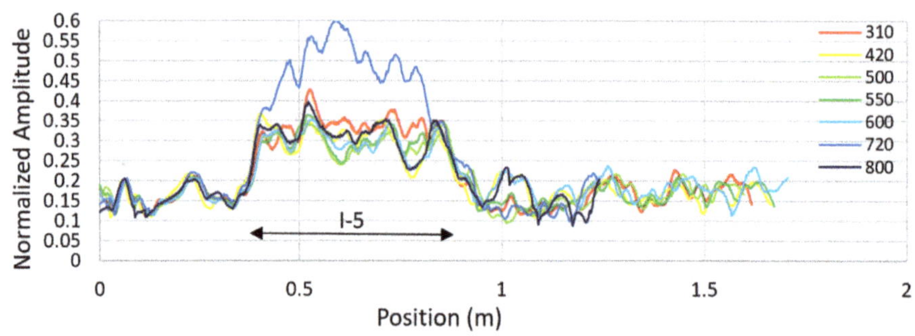

Figure 30. 2.6 GHz Normalized RW amplitudes on defect I-5, in the longitudinal direction vs. K loadings.

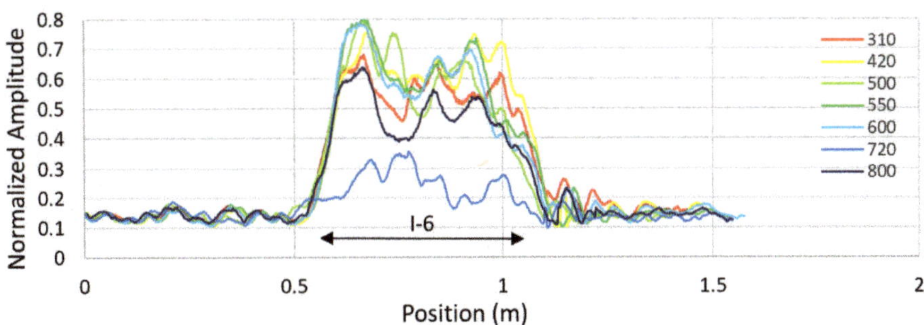

Figure 31. 2.6 GHz Normalized RW amplitudes on defect I-6, in the longitudinal direction vs. K loadings.

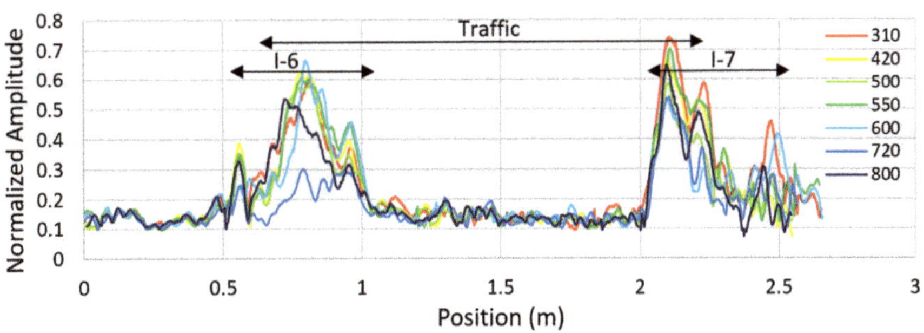

Figure 32. 2.6 GHz Normalized RW amplitudes on defects I-6 and I-7, in the transversal direction vs. K loadings.

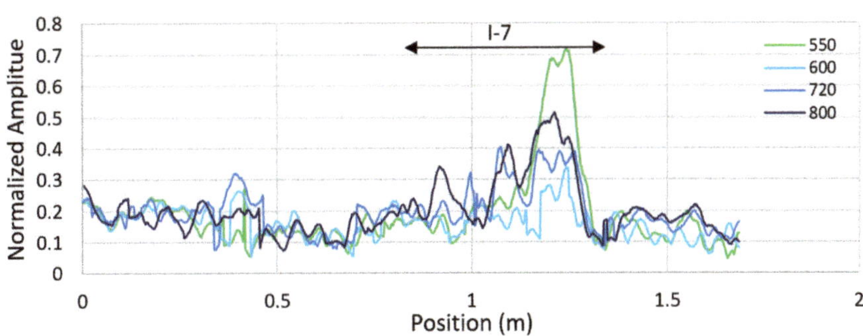

Figure 33. 2.6 GHz Normalized RW amplitudes on defect I-7, in the longitudinal direction vs. K loadings.

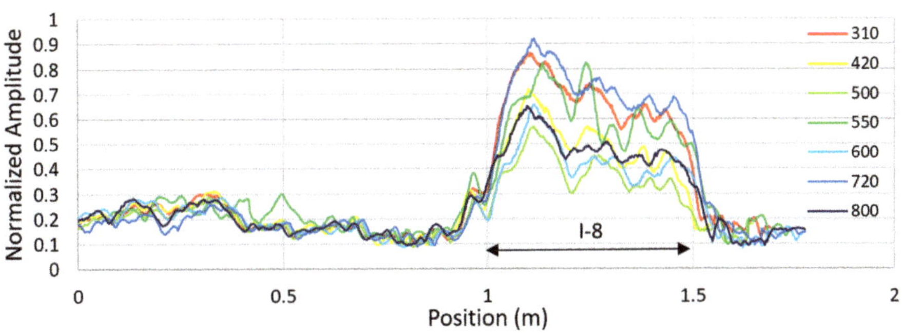

Figure 34. 2.6 GHz Normalized RW amplitudes on defect I-8, in the longitudinal direction vs. K loadings.

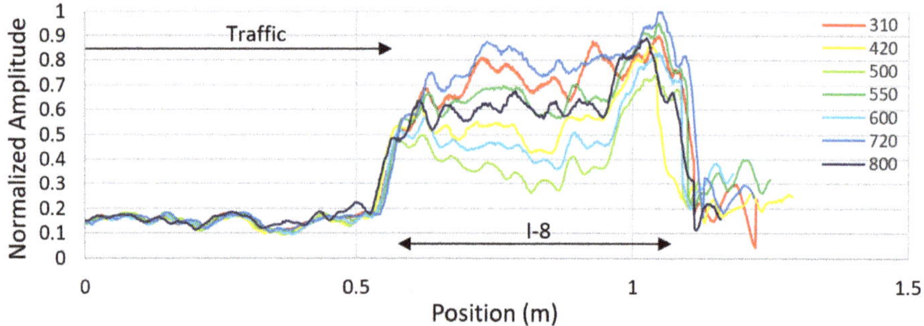

Figure 35. 2.6 GHz Normalized RW amplitudes on defect I-8, in the transversal direction vs. K loadings.

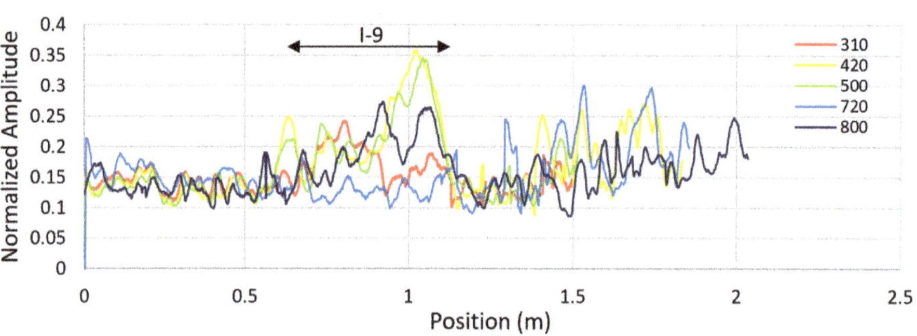

Figure 36. 2.6 GHz Normalized RW amplitudes on defect I-9, in the transversal direction vs. K loadings.

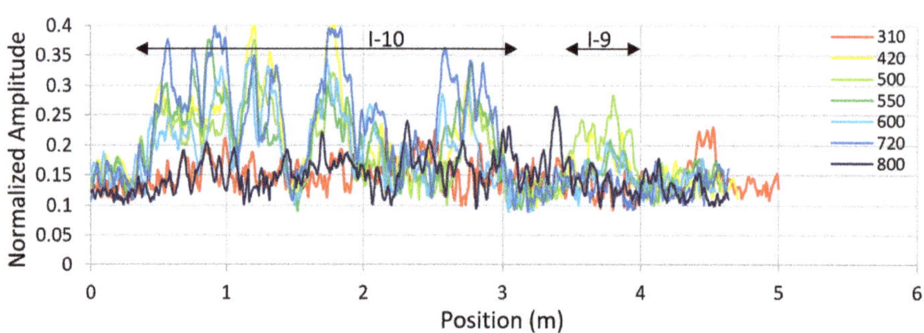

Figure 37. 2.6 GHz Normalized RW amplitudes on defects I-9 and I-10, in the longitudinal direction vs. K loadings.

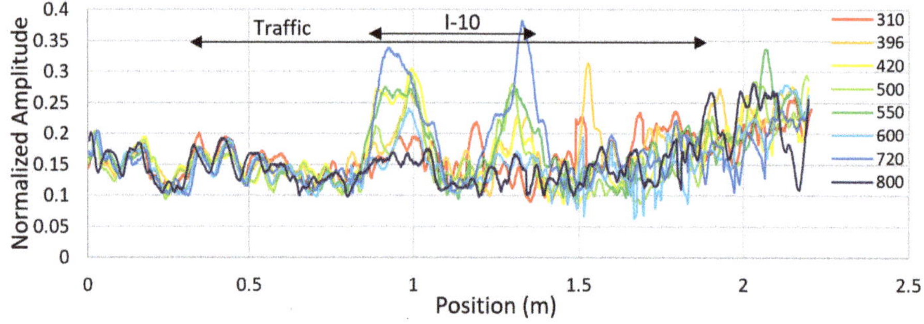

Figure 38. 2.6 GHz Normalized RW amplitudes on defect I-10, in the transversal direction vs. K loadings.

The analysis of measures done on defects I-5 and I-8 can be interesting since these areas reveal the evolution of the pavement structure without traffic function of the months and seasons. Moreover, Figures 30–34, results show a variability of building. Indeed, I-5 appears to be very stable regardless of the traffic step and the period of acquisition (the exception being the 720 K loading step, the only one done in winter), while I-8 shows strong variations, which may be due to variation of water ingress.

Defects I-4, I-9, and I-10, located on the center of the traffic zone, present amplitude values roughly two times lower than the value obtained from defect I-12, with some punctual zones in which amplitude values exceed just the average of neighboring healthy areas. These strong variations of EM contrasts suggest variation of tack coat gluing, inducing possible water ingress.

Most surprising is the shape of I-10 results (Figure 38) showing detection only on the borders of the defect. Amplitudes are growing vs. traffic during the summer 2019 (from 310 K to 550 K loadings) and a slow decrease at 600 K loading step. For these last measurements, the explanation could come from the fact that the heavy traffic, going towards 600 K loadings, was realized several weeks before GPR measurements, letting the structure auto-repair with the high temperatures. We find a similar situation for the 800 K loading step: Low radar amplitudes, measurements done at the end of May, and traffic cycles done numerous weeks before. The exception comes from the 720 K loading steps, as GPR measurements were performed in winter (January 2020) during a cold and rainy period, and showing strong EM contrasts may be due to water ingress. This analysis, done on defect I-10, is representative of almost other defects.

Concerning transversal profiles, we can see the combined effects of lateral traffic density (see Figure 32) and topography (see Figure 35), which can induce inward water migration, on the GPR amplitude.

4. Autopsy of the Road Section

4.1. Extraction of Transversal Blocks

At the end of the loading series and GPR experiments, several trench cores were sampled to get a ground truth of the damaged structure, and first of all, the real thickness of the defects and their lateral length. Six trenches have been realized in the area of the largest defects. Four transverse trenches were located at the center of defects I-10, I-11, I-12, and I-13, at the level of the GPR profiles, and dimensions of approximately as follows: 1.30 × 0.25 × 0.15 m (Figure 39). One transverse trench was done between defects I-11 and I-12 to obtain information of the traffic effect on a healthy zone. The last one was sawn longitudinally from the center of I-12 towards I-13, 1.2 m long.

(a)

(b)

Figure 39. Transverse trenches and corresponding asphalt slabs of (**a**) defect I-10 and (**b**) defects I-12.

During the extraction of the blocks, all but the healthy one broke, as shown in Figure 39a. From the perspective of road experts, this was due to weakness of the top layer from the defective zones, presenting numerous vertical micro-cracks on the side (some of them being attributed to the extraction).

Going into detail (Figure 40), the findings are as follows:

- The geotextile-based defect presents no clear degradation. The geotextile is about 5 mm thick and remains glued to the layer 1. When it is debonded to layer 2, we cannot see its lateral extension.
- Concerning the sand-based defect, a layer of void is visible due to sawing under water, which carried away the sand. The apparent thickness of this debonding is about 3 mm, with a maximum of 11 mm near the inner radius and a minimum around 1.2 mm seen near the center of the traffic. Moreover, it seems that no aggregates were loosened from one of the two asphalt layers.
- The tack-free-based defect presents a narrow debonding (visible by wetting in Figure 40c) all along the defect of sub-millimeter to millimeter thickness, but no loose aggregates.

To conclude, the extracted blocks and the trenches have shown no loosing of material from asphalt layers, but only micro-cracks, mainly vertically oriented.

(a) (b)

(c) (d)

Figure 40. Details of defects visible on the side of asphalt slabs, based on (**a**) sand (I-11), (**b**) geotextile (I-10), (**c**) tack-coat free (I-13), and (**d**) healthy zone.

4.2. Estimation of Void Content and Relative Permittivity

The estimation of void content has been performed on layer 1 of the trenched block, extracted over I-10. It was then tested under gamma radiation in the laboratory several weeks after its extraction. Gammadensimetry, a semi-destructive radiation method (Figure 41a), is viewed as a reliable laboratory testing method and considered as a reference in the field of the bulk density variation measurement [19].

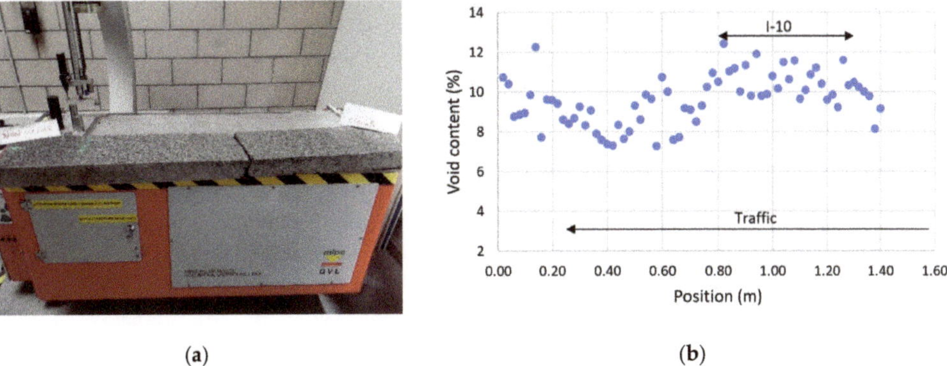

(a)   (b)

Figure 41. (a) Gammadensimetric acquisition on asphalt layer, (b) void content estimation vs. of lateral location.

Measurements were performed on six parallel lines of 1.4 m long. While considering an average bulk density of the asphalt, a void content value is calculated every 2 cm. As the gamma ray is very narrow (>10 mm), an average is done so that values are representative of the mixing vs. location of the traffic (Figure 41b).

Data show that the asphalt presents higher void content (~10.5%) above the defect than in healthy zone under traffic (~8.7%), which can be explained by higher density of micro-cracks.

In parallel, two actions were performed to estimate the permittivity of the asphalt layer 1 along the transversal direction. First, travel time Δt in layer 1 was extracted from the automatic time picking of the 2.6 GHz amplitude Bscan done on I-10 after the 800 K loading step, taking account the ground-coupled bistatic mode (offset X) of acquisition (Figure 42a). While knowing the thickness of the layer 1, this approach is complementary to [20] studying the GPR amplitude vs. compaction.

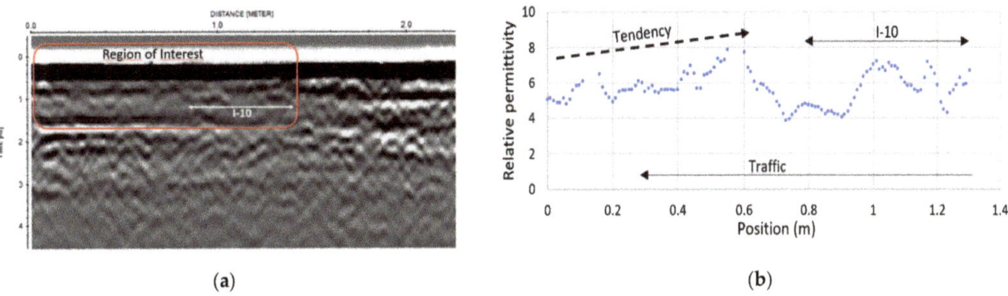

(a)   (b)

Figure 42. (a) 2.6-GHz transversal profile done on defect I-10 at 800 K loadings. (b) Estimation of the relative permittivity of the asphalt layer 1.

Second, 12 Mpx images were taken on the side of the asphalt slab. An operator manually calibrated the images in size and accurately measured the thickness D of layer 1 every 1 cm along the entire length.

From this information, the relative permittivity of the asphalt layer can be calculated as presented in Figure 42b, using the following equation where c is the velocity in the air:

$$\varepsilon_r = \left[\frac{c.\Delta t}{2\sqrt{\left(\frac{X}{2}\right)^2 + D^2} - X} \right]^{\frac{1}{2}} \tag{1}$$

We note an increase of permittivity while going from the inner area toward the center of traffic in the first 0.6 m. Then, picking is biased by the defect I-10. This result is compatible with the ones obtained from 2.6-GHz DW in Figures 22–24, as the increase of permittivity reduces the radiation pattern and then the DW amplitude. A possible interpretation is an increase of porosity (or micro-cracking) in which humidity has penetrated and is kept by capillary effect.

The decrease of relative permittivity of the asphalt slab from gammadensimetry results, being correlated to an increase of porosity or micro-cracking, is explained by an indoor natural drying before measurements.

These results show an opposite tendency to the ones observed on the measurements of the 1.5 GHz Stepped-frequency array system (Figure 11). A possible explanation could come from the fact that the antennas, inserted in a wide plastic box, lie at a small height from the asphalt layer. This could induce a merge of an aerial direct wave and with the direct wave in the surveyed medium, and then an increase of amplitude as the permittivity of the layer decreases.

5. Conclusions

A major GPR experiment was performed on a full-scale accelerated pavement testing at the university Gustave Eiffel, on a section presenting three artificial defects of bonding (tack-free, geotextile, and sand-based). Measurements were done periodically with several GPR systems all along the life-cycle of the road structure in two stages: 300 K loadings in 2012–2013 and 500 K loadings in 2019–2020 leading to a structure considered as strongly degraded.

Four GPR systems were tested, mainly during the second stage of experiment from 1.5 to 5 GHz central frequency. More than 500 GPR profiles were acquired, most of them being processed for this paper. Data processing was focused on the amplitude analysis (amplitude ratio test) of the reflected waves at the interface of the defects and on the direct waves.

The major results are the following:

- Defects are detected almost all the time (with exceptions), due to sufficient dielectric contrasts. The geotextile defect presents the strongest EM contrast followed by the sand-based defect, with the tack-free defect being nevertheless detectable.
- Debondings did not show lateral expansions during the life-cycle of the tested structure.
- Variations of normalized reflected amplitude of the interface echoes did not show a reliable correlation with heavy traffic. Amplitudes appeared to be more sensitive to the meteorology (humidity, water ingress from rainy periods, temperature) and the time delay between the end of traffic step and GPR measurements. Indeed, for the first point, the appearance of micro-cracks allowed increased moisture content inside the defect due to the weather conditions from the previous days. For the second point, under high temperatures, asphalt layers with visco-elastic materials could slightly auto-repair, therefore reducing the EM contrasts.
- When comparing the central frequencies for the detectability of debonding defects, considered as very thin layers, 2.6 GHz corresponds to the best frequency for debonding detection, compared to the two others. This central frequency, higher than 1.5 GHz, shows better sensitivity to very thin layer as debonding, and is not attenuated as

at 5 GHz. Indeed, at such very high frequencies, EM waves are scattered by the biggest aggregates (diam. = 10 mm) inducing an attenuation that counterbalances their sensitivity to very-thin-layer detection. This phenomenon is not visible on the B-scans as the beamwidth of the air-coupled antennas induces an average on this scattering effect.

- Concerning the study of GPR direct waves, we observed a general tendency showing that degradation of the surface layer is visible by GPR measurements. A general tendency of decreaing DW amplitude vs. traffic is associated with the increase in micro-cracks, and then of water content of the medium.
- At last, we observe that overall, the results are similar to those from SHRP-2 [9] if we do not consider the influence of the traffic. In this report, GPR results on full-scale road sections, from five radar systems, have shown that degraded zones (no bond and stripping, similar to defects I-13 and I-11) were not always detected, with the exception of after water ingress. Moreover, the choice of time-depth slices as a detection parameter is not optimal as thicknesses of layers are not perfectly constant and thus the extracted amplitudes may not correspond to a maximum of EM contrast. This problem is removed when picking the echoes at an interface (problem solved for the 3D-radar system in 2019, see Figures 10 and 11).

The database is available to the GPR community and may serve as a reference benchmark for both developing and testing the performance of various processing or monitoring purposes of debonded areas of pavement structures under full-scale controlled traffic [16,17].

Author Contributions: Conceptualization, X.D., V.B., and J.-M.S.; methodology, X.D., V.B., and J.-M.S.; software, X.D., V.B., and C.N.; validation, X.D., V.B., and J.-M.S.; formal analysis, X.D., V.B., S.S.T., C.N., and H.-Y.H.; investigation, X.D., V.B., C.N., and H.-Y.H.; resources, X.D., V.B., and J.-M.S.; data curation, X.D., V.B., and J.-M.S.; writing—original draft preparation, X.D.; writing—review and editing, X.D., V.B., and J.-M.S.; funding acquisition, X.D., V.B., and J.-M.S. All authors have read and agreed to the published version of the manuscript.

Funding: The data collection in 2019 has been supported by the French Research Agency (grant ANR-18-CE22-0020), the FEREC foundation (grant CINC-RSF), and the Ministry of Ecological and Solidarity Transition (grant MTES N° 18/402).

Institutional Review Board Statement: Not applicable.

Informed Consent Statement: Not applicable.

Data Availability Statement: Database of GPR B-scans are available at: https://doi.org/10.255 78/N0RXSQ (accessed on 4 March 2021) and, https://doi.org/10.25578/MM9Q0R (accessed on 4 March 2021).

Acknowledgments: The authors would like to thank the technical staff of univ. Gustave Eiffel Carrousel (S. Trichet, T. Gouy, G. Coirier, Y. Baudry, M. L. Nguyen et J. Blanc), the one of GeoEND laboratory (O. Durand, G. Gugole and T. Devie) and J.M. Moliard from S2I laboratory. The authors also thank P.H.G. Ching for experimental help in the frame of a collaboration with Polytechnic university of Hong Kong. The first part of the data collection in 2012, has been a contribution of the RILEM technical committee TC241-MCD ("Mechanisms of cracking and debonding in asphalt and composite pavements", www.rilem.org (accessed on 4 March 2021)) and the COST action TU1208 ("Civil engineering applications of ground-penetrating radar").

Conflicts of Interest: The authors declare no conflict of interest. The funders, cited in the Acknowledgments section, had no role in the design of the study; in the collection, analyses, or interpretation of data; in the writing of the manuscript, or in the decision to publish the results.

Appendix A

Presentation of longitudinal B-scans performed at about 396 K and 720 K loadings on the defect I-12.

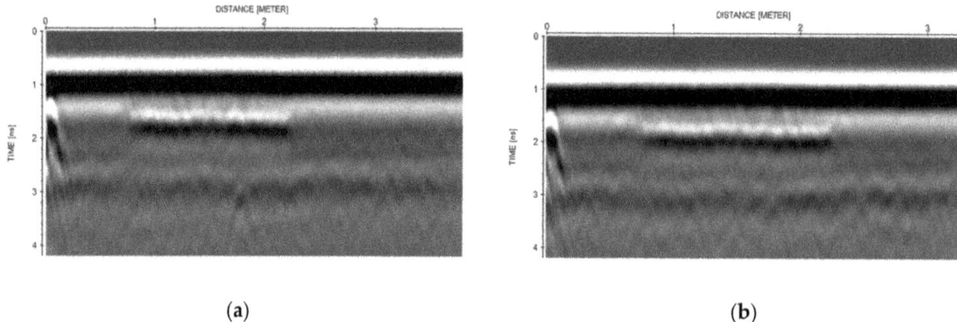

(a) (b)

Figure A1. 1.5 GHz impulse B-scans performed on geotextile (I-12) at (**a**) 396 K and (**b**) 720 K loadings.

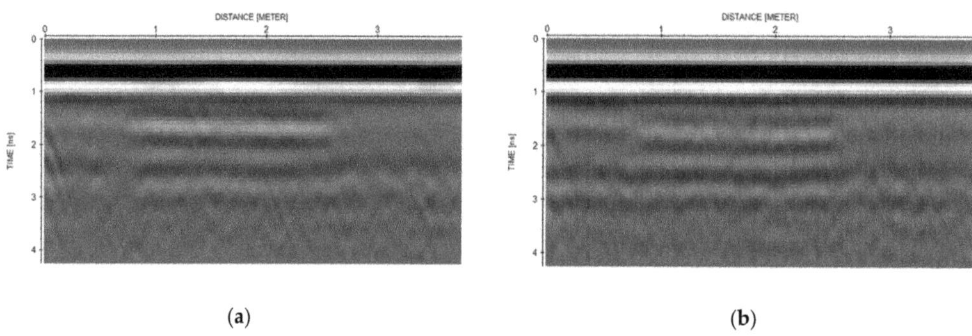

(a) (b)

Figure A2. 1.5 GHz stepped-frequency B-scans performed on geotextile (I-12) at (**a**) 396 K and (**b**) 720 K loadings.

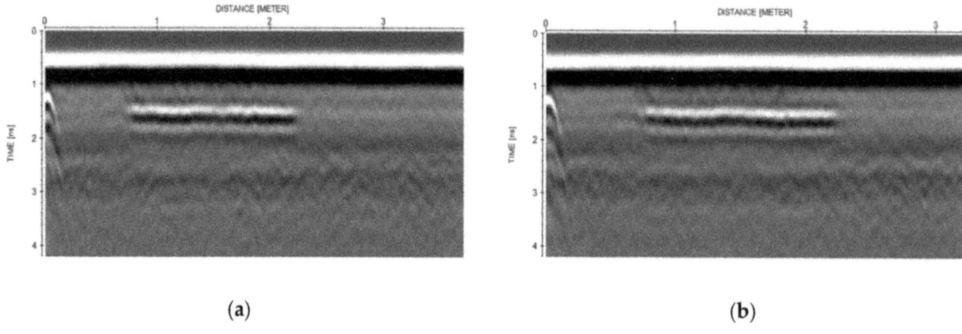

(a) (b)

Figure A3. 2.6 GHz impulse B-scans performed on geotextile (I-12) at (**a**) 396 K and (**b**) 720 K loadings.

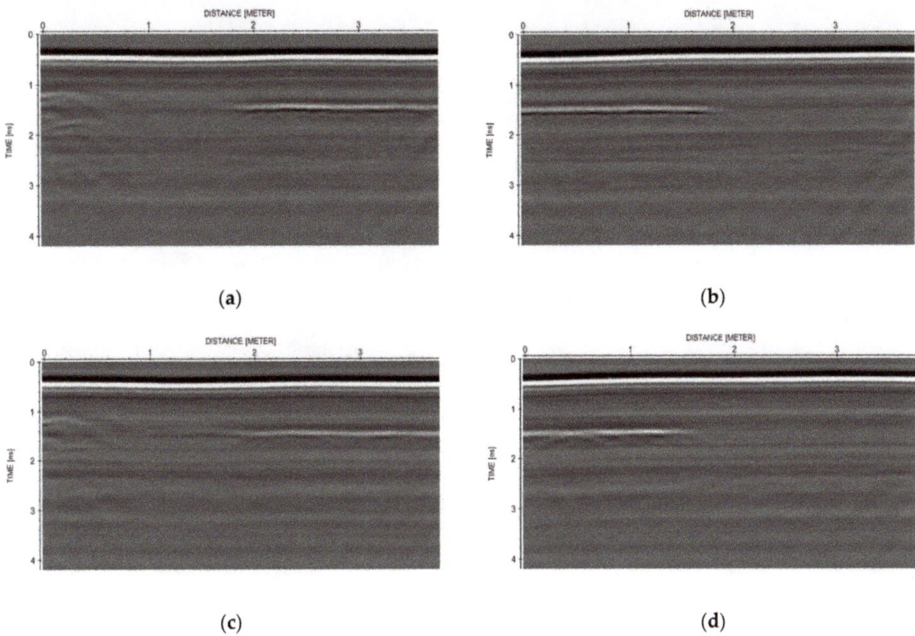

(**a**) (**b**)

(**c**) (**d**)

Figure A4. 5 GHz stepped-frequency B-scans performed on geotextile (I-12) in two steps at (**a**,**b**) 420 K and (**c**,**d**) 720 K loadings.

Appendix B

Presentation of normalized RW amplitudes of 2.6 GHz GPR profiles done on defects I-11 to I-13 during the 2012 series.

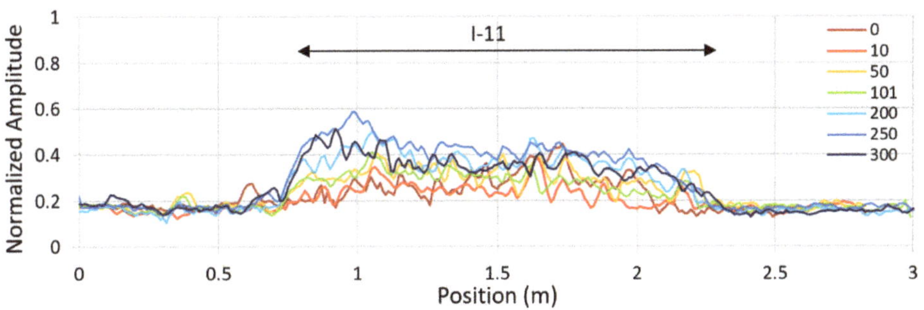

Figure A5. 2.6 GHz normalized RW amplitudes on sand (I-11) in the longitudinal direction vs. K loadings.

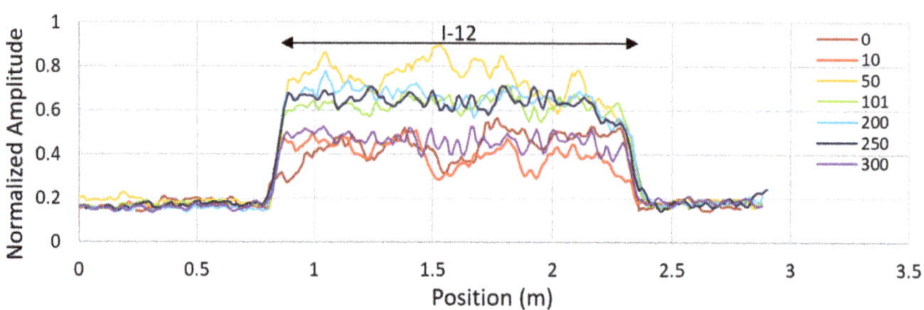

Figure A6. 2.6 GHz normalized RW amplitudes on geotextile (I-12) in the longitudinal direction vs. K loadings.

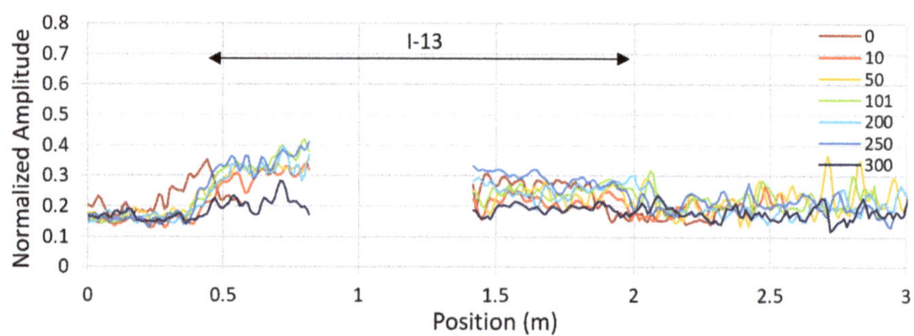

Figure A7. 2.6 GHz normalized RW amplitudes on tack-coat free (I-13) in the long. direction vs. K loadings.

Appendix C

Presentation of normalized DW amplitudes of 1.5 GHz GPR profiles done on defects I-11 to I-13 during the 2019–2020 series.

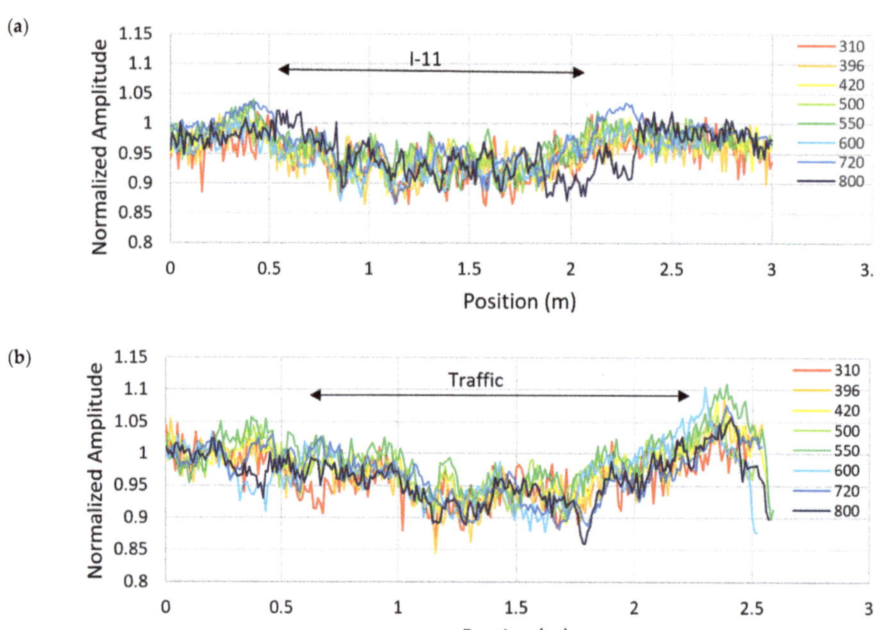

Figure A8. 1.5 GHz Norm. DW amplitudes on sand (I-11) in the (**a**) long. and (**b**) transv. direction vs. K loadings.

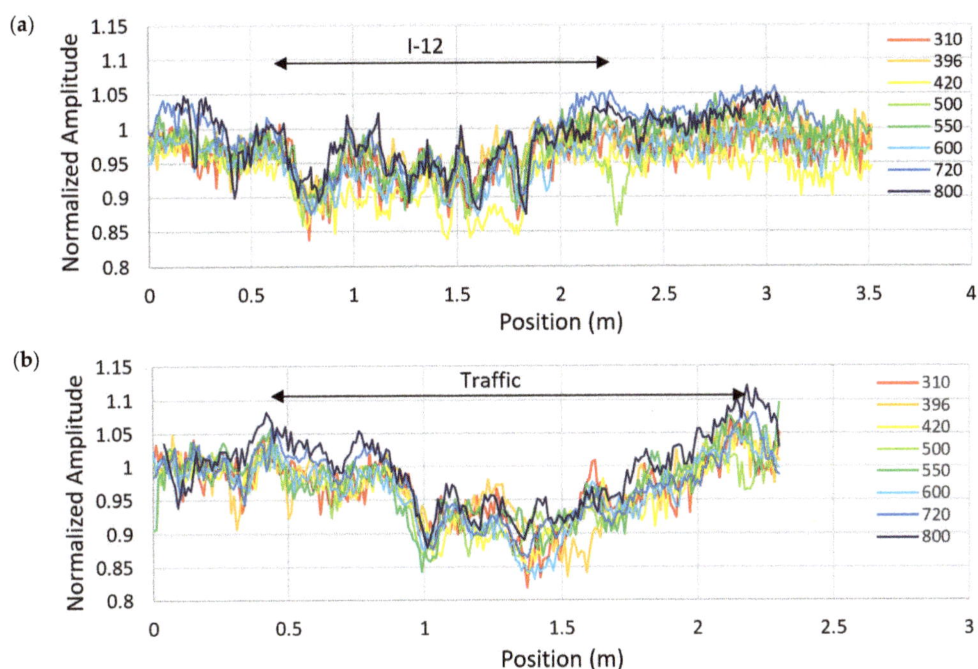

Figure A9. 1.5 GHz Norm. DW amplitudes on geotextile (I-12) in the (**a**) long. and (**b**) transv. direction vs. K loadings.

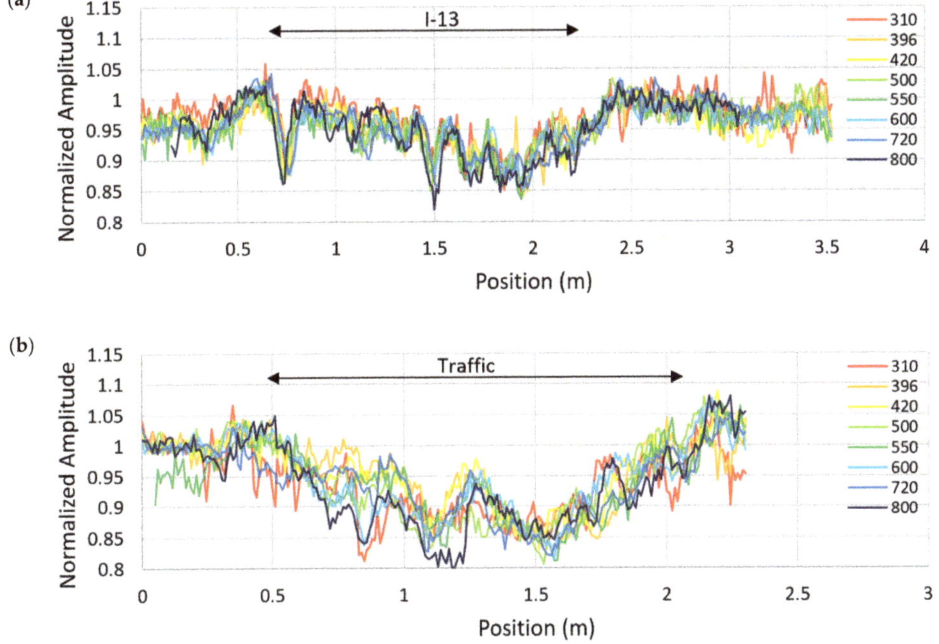

Figure A10. 1.5 GHz Norm. DW amplitudes on tack-coat free (I-13) in the (**a**) long. and (**b**) transverse direction vs. K loadings.

Appendix D

Presentation of DW amplitudes of 2.6 GHz GPR profiles done on defects I-11 to I-13 during the 2012 series.

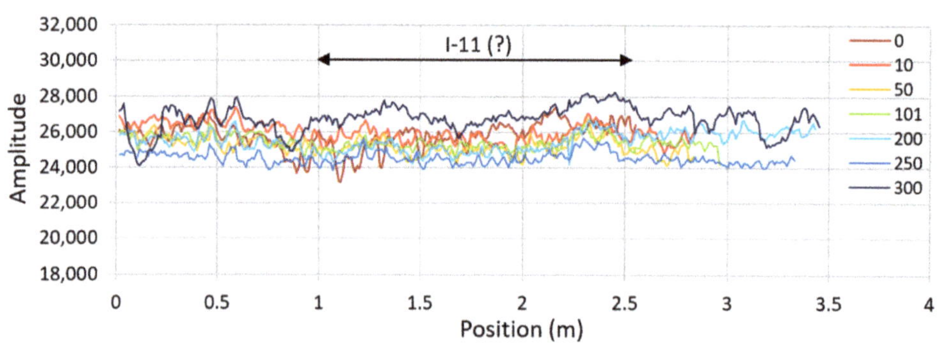

Figure A11. 2.6 GHz DW amplitudes on sand (I-11) in the longitudinal direction vs. K loadings.

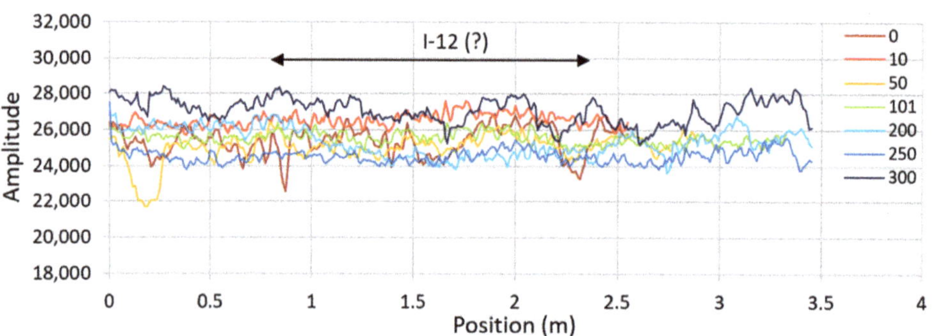

Figure A12. 2.6 GHz DW amplitudes on geotextile (I-12) in the longitudinal direction vs. K loadings.

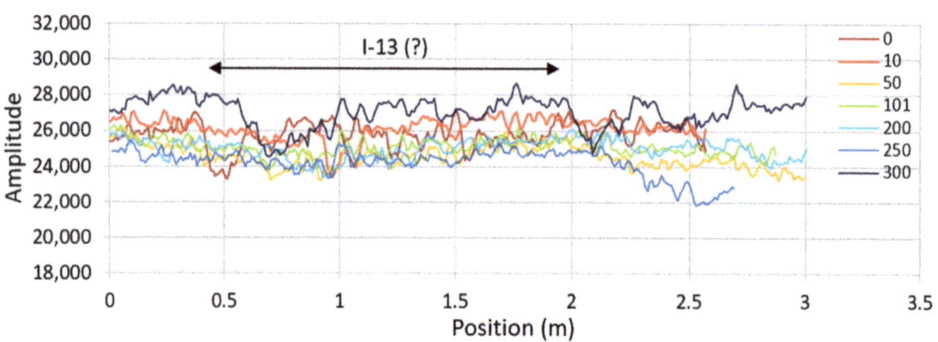

Figure A13. 2.6 GHz DW amplitudes on tack-coat free (I-13) in the longitudinal direction vs. K loadings.

References

1. Baltazart, V.; Moliard, J.-M.; Amhaz, R.; Cottineau, L.-M.; Wright, A.; Wrigh, D.; Jethwa, M. Automatic Crack Detection on Pavement Images for Monitoring Road Surface Conditions—Some Results from the Collaborative FP7 TRIMM Project. In *8th*

RILEM *International Conference on Mechanisms of Cracking and Debonding in Pavements*; Springer: Dordrecht, The Netherlands, 2016; Volume 13, pp. 719–724. [CrossRef]

2. Kruncheva, M.R.; Collop, A.C.; Thrn, N.H. Effect of bond condition on flexible pavement performance. *J. Transp. Eng.* **2005**, *131*, 880–888. [CrossRef]
3. Gharbi, M.; Nguyen, M.L.; Chabot, A. Characterization of debonding at the interface between layers of heterogeneous materials coming from roads. In Proceedings of the Congrès Français de Mécanique, Lille, France, 28 August–1 September 2017.
4. Saarenketo, T.; Scullion, T. Road evaluation with ground penetrating radar. *J. Appl Geophys.* **2000**, *43*, 119–139. [CrossRef]
5. Plati, C.; Georgouli, K.; Loizos, A. A Review of NDT assessment of road pavements using GPR. *RILEM Books* **2012**, *6*, 855–860.
6. Solla, M.; Lagüela, S.; Gonzalez, H.; Arias, P. Approach to identify cracking in asphalt pavement using GPR and infrared thermographic methods: Preliminary findings. *NDT E Int.* **2014**, *62*, 55–65.
7. Marecos, V.; Solla, M.; Fontul, S.; Antunes, V. Assessing the pavement subgrade by combining different non-destructive methods. *J. Constr. Build. Mat.* **2017**, *135*, 76–85. [CrossRef]
8. Lai, W.L.; Dérobert, X.; Annan, A.P. A review of ground penetrating radar application in civil engineering: A 30-year journey from locating, testing and evaluation to imaging and diagnosis. *NDT E Int.* **2018**, *96*, 58–78.
9. *Nondestructive Testing to Identify Delaminations between HMA Layers*; SHRP-2 report S2-R06D-RW-3; Transportation Research Board: Washington, DC, USA, 2013; Volume 3, p. 146.
10. *Nondestructive Testing to Identify Delaminations between HMA Layers*; SHRP-2 report S2-R06D-RW-4; Transportation Research Board: Washington, DC, USA, 2013; Volume 4, p. 158.
11. Buttlar, W.; Chabot, A.; Dave, E.V.; Petit, C.; Tebaldi, G. RILEM State Art Reports. In *Mechanisms of Cracking and Debonding in Asphalt and Composite Pavements*; Springer: Cham, Germany, 2018; Volume 28, p. 237.
12. Simonin, J.M.; Baltazart, V.; Hornych, P.; Kerzrého, J.P.; Dérobert, X.; Trichet, T.; Durand, O.; Alexandre, J.; Joubert, A. Detection of debonding and vertical cracks with ND techniques during accelerated tests. In Proceedings of the Accelerated Pavt Testing (APT) Conference, Davis, CA, USA, 19–21 September 2012.
13. Simonin, J.M.; Hornych, P.; Baltazart, V.; Dérobert, X.; Thibaut, E.; Sala, J.; Utsi, V. Case study of detection of artificial defects in an experimental pavement structure using 3D GPR. In Proceedings of the GPR'2014 Congress, Brussels, Belgium, 30 June–4 July 2014.
14. Simonin, J.M.; Baltazart, V.; Le Bastard, C.; Dérobert, X. Progress in monitoring the debonding within pavement structures during accelerated pavement testing on the Ifsttar's fatigue carousel. In Proceedings of the 8th International RILEM Conference, Nantes, France, 7–9 June 2016.
15. Todkar, S.; Le Bastard, C.; Baltazart, V.; Ihamouten, A.; Dérobert, X. Performance assessment of SVM-based classification techniques for the detection of artificial debondings within pavement structures from step-frequency A-scan radar data. *NDT E Int.* **2019**, *107*, 102128. [CrossRef]
16. Derobert, X.; Baltazart, V.; Simonin, J.-M.; Durand, O.; Norgeot, C.; Doué, S. Radar database over large debonded areas. *IFSTTAR* **2020**. [CrossRef]
17. Derobert, X.; Baltazart, V.; Simonin, J.-M.; Hui, H.Y. *Radar Database over Narrow Debonded Areas*; Université Gustave Eiffel: Marne-la-Vallée, France, 2020. [CrossRef]
18. Norgeot, C.; Dérobert, X.; Simonin, J.-M.; Doué, S.; Baltazart, V.; Hui, H.Y. 3D GPR monitoring of artificial debonded pavement structures during accelerated tests. In Proceedings of the GPR2020 Conference, Boulder, CO, USA, 14–19 June 2020.
19. Villain, G.; Thiery, M. Gammadensimmetry: A method to determine drying. *NDT E Int.* **2006**, *39*, 328–337. [CrossRef]
20. Plati, C.; Loizos, A. Estimation of in-situ density and moisture content in HMA pavements based on GPR trace reflection amplitude using different frequencies. *J. Appl. Geophys.* **2013**, *97*, 3–10. [CrossRef]

 *remote sensing*

Article

GPR Spectra for Monitoring Asphalt Pavements

Josep Pedret Rodés [1], Adriana Martínez Reguero [1] and Vega Pérez-Gracia [2,*]

[1]   Department of Civil & Environmental Engineering, Universitat Politècnica de Catalunya, Campus Nord,
      Jordi Girona 1-3, 08034 Barcelona, Spain; josep.pedret@upc.edu (J.P.R.); adriana.martinez@upc.edu (A.M.R.)
[2]   Department of Strength of Materials and Structural Engineering, Universitat Politècnica de Catalunya,
      Campus Diagonal Besós, Barcelona East School of Engineering, EEBE, Av. Eduard Maristany, 16,
      08019 Barcelona, Spain
*   Correspondence: vega.perez@upc.edu

Received: 8 April 2020; Accepted: 25 May 2020; Published: 29 May 2020

Abstract: Ground Penetrating Radar (GPR) is a prospecting method frequently used in monitoring asphalt pavements, especially as an optimal complement to the defection test that is commonly used for determining the structural condition of the pavements. Its application is supported by studies that demonstrate the existence of a relationship between the parameters determined in GPR data (usually travel time and wave amplitude) and the preservation conditions of the structure. However, the analysis of frequencies is rarely applied in pavement assessment. Nevertheless, spectral analysis is widespread in other fields such as medicine or dynamic analysis, being one the most common analytical methods in wave processing through use of the Fourier transform. Nevertheless, spectral analysis has not been thoroughly applied and evaluated in GPR surveys, specifically in the field of pavement structures. This work is focused on analyzing the behavior of the GPR data spectra as a consequence of different problems affecting the pavement. The study focuses on the determination of areas with failures in bituminous pavement structures. Results epitomize the sensitivity of frequencies to the materials and, in some cases, to the damage.

Keywords: Ground Penetrating Radar; pavement monitoring; spectral analysis; NDT

---

1. Introduction

1.1. Detection of Failures in Pavements

Traditionally, the presence of defects that affect pavement preservation to the greatest extent (moisture in the lower layers and lack of adhesion in the top pavement layers) is usually detected with deflection tests, although the study of the evolution of standardized indicators is also used, such as the International Roughness Index (IRI).

The basic parameter measured in the deflection tests is the vertical displacement produced in the pavement after applying a load. The result is the response of all the layers that make up the road structure, including the sub-grade. In addition, some measuring equipment, such as the Falling Weight Deflectometer (FWD) and the curviameter allow the deflection bowl generated by the load applied during the test to be interpreted, and the modulus of elasticity of the various layers that make up the pavement to be determined by means of back calculation [1].

Studies, Chea, and Martínez [2] for example analyse the adhesion between layers in a semi-rigid pavement showed that the deflection curve did not vary significantly, but that its first derivative and the radius of curvature under the loaded wheel could be used as a lack of adhesion indicator.

Another common problem, excessive moisture in the pavement layers, can affect its strength and thus reduce its useful life [3–5]. Recently, a study carried out in Torpsbruk (Sweden) demonstrated the applicability of the impact deflectometer in evaluating the effect of moisture in the unbound layers on the bearing capacity of a flexible pavement [6]. However, among its conclusions, the need to carry out a more intensive investigation that would allow the interpretation of the results to be improved was highlighted, so that environmental factors that affect the pavement response could be understood.

The work of Gedafa et al. [7] proposed a methodology based on the impact deflectometer using measurements made over eight years, with the aim of estimating the remaining life of a pavement depending on the deflections. Using a non-linear regression procedure, a very good fit sigmoidal relationship is determined that allows its remaining life to be predicted.

Various researchers have developed indices that consider the deflection value, the Structural Number (SN) being the most well-known [8,9]. Other indices, such as Surface Curvature (SCI), Base Damage (BDI), and the Base Curvature (BCI) were used to establish evaluation criteria for the granular layers treated with cement and for the bituminous layers [10–12].

The IRI test does not allow for the structural condition of a road to be defined, but its evolutionary analysis over time does allow its evolution to be studied [13]. Some researchers have developed models for predicting the Pavement Condition Index (PCI) based on the IRI. For example, Arhin et al. [14], through a functional classification (type of road: local, collector, arterial) and the type of pavement (flexible and rigid pavements), and based on data obtained in the district of Columbia over two years, found statistically significant results with a 5% margin of error. Park et al. [13] established a logarithmic relationship between the IRI and the PCI based on the results obtained from the acoustic monitoring of roads in nine North American states for a period of nine years. Dewan and Smith [15] related both indices for the streets of the San Francisco Bay Area, and developed a model that can be used to estimate vehicle operating costs directly based on deterioration identified in the pavement.

1.2. Geophysical Surveys in Pavement Assessment

As complementary evaluation of roads, geophysical surveys were tested and applied in many cases. These surveys provide a non-destructive analysis of the medium, based on the measurement of physical properties on its surface. The analysis of the values obtained for each one of those physical properties can be associated with models of the inner medium. Therefore, the properties and characteristics of the internal medium are deduced from indirect measurements. This type of studies presents benefits and limits. The benefits are mainly the quick data acquisition (in many cases data is acquired at a usual traffic speed) and the non-destructive character of the surveys, i.e., data acquired without damage the pavement. The limits are consequence of the indirect measurements, because the values of the measurement could correspond to different models of the medium. Therefore, in many cases, non-destructive techniques are applied combined with punctual drills or combining different NDT methodologies to avoid the vagueness and to obtain more accurate models. Several authors investigate the improvement of the results in case of different combinations of techniques, both in laboratory or in field tests. Capozzoli and Rizzo [16] compared GPR with resistivity tomography (ERT) and infrared thermography (IRT), obtaining promise results in the study of concrete, highlighting the different data obtained with each one of the techniques. The work of Lagüela et al. [17] demonstrated the ability of different techniques: GPR, IRT and laser terrestrial scanning (LTS), in the assessment of paving, showing the different damage observed with each one of those methods. A revision of the state-of-the-art in the assessment of pavement structural conditions, comparing the results in the estimation of thickness and moduli of different layers can be found in [18]. In many cases, the analysis of the pavement conditions are evaluated by using GPR and falling weight deflectometer (FWD) [19–21]. A wide revision of fundamentals and applications of NDT and geophysical surveys in the assessment of structures and infrastructures can be found in [22,23]. The most widely applied techniques are seismic methods, IRT, ultrasonic tomography, and GPR. Seismic surveys are widely used in pavement assessment. The method consists of the measurements of the pavement response to vibrations that are produced by falling weights. Other methodology used in the pavement tests is thermography, based on the detection of infrared radiation. Several studies indicate that this technique is appropriate to detect pavement defects, differences in compaction, cracking, and delamination [24–26]. Ultrasonic tomography has been also applied to detect cracks in concrete pavements [27,28]. The shallowest layers (asphalt binders) have been analyzed in some cases by using acoustic emission in order to detect incipient cracking [29,30]). GPR is other widely used technique in the assessment of pavements. Many of those techniques are applied in combination with other methodologies and, in all cases, with coring in specific points.

The GPR technique consists of the emission and reception of electromagnetic ultra-wideband frequency waves, in the range of microwaves and radiofrequencies. A transmitter antenna generates and transmit the signal that propagates through the medium. The existence of electromagnetic discontinuities produces partial reflection of the energy that propagates back to the surface of the medium. A receiver antenna detects that reflected energy (Figure 1). Each one of the recorded pulses is called A-scan, and the collection of these individual traces is the B-scan. In many road assessment, GPR usually collects one trace each 20 or 25 cm.

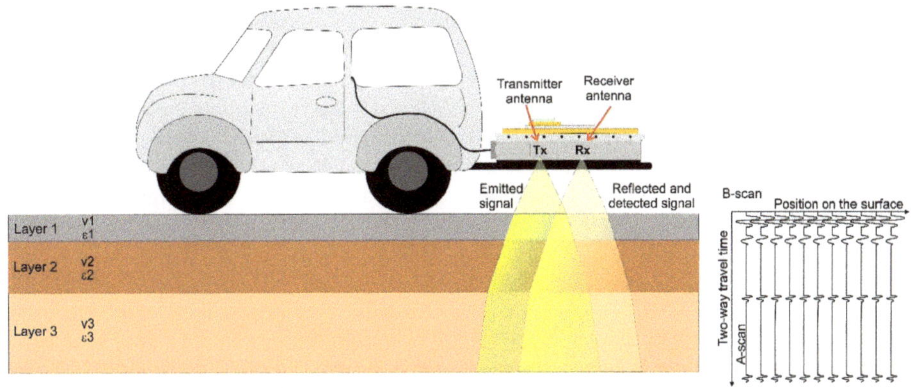

Figure 1. Scheme of GPR assessment of pavements. The antennas are usually mounted in a platform in a car. The existence of changes in the electromagnetic parameters produce the reflection of the energy that is detected by the receiver antenna. The result is the B-scan that includes the amplitude of the signals and the two-way travel time, in front of the position of each individual A-scan.

The first applications of GPR in pavement assessment were focused on the detection of layers, determining their thickness [31–34]. Later, the technique was applied to the detection of cracks [34–37] and to the analysis of properties and parameters of the asphalt and other layers, based on the estimation of their dielectric permittivity [38]. The method was also applied for moisture and infiltration control [39,40]). Even though GPR is applied to determine damage in pavement, distinguishing between specific details or causes is difficult because diverse anomalies could produce similar response in the GPR signals. For example, some researchers showed that non-destructive technologies are not able to identify debonding between asphalt layers because of an inadequate tack coat execution. However, GPR can be used to detect this distress when moisture is trapped in the interface [41,42]. In addition, some researchers discuss the ability of non-destructive technologies to detect an inadequate tack coat execution [41]. However, it was shown that GPR can be used to detect the distress caused when pavement layers are debonded, which can be a consequence of a deficient tack coat application [43]. Moreover, the document published by RILEM Technical Committee 241-MCD (RILEM State-of-the-Art Reports) considers this option: "techniques for detecting debonding using Ground Penetrating Radar have also shown promise" [44]. Also an investigation about the use of GPR to characterize changes in geometric and dielectric properties of the tack coats has been carried out and results prove the suitability of the technique proposed [45].

1.3. Application of GPR in the Detection of Road Failures

Currently, many pavement studies are complemented with non-destructive tests (NDT), being GPR one of the most extended methods (e.g., [46]), in many cases applied in combination with other NDT techniques such as thermography. GPR assessment provides an evaluation of defects that are located by analyzing the anomalies in the images. This type of analysis started to be applied in the 1980s to locate voids under the pavement layers [47], with varying degrees of success (e.g., [48,49]).

The technique is usually based on the determination of the electrical permittivity in the various stretches of road. This parameter depends on the materials that make up each medium, and it substantially determines the speed of propagation of the electromagnetic wave. In dielectric media, the relationship between the signal speed (v) and the relative dielectric permittivity (ε_r) is determined by means of Equation (1).

$$v = \frac{c}{\sqrt{\varepsilon_r}} \tag{1}$$

where c is the speed of propagation of the electromagnetic wave in free space (approximately 30 cm/ns). The limits between the different materials are detected from the reflection of the signal that occurs when there is a significant contrast in the dielectric constant of the media. The two extreme values considered for ε_r are those for water (approximately 81) and those for air (approximately 1), while the other materials that usually form part of the studied media have constants between 2 and 30 [50]. In particular, composite materials with bituminous binders usually have values that vary between 2 and 12 [51,52]. The total thickness of each layer (h) is calculated as the product of the propagation time of the wave reflected at the base of the layer (Δt) divided by two, and its average propagation speed (v), as defined in Equation (2).

$$h = v \cdot \frac{\Delta t}{2} \tag{2}$$

Equation (2) shows that the calculation of h includes an estimation of ε_r, so the speed of propagation of the signal can be determined. Although sometimes values already defined in the existing literature are considered, in other cases they are determined experimentally by analyzing specific cores and comparing them with the propagation times recorded in the area. Occasionally, a comparison will also be used between recorded amplitudes and signal amplitudes obtained in a prior calibration on metal plates. The entire energy incident on a metal plate is reflected; therefore the reflection coefficient in this case is one. By comparing the reflection coefficients in the ideal case (calibration on metal) and for a real surface, a permittivity value can be defined for the outermost surface of the layer by means of a relationship between the amplitudes (Equation (3)).

$$\sqrt{\varepsilon} = \frac{1 + \frac{A_0}{A_m}}{1 - \frac{A_0}{A_m}} \tag{3}$$

where A_0 is the amplitude of the wave reflected from the top of the pavement, and. A_m is the amplitude of the wave reflected on a metal plate located at the same distance from the antenna as the top of the pavement.

Test standard ASTM D4748 [50] states that the resolution of the GPR is sufficient to measure a minimum thickness of 40 mm with an accuracy of 5 mm. Some researchers have verified that the error made on measuring thicknesses is comparable to that made on direct measurements on cores (e.g., [32,52–54]).

Additionally, several studies demonstrate the ability of GPR to detect voids in pavements [55–57] and cracking [25,37]. In 2012, Saarenketo [56] developed a relationship that allowed the content of voids in a bituminous layer to be evaluated according to the average dielectric constant [58].

The method can also be used to analyse the presence of moisture due to the difference between the relative dielectric permittivity of the water and air [45]. A significant effect of moisture on the dielectric constant of bituminous mixes was observed [40,58–62].

1.4. The Frequency Spectrum

There are few works based on the analysis of the frequency spectrum of the GPR reflected signals. Some of the first studies used the frequency only to distinguish the antenna, determining differences in wave velocity [63] and in the absorption [64] depending on the antenna frequency. Frequency analysis was also included in laboratory resolution studies [65], comparing the frequency spectrum in air and in different media, which reduce the center frequency, the amplitude, and the bandwidth. Some studies focused on the detection of changes in water content in concrete determine also that the center frequency and the bandwidth decreases as water content increases [66,67], even though the main objective was the analysis of the wave amplitude, analyzing in some cases the spectra amplitude attenuation depending on the water content [68]. The analysis of the spectrum behavior combined with backscattering was also applied in the study of shallow geology to detect seasonal subterranean streams, differentiating between active and non-active streams [69] and in the study of compaction and moisture in sandy loam [70]. Further evaluations studied the propagation of GPR signals in unsaturated ground using the Rayleigh dispersion and confirmed that the frequency of the waves changed depending on the moisture content: the maximum amplitude observed moved to lower frequency values as the water content increased [61]. The changes in the GPR spectrum were also observed in studies evaluating the hydration phenomena of Portland cement as it passed from the fresh to hardened state by measuring the changes in the GPR signal spectrum over 90 days and recording the variations in the maximum amplitude values of the frequency spectrum [62]. It was also observed that the amplitude increased with the age of the concrete, confirming the relationship observed in other materials. Laboratory experiments with soils denote also that the clay content affected the shift and peaks of GPR frequency spectra, obtaining peaks at lower values as clay content increases [71,72].

Similar line of analysis, applied to the assessment of the pavement base, was focused on the study of water content, providing promising results and showing that the shift and peaks of the spectrum could most likely be and indicator that help in the mapping of spatial soil moisture variability [60,73]. The spectrum is also sensible to clay content in the pavement base, and several studies point to the possibility of using the peaks displacement to detect changes in clay content [74]. The work of Pedret et al. [42] analyses a section of flexible pavement by evaluating the changes in the spectrum according to the thickness of the bituminous mixture layers, the moisture and the detachment between layers. Based on the variations observed in the bandwidth and in the amplitude of the frequency spectrum maximums, the use of GPR is proposed to define stretches of road in accordance with the parameters observed in the frequency spectrum.

1.5. Study Objectives

The preliminary tests in selected areas show an interesting correlation between the form of the amplitude spectrum and the structural condition of the pavement [42]. However, in these first analyses, the records were studied at certain points by evaluating individual traces. The analysis of individual traces could introduce errors as on occasions sporadic alterations can occur in a single trace or a few traces due to factors external to the study. For this reason, and based on the results obtained, the possibility was considered of analyzing different stretches by characterizing them with an average spectrum for each one. This would reduce the effect of anomalous traces and produce a more accurate and reliable result. That previous work [42] demonstrated the existence of changes on the frequency pattern associated with the thickness of the layers, being the peaks of the spectrum moved to the lower frequencies in the case of higher thickness. Those results pointed to a relation between the different zones of the spectrum and the contacts between materials and, in some cases, with debonding between layers.

These results were complemented with more detailed tests and further processing, including the average of traces in the same place. Therefore, the purpose of this work is, therefore, to develop and analyze a possible methodology that can be applied to the analysis of the pavement which is based on the study of the frequencies of the GPR records, based on a comparative analysis of the frequencies that would enable possible structural changes and the existence of damage to be identified. For this, the results of an exhaustive prospection on a pavement are studied. The prospection was focused on determining changes in pavement sections and on the effect due to voids and moisture.

2. Methodology

To achieve the objectives of the work, a section of motorway was selected with known pavement design cross-sections and conditions. A prospection campaign with GPR was carried out in this sector. Previous deflection and surface roughness test results are available for the same sector, as well as information on the traffic carried.

The GPR records are used to carry out a comparative study between the spectra of the recorded signals measured in various areas of the motorway. This comparison allows the small alterations observed in the form of the spectrum to be evaluated with respect to a reference spectrum.

The motorway chosen for the study is one of the main accesses from the north of the metropolitan area of Barcelona with an annual average daily traffic (AADT) of 15,000 with 10% heavy vehicles, it was put into service in 2006. It was chosen because in a short stretch (about 7800 m) three types of well-defined and known design cross-sections coexist. Each one of the cross-sections represents one of the best-known pavement typologies [75].

A total of 7800 m is studied which ensures the geographical proximity of each one of the records. Therefore, it is considered that there are no local climate or traffic density changes that affect the test, as the length of the analyzed stretch is short.

The sectors are defined taking into consideration the three existing design cross-sections over the total analyzed stretch. For each cross-section, three subsectors of length between 700 m and 1000 m are considered depending on the state of preservation of the pavement. The information used to define the subsectors was obtained from the previous deflection and IRI tests.

The analysis is carried out by comparing the frequency spectra of the radar records, which are characteristic of each subsector, and evaluating the influence of the design and the state of preservation of the pavement.

The standardized CDA (Cumulative Difference Approach) segmentation method has been used, as set out in the AASHTO (1986) [76] guide, for delimiting the subsectors. The criteria used consist of delimiting the changes of slope of the cumulative deviations of the set of values recorded in the deflection measurements. This ensures sufficient homogeneity in the characteristics of each one of the subsectors. Table 1 shows the three sectors and the nine subsectors considered, their length and segmentation in accordance with the deflection and roughness test results.

Table 1. Stretches studied with the deflection and surface roughness (IRI) test results. Being: **AC**, Asphalt concrete; **GB**, Granular Base; and **CTB**, Cement-treated Base. **D1**, **D2** and **D3** are the classification of zones; the increasing numbers indicate increasing deflection ranges.

Section	Design Section	Condition	Initial K.P. (km)	Final K.P. (km)	Tested Length (m)	Deflection (μm)	IRI (m/km)
		D1	60 + 400	61 + 100	700	82	0.9
		D2	61 + 100	61 + 800	700	94	1
		D3	61 + 800	62 + 500	700	103	1
	AC (25 cm)	D1	71 + 600	72 + 600	1000	53	0.9
		D2	66 + 000	67 + 000	1000	90	0.9
		D3	68 + 200	69 + 200	1000	130	1.7
	AC (20 cm)	D1	77 + 000	78 + 000	1000	26	0.7
		D2	78 + 000	79 + 000	1000	24	1
		D3	80 + 300	81 + 300	700	82	1.3

2.1. Design Cross-Sections Considered

The three design cross-sections of the stretch of road studied (called A, B, and C) are shown in Figure 2. Cross-section A shows a full depth type pavement with a very thick layer of bituminous mix on a base of compacted soil forming the sub-grade. The design cross-section of the pavement is made up of different layers of asphalt mix with a total thickness of 35 cm (Figure 2a). Cross-section B is a flexible pavement. The design cross-section is made up of several bituminous mix layers of 25 cm thickness in total, on an untreated granular base of 25 cm (Figure 2b). Cross-section C is a semi-rigid pavement. The design cross-section is made up of a series of bituminous mix layers with a total thickness of 20 cm on a cement-treated granular base of 25 cm (Figure 2c).

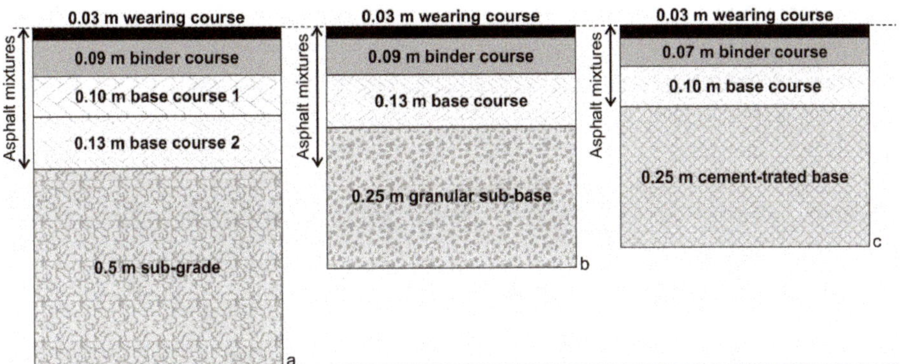

Figure 2. Design cross-sections, type A full depth asphalt (**a**), B flexible (**b**) and C semi-rigid pavement with treated cement base (**c**).

2.2. State of Preservation of the Studied Sectors

The initial definition of the nine subsectors of the studied stretch of road, taking into consideration their construction characteristics and state of preservation, was made based on the results of the deflection and roughness tests using cumulative frequency histograms. The presented values correspond to all the measurements carried out on each one of them.

Figure 2 shows the resultant values of the deflection measured with the Lacroix deflectometer device [77] in each one of the considered subsectors. The continuous line and discontinuous line graphs in Figure 2 show the subsectors in best and intermediate conditions, respectively. The dotted line graphs in Figure 2 show the subsectors in the worst conditions. The deflections obtained in cross-section A (Figures 2 and 3) showed low values due to the thickness of the combined bituminous layers which gives the system a high rigidity.

The deflections obtained in cross-section B (Figures 2 and 3) show a large range of conditions. A subsector can be seen in very good condition with low deflections, a second subsector in reasonable condition, and a third subsector that has very few areas in good condition. In the subsector in worst condition, there are practically no points under the 100 μm value.

In the deflections obtained in cross-section C (Figures 2 and 3), two of the subsectors are shown to be in very good condition while the third is in worse condition. The different zones, depending on the deflection results were called D1, D2, and D3, indicating increasing levels of deflection, although the thresholds are low in all three cases

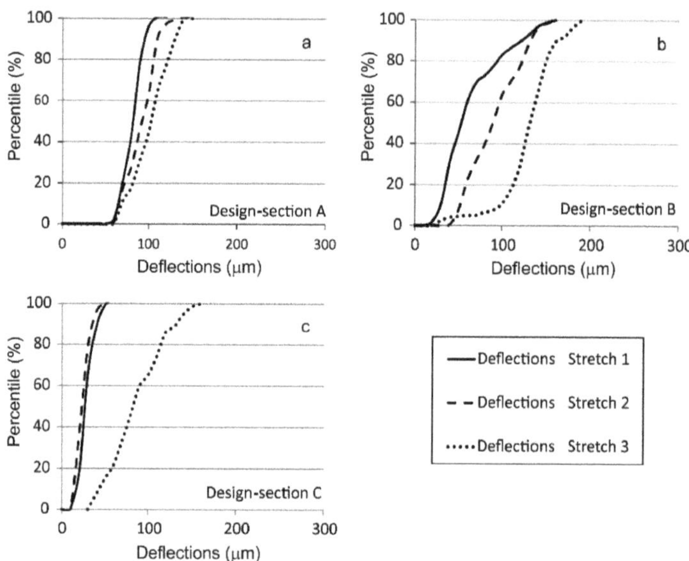

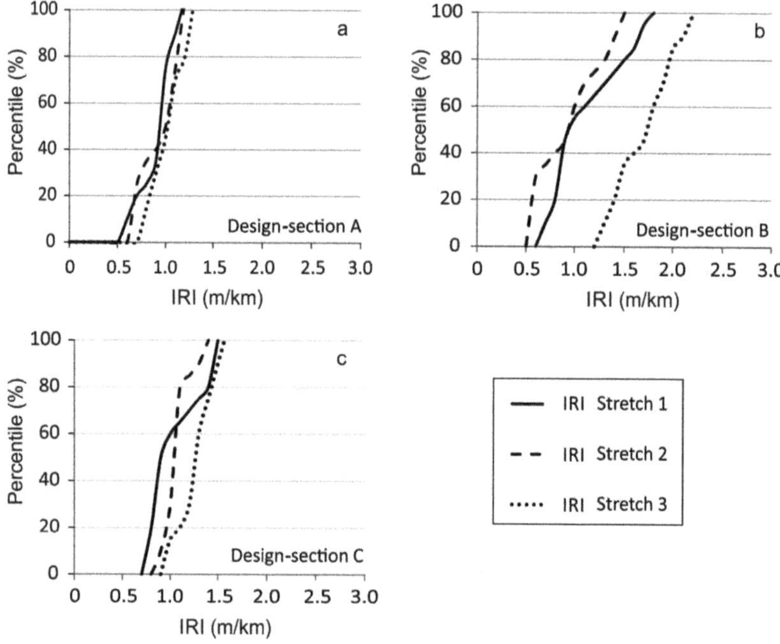

Figure 3. Deflection test cumulative frequency histograms in cross-section type A (**a**), type B (**b**) and type C (**c**).

Figure 4 shows the pavement roughness results, measured with the IRI index, for the three pavement cross-sections.

Figure 4. Roughness test cumulative frequency histograms (IRI) in cross-section type A (**a**), type B (**b**) and type C (**c**).

Considering the 50 percentile of the set of values, a consistency can be seen in the deflection results, that is to say, the stretches in best condition also usually show a better IRI, even though there is no direct correlation of values.

The surface roughness test has only been taken into consideration in this study as a criterion for assessing the preservation condition in stretches where it is considered that the deflection test is not sufficient to provide the classification, as values are obtained within the same order of magnitude.

2.3. Methodology Applied for Evaluation by GPR.

Once the subsectors into which each sector is divided are defined (by a characteristic structural cross-section), the inspection campaign with GPR was carried out on each one of them continuously and at an adequate speed (between 60 and 80 km/h) so as not hold up the traffic, a trace was taken every 25 cm with a sampling frequency of 15,000 MHz.

The profile data acquisition was carried out continuously in the center of the right-hand lane which carries mainly heavy vehicles (Figure 5). For the test, an antenna with a central frequency of 900 MHz was used after prior calibration. For this, the direct and reflected wave were analyzed after it was propagated through the air.

Figure 5. Radar data acquisition with a 900 MHz center frequency antenna.

The calibration results are shown in Figure 6, which shows its pulse form and its frequency spectrum. It can be seen that the wave has a central frequency (f) at around 950 MHz, a bandwidth (BW) of about 350 MHz measured at −3 dB and a period of approximately 1.2 ns.

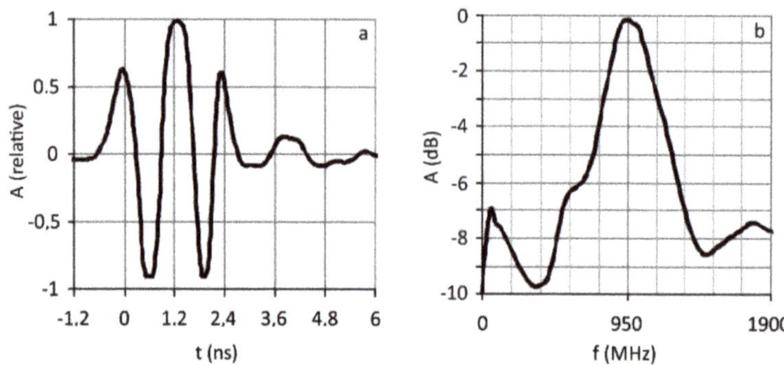

Figure 6. Characteristics of the antenna used. Recorded wave (**a**) and frequency spectrum (**b**).

2.4. Calibration of the Signals in Known Media

Before carrying out the signal acquisition, GPR data was acquired in specific and known zones, and the results were analyzed in order to determine propagation velocities and possible correlations between the form of the frequency spectrum and the different characteristics and pathologies of the pavement. The results of those preliminary tests are analyzed and discussed in [42]. Further contrast images were also acquired in the motorway studied in this work, always in points at which cores were obtained, in parts without visible damage. Figure 7 shows the core and the radar images obtained in one of these contrast tests, in kilometric points (KP) without construction defects and visible damage.

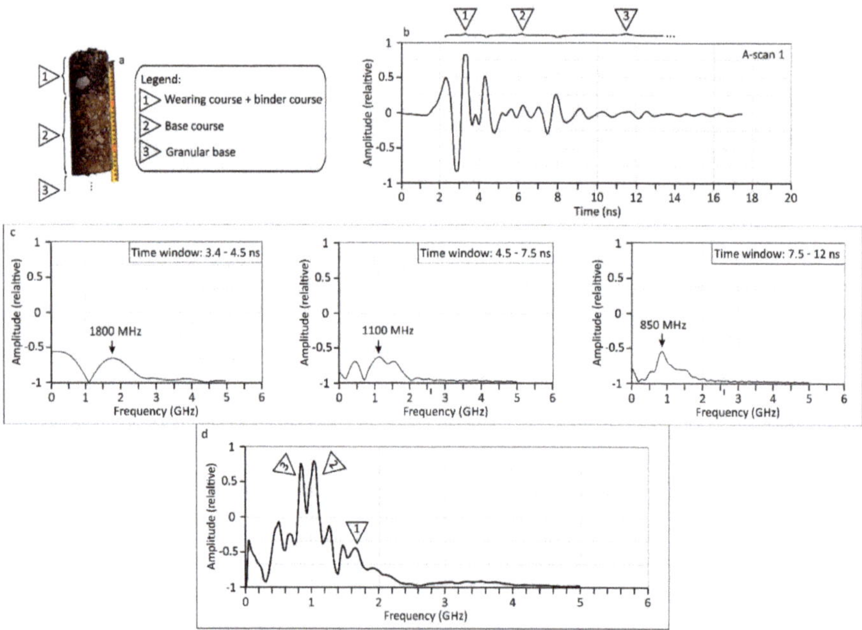

Figure 7. Results obtained in a cross-section without defects. (**a**) Core showing the structural layers. (**b**) A-scan, in the time domain. (**c**) Frequency content of the signal considering three time windows that are defined considering the core layers and the A-scan. (**d**) Frequency spectrum of the A-scan 1. The contribution determined in the three time windows define the peaks observed in the spectrum.

The road cross-section observed in the core (Figure 7a) presents three main layers: an upper part composed by wearing and binder courses, a second layer corresponding to the base course, and the end of the core is placed in the contact with the granular base. Therefore, the structural cross-section of the core shows two asphalt layers and the contact with the granular base. The top asphalt layer comprises both the wearing and binder course layers. Underneath there is a bituminous base layer spread over a granular base.

The wave velocity was obtained at each layer by comparing the two-way travel time in the A-scan to the thickness observed in the core (Figure 7b). This trace shows the correspondence between the number of transitions between layers and the number of amplitude maximums of the wave due to the reflection of the signal at each discontinuity. The maximum located at a time around 4.3 ns corresponds to the transition between the bituminous base and the granular base. The peak located around 1.5 ns corresponds to the transition between the two bituminous mix layers, and the initial peak located at 0 ns corresponds to the transition between the air and the pavement. Comparing those times with the core layers thickness, an average velocity of about 11.4 cm/ns is obtained, corresponding to the shallowest layer (wearing and binder courses), and a wave velocity of about 8.7 cm/ns in the base course.

The spectrum of the A-scan was compared to the spectrum obtained in air. A detailed analysis of the frequency of the received signal also shows variations in its maximum amplitude frequencies that appear to correspond with the different structural discontinuities of the pavement [42]. In general, the amplitude of the spectrum diminishes, and several peaks appear at frequencies of about 850 MHz, 1.1 GHz, and 1.3 GHz (Figure 5). These three maximum values are associated with three layers observed in the core and in the A-scan. This spectrum represents the frequency distribution of the received signal. The graph shows that the two main peaks are placed around 950 MHz. Each one of these corresponds to the two predominant materials in the environment: The asphalt mix of the base (frequency higher than 950 MHz) and the granular base (frequency lower than 950 MHz). Those results are considered to be contrast data to be compared with the radar data acquired in different sectors with a different damage degree.

The material of the medium through which the wave propagates, acts as a low pass filter. For this reason, the result obtained is logical as the layers located nearest the surface appear to be represented by the maximum amplitudes in the frequency spectra. In the spectrum shown in Figure 7c, the maximum indicated as 1, located at about 1200 MHz, represents the combined wearing and binder courses, which are the uppermost layers. The second maximum (2), at about 1000 MHz, represents the base mix. Finally, the maximum indicated as 3, located at about 900 MHz, corresponds to the granular base.

2.5. Signal Processing

(a) Processing A-scan—Signals recorded in each one of the stretches were processed to correct baseline deviations of the records, moving the zero offset of each A-scan with a dewow filter [78]. Other filters were not applied to the traces. This processing is usual in pavements GPR surveys to remove the low frequency and the down-shifting, but including in many cases a gain function application (e.g., [25,34,79]). Those filters were not used in this case because the study was focused on the spectrum analysis. The frequency spectra were obtained by means of a Fourier transform. Subsequently, they were smoothed by using a low pass filter with a cutoff frequency of 6 GHz, so the maximum values obtained could be more clearly seen.

(b) Processing B-scan—The analysis of individual trace spectra can provide results that are unrepresentative of the pavement studied when the study is carried out on randomly selected A-scans chosen from B-scans. An anomalous element on the surface, such as an irregularity in the ground or a waterlogged area, can affect a trace or a small and limited set of traces. If one of them is selected in the pavement analysis, the result may be due to external elements and will not show the average characteristics of the analyzed stretch of road.

For this reason, all the A-scans obtained over the same subsector (previously determined based on IRI and FWD data) have been averaged. In the resultant average trace, the effects due to small elements and surface factors are minimized, while those elements that are nearly constant throughout the subsector are represented. In this way, a record is obtained with very little noise that can be considered characteristic of the stretch. After applying the Fourier transform to this averaged trace, the resultant spectrum characteristics depends on different more general aspects of the pavement, such us number of layers, thickness of the layers, state of preservation and moisture. The shape and characteristics of the spectrum are determined by some of those aspects of by the combination of several of them. It is not possible to distinguish the causes of the changes. Hence, it is crucial to differentiate between types of pavement and state of preservation in order to associate with less uncertainty each GPR anomaly to the different pavement characteristics. The previous inspection is based on IRI and FWD, and determines each one of the subsectors identifying their constructive typology and by their degree of preservation.

3. Results Obtained

To check the sensitivity of the proposed frequency analysis-based method, a comparative study of the obtained frequency spectra is carried out, in accordance with the considered stretches of the road. Therefore, the study compares possible variations in the spectrum by averaging the results obtained in a particular road subsector. The comparison of these spectra with the IRI and FWD data and with specific cores shows the effect due to the typology of the construction cross-section and, for a same cross-section typology that is due to the different integrity levels.

3.1. Effects of Construction Cross-Section Variations

In the first test, the behavior is studied of the signal frequency spectrum in the case of variations in the pavement structure.

To avoid the effect due to the state of preservation of the pavement on the spectrum, a comparison is only made on those stretches in best condition. The known stretches with different cross-sections are analyzed (A, B, and C in Figure 1). Figure 6 shows the results obtained, averaged for each selected subsector. Standard deviation obtained for each one of the averaged spectra in the zone of interest (near 950 MHz) is approximately between 10% and 20%.

The graphs in Figure 8 are characterized by a minimum or an inflection at 950 MHz. The spectrum that has a minimum at these frequencies shows behavior similar to that of the spectrum in Figure 5. However, on comparing the three cross-sections in Figure 6, a notable change in this behavior is apparent for each section. Depending on the section, the minimum becomes a change in the curve inflection.

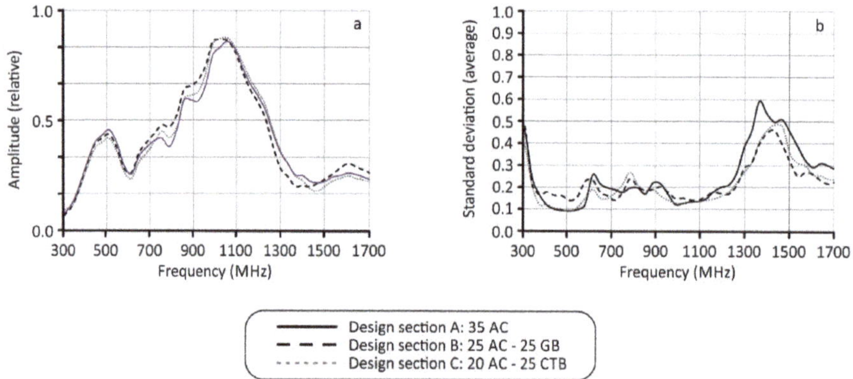

Figure 8. Frequency spectra of the construction cross-sections. Averaged results for each subsector (**a**) and standard deviation of the averaged spectra (**b**).

The maximum corresponding to the frequency above 950 MHz, associated with the combined bituminous mix layers, remains almost unchanged in the three design cross-sections. However, the same is not true of the maximum associated with the granular and cement-treated bases, located at about 900 MHz. At a lower frequency, the amplitude of this maximum reduces considerably in the type A structural cross-section, the amplitude is higher in the type B cross-section and higher still in the type C cross-section.

The amplitude differences of the maximums of the spectra associated with the bases in cross-section A and B are small, and could be due to the depth of each layer, as the depth of the granular base in cross-section A is, on average, at about 35 cm, while in the case of B, it is at about 25 cm. As B is closest to the top surface, it is logical that the amplitude of the maximum associated with the granular base is greater, as the high frequencies have not been attenuated to the same extent by the effect of the media.

The case of cross-section C is different, as the maximum that corresponds to the cement-treated base layer is very pronounced, and the difference of depth of this layer with respect to cross-section B is small (on average, about 5 cm). Therefore, a possible explanation of that observed in the record is the effect of the cement-treated base layer on the signal. This would indicate that this layer has electromagnetic characteristics that are sufficiently different from the other materials, so that changes in the frequency content of the records can be detected.

Although a similar case was analyzed [35] in preliminary studies, there are no exhaustive studies on the frequency behavior of GPR signals obtained in a pavement with a cement-treated base layer.

The results can be evaluated based on the composition of these types of layers. Table 2 shows the physical characteristics (relative dielectric permittivity and electrical resistivity) of asphalt mixtures, conventional granular bases, and cement-treated bases [80]. This data is obtained measuring by a 900 MHz shielded antenna and a distance of 20 cm above the surface pavement.

Table 2. Characteristics of layers AC, GB and CTB [80].

Layer Type	Relative Dielectric Permittivity (ε_r)	Electrical Resistivity (σ) (Ωm)
AC	7.6–8.2	0.001
GB	4.5–4.8	0.001
CTB	15.9	0.1

The dielectric permittivity and electrical resistivity of the asphalt mixtures are more similar to those of a conventional granular base than those of a cement-treated base.

3.2. Effect of Variations of the Pavement Integrity Condition

The analysis was carried out by comparing averages of the frequency spectrum of records obtained on a stretch of road. This stretch was divided into sectors (in accordance with the structural topology) with a length of approximately 700 m. The results obtained in a specific sector were compared, taking into consideration the three pavement preservation levels: D1, D2, and D3. This classification allows subsections to be defined for each structural cross-section.

Figures 8–11 show the results obtained for each sector according to their preservation condition. All of them show the wave trace diagrams (A-scans) in their time domain and the frequency spectra. These all correspond to the average of all the measurements carried out in a specific sector.

For each design cross-section, three stretches are considered that are represented in the following way: by a continuous line: the subsector in the best structural condition (D1); by a discontinuous line: the subsector in the intermediate condition (D2); finally, by a dotted line: the subsector that in the worst condition (D3). The results obtained from the type A pavement (Figure 2) are summarized in Figure 9.

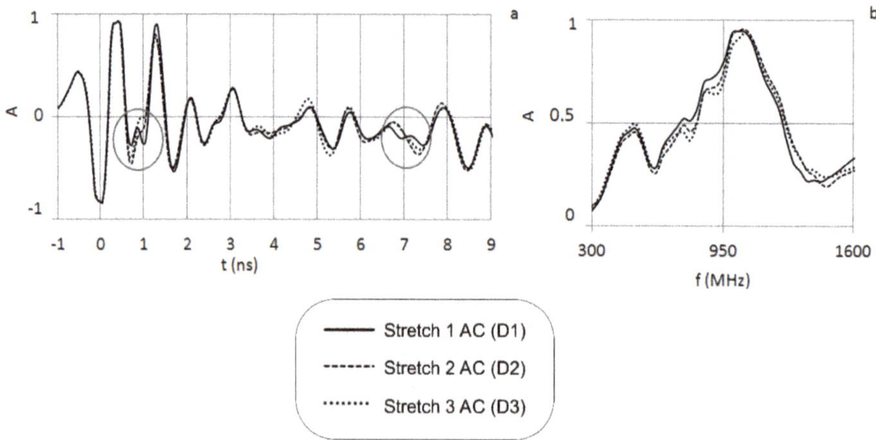

Figure 9. Time domain (**a**) and frequency spectra (**b**) in the three subsectors of the type A cross-section.

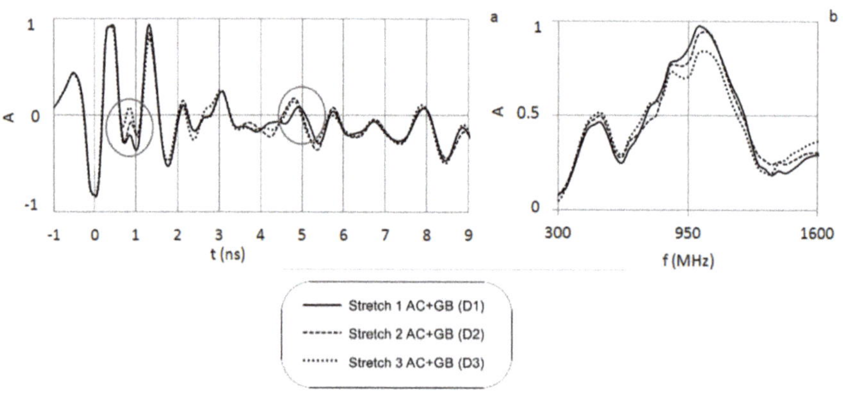

Figure 10. Time domain (**a**) and frequency spectra (**b**) in the three subsectors of the type B cross-section.

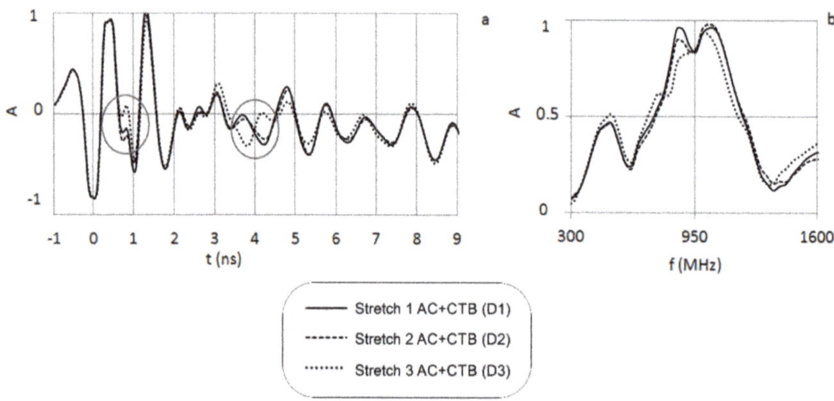

Figure 11. Time domain (**a**) and frequency spectra (**b**) in the three subsectors of the type C cross-section.

In the left-hand diagram (Figure 9a), the A-scans of the three subsectors determined by their different degree of preservation are compared. It can be seen that the record pattern is the same in the three cases, except in the two specific areas that are marked with circles in Figure 9a. In these two areas, changes of amplitude occur corresponding to the signal reflections. These changes show discrepancies between the three averages. They are located around 0.8 ns and 7 ns, which correspond to the contact surfaces between wearing and binder courses, and base course and sub-grade. Assuming an average wave speed for a standard asphalt concrete of 10 cm/ns approximately, the signal reflections are located around 3 cm and 35 cm depth from the surface of pavement.

On analyzing the results for the same pavement typology with different preservation conditions (D1, D2, and D3), no relationship is observed between the double propagation time of the reflected wave and the structural condition of the pavement; therefore, it is deduced that the propagation time analysis does not provide relevant information with regard to the preservation condition of the pavement.

The diagram in Figure 9b shows the variation in the frequency spectrum for the three deterioration levels considered for the pavement. The maximum recorded for the highest frequencies associated with the combined bituminous mix layers remains virtually unchanged in the three spectra. However, in the case of the maximum peak associated with the sub-grades, located at about 900 MHz, corresponding to the second maximum, the higher amplitude is obtained for the signal obtained in the stretch in the best conditions (D1), and the maximum loses amplitude as the structure deteriorates.

These results suggest that the greatest deterioration of the structure would be found in the sub-grade, under the bituminous mix structure. A possible cause may be the debonding between layers or the breaking up of material in the sub-grade, which generates an energy dispersion effect and therefore a loss of energy.

Figure 10 shows the results obtained in the studied subsector, characterized by a type B design cross-section.

In the diagram of Figure 10a, two amplitude changes can be seen in the trace again. They are located around 0.8 ns and 5 ns, which correspond to the interface between wearing and binder courses, and bituminous base course and granular base, respectively. Assuming an average wave speed for standard asphalt concrete, the signal reflections are located around 3 cm and 25 cm depth from the surface of pavement.

Stretches 2 and 3 show higher amplitude in the areas identified with circles. This phenomenon could be associated with a debonding process between layers [81].

The diagram in Figure 10b also shows the variation of the signal frequency spectrum in each one of the three subsectors.

In this case, the form of the spectrum and the location of its maximums, both that related to the combined bituminous mix layers and that related to the granular bases, change following the same pattern, i.e., the amplitude and bandwidth is maintained in the stretch in best condition, and they diminish as the structure deteriorates.

This result could indicate that in contrast to that occurring in the type A cross-section, the structural failure could take place in the system as a whole, i.e., in the AC and GB layers.

The graphs obtained in the study of the type C cross-section stretches are shown in Figure 11. In the diagram in Figure 11a, it can be seen that there are again two changes in the amplitude associated with the contacts between wearing course and binder course, and bituminous base course and cement-treated base. The second change is located around 4.2 ns for the best and intermediate stretch, and at 3.7 ns for the worst section (D3). An analysis similar to that carried out in the other tests gives a thickness of base course about 20 cm in the case of the best (D1) and intermediate (D2) subsectors, and 17 cm for the worst subsector. This thickness difference reveals the significant effect of the base course over the cement-treated base in the behavior of the structure (higher deflections).

In the diagram in Figure 11b, it can also be seen how the amplitude of the spectrum and the bandwidth decrease in the case of the graph that corresponds with the subsector in the worst conditions.

This data suggests that the failure in the structure could occur, among other reasons, due to the reduced thickness of the bituminous mix in this stretch of road.

Table 3 summarizes the main results, comparing in the three types of pavement, the three degrees of state of conservation (D1, D2, and D3).

Table 3. Main results, highlighting the changes in the spectra as consequence of the deflection ranges D1, D2 and D3, for each one of the three types of pavement.

Stretch	Spectrum		A-Scan	Possible Damage in Case D3
	Peak 1 (900 MHz)	**Peak 2 (1100 MHz)**		
AC	Maximum amplitude in D1 and minimum amplitude in D2	Same amplitude approximately in D1, D2 and D3	Small changes of amplitude in contacts when comparing D1, D2 and D3	Possible damage under bituminous layer (AC): possible debonding or damage in sub-grade
AC + GB	Amplitude diminish as structure deteriorates (maximum amplitude in D1 and minimum in D3)	Amplitude diminish as structure deteriorates (maximum amplitude in D1 and minimum in D3), with an important decrease in the case D3	Clear changes of amplitude, being higher the amplitude in the contacts in the case of worst pavement conditions (D3)	Damage in both AC and Gb layers. Possible debonding
AC + CTB	Amplitude diminish as structure deteriorates (maximum amplitude in D1 and minimum in D3)	No evident changes of amplitude	Change of amplitude and phase in the second contact	Possible damage in the bituminous layer AC and in layer CTB

4. Discussion and Conclusions

The GPR test is included in various pavement design guides in different countries. However, its use is usually restricted to propagation time analysis, there being few studies based on the frequency domain of the records.

The objective of the work was to analyse the frequency content of the spectrum of the signals recorded in various sectors of a stretch of road differentiated by their construction cross-sections. Each one of the sectors was subdivided into subsectors which were selected in a way that ensured their homogeneity, and within the criteria set out in the AASHTO pavement design guide. These subsectors were selected and classified in accordance with the results of the deflection and roughness tests carried out to find out their preservation condition.

It was the intention, therefore, to analyze the sensitivity of the frequency spectrum of the GPR records to changes in the construction typology and their state of preservation. The study is a first analysis to determine the capability of the proposed method; therefore, quantitative relationships are not determined between its sensitivity and the studied parameters. However, the results are promising and indicate the way in which these studies can be applied systematically.

The proposed methodology comprised a statistical analysis of a set of deflection values, IRI data and GPR records selected after deciding on some common criteria. Based on this selection, a results diagram was obtained in each case for evaluating the typology and the state of preservation of the studied pavement stretches. In the case of the GPR tests, diagrams averaged in the time and frequency domain are shown, and it was endeavored to ensure that they are sufficiently representative of each stretch. The results have led to the following conclusions:

(1) The frequency spectrum is sensitive to the typology of material used in a pavement. Tests were carried out on stretches where different materials were used for the base: bituminous mixes, granular and cement-treated materials. In the comparative study, changes were observed in the amplitude maximums of the spectrum within the frequency range associated with that layer.

(2) A statistical study of the wave transmitted to each medium allowed the amplitude changes recorded in the time domain to be determined, as a consequence of the reflections that occurred at the contacts or in the transition zones between the various layers that make up the pavement. This allowed the approximate average thicknesses of each one of the layers to be defined.

(3)　On obtaining the frequency spectrum of the time signals by means of the Fourier transform, their variations could be analyzed when the conditions of the media changed, which were studied from the time record. It can be seen that the frequency spectrum is sensitive to the structural condition of a pavement (this condition was previously defined based on a standardized deflection test). In all the analyzed cases, as the condition becomes worse, the amplitude of the frequency spectrum decreases as its bandwidth reduces. The tests carried out appear to indicate that the cause of these phenomena may be the debonding between layers or the breakup of the material in the sub-grade. This second problem would produce an increase in energy dispersion, thus attenuating the signal. The result would be a decrease in the energy recorded in the frequency range associated with the affected material and, therefore, a smaller amplitude in the area of the spectrum associated with that range.

(4)　Although none of the NDT technologies is capable of identifying partial or no bond due to inadequate tack coat during construction. GPR can identify variations in the pavement, isolate the depth of a discontinuity in the pavement, and provide a relative degree of severity. Severe conditions, such as stripping, can be observed with conventional analysis software. Detecting debonding between asphalt layers is only possible when there is moisture trapped in the debonded area between the layers using current analysis methodology, although the results indicates possible detection of those damages.

The results shows that GPR is sensitive to the different layers that make up an asphalt pavement. Furthermore, through the study of the response wave in its frequency domain it is possible to know whether the granular base is treated with cement or not. On the other hand, the amplitude generated by each frequency peak is also sensitive to its integrity condition. However, the tests have been carried out on a known structure with homogeneous thicknesses, which are also known. For this reason, it would be of great interest to be able to check whether the behavior of the response wave in its frequency domain is similar in pavements composed of an asphalt layer and a granular base or cement-treated base with different thicknesses. Should this be the case, the study could allow in the future to approximate a deflection value only by studying the shape of each peak generated by each pavement layer. Given that the GPR test is carried out at high speeds with a very high density of sampling (more data can be obtained per length unit with GPR than with deflection surveys), obtaining results in a very short time, it would be reasonable to use this method to define homogenous sections of a flexible pavement according to its characteristics, also approximating its maintenance condition.

Although the promising results showing the sensitivity of the spectrum to several characteristics and conditions of the pavement, further research is still needed to prove and confirm the conclusions and to consolidate the results. Future works might be focused on laboratory tests and numerical simulation of the spectrum behavior associate to changes in water content, debonding and number of layers. In addition, a statistical analysis based on field surveys could also be helpful to corroborate the results obtained in this work and to associate the different changes in the frequency content to particular features.

Author Contributions: Conceptualization, A.M.R., V.P.-G.; methodology, V.P.-G., J.P.R.; validation, A.M.R., V.P.-G.; formal analysis, A.M.R., V.P.-G.; investigation, J.P.R.; data acquisition, J.P.R., data processing, J.P.R.; data curation, J.P.R., A.M.R., V.P.-G.; writing—original draft preparation, J.P.R., A.M.R., V.P.-G.; writing—review and editing, A.M.R., V.P.-G.; supervision, A.M.R., V.P.-G. All authors have read and agreed to the published version of the manuscript.

Funding: This research has been partially funded by the Spanish Ministry of Economy and Competitiveness (MINECO) of the Spanish Government and by the European Regional Development Fund (FEDER) of the European Union (UE) through projects referenced as CGL2011-23621 and CGL2015-65913-P (MINECO/FEDER, UE). The study is also a contribution to the EU Action COST TU1208 "Civil Engineering Applications of Ground Penetrating Radar" financed by the European Union.

Acknowledgments: The authors would like to acknowledge the contribution of all those people and institutions that have helped and collaborated in the radar data acquisition tasks and in the various experimental tests.

Conflicts of Interest: The authors declare no conflict of interest.

References

1. Stubstad, R.; Jiang, Y.; Lukanen, E. Forward calculation of pavement moduli with load-deflection data. *Transp. Res. Rec. J. Transp. Res. Board* **2007**, *2005*, 104–111. [CrossRef]
2. Chea, S.; Martinez, J. Using Surface Deflection for Detection of Interface Damage between Pavement Layers. *Road Mater. Pavement Des.* **2008**, *9*, 359–372. [CrossRef]
3. Lekarp, F.; Isacsson, U.; Dawson, A. State of the art. I: Resilient response of unbound aggregates. *J. Transp. Eng.* **2000**, *126*, 66–75. [CrossRef]
4. Erlingsson, S. Impact of water on the response and performance of a pavement structure in an accelerated test. *Road Mater. Pavement Des.* **2010**, *11*, 863–880. [CrossRef]
5. Cary, C.E.; Zapata, C.E. Resilient modulus for unsaturated unbound materials. *Road Mater. Pavement Des.* **2011**, *12*, 615–638. [CrossRef]
6. Salour, F.; Erlingsson, S. *Impact of Groundwater Level on the Mechanical Response of a Flexible Pavement Structure: A Case Study at the Torpsbruk Test Section along County Road 126 Using Falling Weight Deflectometer*; VTI: Linköping, Sweden, 2014.
7. Gedafa, D.S.; Hossain, M.; Miller, R.; Van, T. Estimation of remaining service life of flexible pavements from surface deflections. *J. Transp. Eng.* **2009**, *136*, 342–352. [CrossRef]
8. National Cooperative Highway Research Program. *AASHTO Guide for Design of Pavement Structures*; The American Association of State Highway and Transportation Officials: Washington, DC, USA, 1993.
9. Crook, A.; Montgomery, S.; Guthrie, W. Use of Falling Weight Deflectometer Data for Network-Level Flexible Pavement Management. *Transp. Res. Rec. J. Transp. Res. Board* **2012**, *2304*, 75–85. [CrossRef]
10. Horak, E. The use of surface deflection basin measurements in the mechanistic analysis of flexible pavements. In *Sixth International Conference, Structural Design of Asphalt Pavements, Proceedings, University of Michigan, 13–17 July 1987*; Michigan University: Ann Arbor, MI, USA, 1987; Volume I.
11. Horak, E. Benchmarking the structural condition of flexible pavements with deflection bowl parameters. *J. S. Afr. Inst. Civ. Eng.* **2008**, *50*, 2–9.
12. Donovan, P.; Tutumluer, E. Falling weight deflectometer testing to determine relative damage in asphalt pavement unbound aggregate layers. *Transp. Res. Rec. J. Transp. Res. Board* **2009**, *2104*, 12–23. [CrossRef]
13. Park, K.; Thomas, N.E.; Wayne Lee, K. Applicability of the International Roughness Index as a Predictor of Asphalt Pavement Condition 1. *J. Transp. Eng.* **2007**, *133*, 706–709. [CrossRef]
14. Arhin, S.A.; Williams, L.N.; Ribbiso, A.; Anderson, M.F. Predicting Pavement Condition Index using International Roughness Index in a Dense Urban Area. *J. Civ. Eng. Res.* **2015**, *5*, 10–17.
15. Dewan, S.A.; Smith, R.E. Estimating IRI from pavement distresses to calculate vehicle operating costs for the cities and counties of San Francisco Bay area. *Transp. Res. Rec.* **2002**, *1816*, 65–72. [CrossRef]
16. Capozzoli, L.; Rizzo, E. Combined NDT techniques in civil engineering applications: Laboratory and real test. *Constr. Build. Mater.* **2017**, *154*, 1139–1150. [CrossRef]
17. Lagüela, S.; Solla, M.; Puente, I.; Prego, F.J. Joint use of GPR, IRT and TLS techniques for the integral damage detection in paving. *Constr. Build. Mater.* **2018**, *174*, 749–760. [CrossRef]
18. Goel, A.; Das, A. Nondestructive testing of asphalt pavements for structural condition evaluation: A state of the art. *Nondestruct. Test. Eval.* **2008**, *23*, 121–140. [CrossRef]
19. Marecos, V.; Fontul, S.; de Lurdes Antunes, M.; Solla, M. Evaluation of a highway pavement using non-destructive tests: Falling Weight Deflectometer and Ground Penetrating Radar. *Constr. Build. Mater.* **2017**, *154*, 1164–1172. [CrossRef]
20. Borecky, V.; Haburaj, F.; Artagan, S.S.; Routil, L. Analysis of GPR and FWD data dependency based on road test field surveys. *Mater. Eval.* **2019**, *77*, 214–225.
21. Plati, C.; Loizos, A.; Gkyrtis, K. Integration of non-destructive testing methods to assess asphalt pavement thickness. *NDT E Int.* **2020**, 102292. [CrossRef]
22. Pérez-Gracia, V.; Fontul, S.; Santos-Assunção, S.; Marecos, V. Geophysics: Fundamentals and Applications in Structures and Infrastructure. *Non-Destr. Tech. Eval. Struct. Infrastruct.* **2016**, *11*, 59.
23. Solla, M.; Lorenzo, H.; Martínez-Sánchez, J.; Pérez-Gracia, V. Applications of GPR in association with other non-destructive testing methods in surveying of transport infrastructures. In *Civil Engineering Applications of Ground Penetrating Radar*; Springer: Cham, Switzerland, 2015; pp. 327–342.

24. Moropoulou, A.; Avdelidis, N.P.; Koui, M.; Kakaras, K. An application of thermography for detection of delaminations in airport pavements. *NDT E Int.* **2001**, *34*, 329–335. [CrossRef]

25. Solla, M.; Lagüela, S.; González-Jorge, H.; Arias, P. Approach to identify cracking in asphalt pavement using GPR and infrared thermographic methods: Preliminary findings. *NDT E Int.* **2014**, *62*, 55–65. [CrossRef]

26. Plati, C.; Georgiou, P.; Loizos, A. Use of infrared thermography for assessing HMA paving and compaction. *Transp. Res. Part C Emerg. Technol.* **2014**, *46*, 192–208. [CrossRef]

27. Hoegh, K.; Khazanovich, L.; Yu, H.T. Ultrasonic tomography for evaluation of concrete pavements. *Transp. Res. Rec.* **2011**, *2232*, 85–94. [CrossRef]

28. Khazanovich, L.; Velasquez, R.; Nesvijski, E.G. Evaluation of top-down cracks in asphalt pavements by using a self-calibrating ultrasonic technique. *Transp. Res. Rec.* **2005**, *1940*, 63–68. [CrossRef]

29. Behnia, B.; Buttlar, W.; Reis, H. Evaluation of low-temperature cracking performance of asphalt pavements using acoustic emission: A review. *Appl. Sci.* **2018**, *8*, 306. [CrossRef]

30. Apeagyei, A.K.; Buttlar, W.G.; Reis, H. Assessment of low-temperature embrittlement of asphalt binders using an acoustic emission approach. *Insight-Non-Destr. Test. Cond. Monit.* **2009**, *51*, 129–136. [CrossRef]

31. Al-Qadi, I.L.; Lahouar, S. Measuring layer thicknesses with GPR–Theory to practice. *Constr. Build. Mater.* **2005**, *19*, 763–772. [CrossRef]

32. Lahouar, S.; Al-Qadi, I.L. Automatic detection of multiple pavement layers from GPR data. *NDT E Int.* **2008**, *41*, 69–81. [CrossRef]

33. Le Bastard, C.; Baltazart, V.; Wang, Y.; Saillard, J. Thin-pavement thickness estimation using GPR with high-resolution and superresolution methods. *IEEE Trans. Geosci. Remote Sens.* **2007**, *45*, 2511–2519. [CrossRef]

34. Diamanti, N.; Redman, D. Field observations and numerical models of GPR response from vertical pavement cracks. *J. Appl. Geophys.* **2012**, *81*, 106–116. [CrossRef]

35. Krysiński, L.; Sudyka, J. GPR abilities in investigation of the pavement transversal cracks. *J. Appl. Geophys.* **2013**, *97*, 27–36. [CrossRef]

36. Rasol, M.A.; Pérez-Gracia, V.; Solla, M.; Pais, J.C.; Fernandes, F.M.; Santos, C. An experimental and numerical approach to combine Ground Penetrating Radar and computational modelling for the identification of early cracking in cement concrete pavements. *NDT E Int.* **2020**, 102293. [CrossRef]

37. Rasol, M.A.; Pérez-Gracia, V.; Fernandes, F.M.; Pais, J.C.; Santos-Assunçao, S.; Santos, C.; Sossa, V. GPR laboratory tests and numerical models to characterize cracks in cement concrete specimens, exemplifying damage in rigid pavement. *Measurement* **2020**, *158*, 107662. [CrossRef]

38. Liu, H.; Sato, M. In situ measurement of pavement thickness and dielectric permittivity by GPR using an antenna array. *NDT E Int.* **2014**, *64*, 65–71. [CrossRef]

39. Grote, K.; Hubbard, S.; Harvey, J.; Rubin, Y. Evaluation of infiltration in layered pavements using surface GPR reflection techniques. *J. Appl. Geophys.* **2005**, *57*, 129–153. [CrossRef]

40. Plati, C.; Loizos, A. Estimation of in-situ density and moisture content in HMA pavements based on GPR trace reflection amplitude using different frequencies. *J. Appl. Geophys.* **2013**, *97*, 3–10. [CrossRef]

41. Heitzman, M.; Maser, K.; Tran, N.H.; Brown, R.; Bell, H.; Holland, S.; Hiltunen, D. *Nondestructive Testing to Identify Delaminations between HMA Layers*; Transportation Research Board: Washington, DC, USA, 2013.

42. Pedret Rodes, J.; Pérez-Gracia, V.; Martínez-Reguero, A. Evaluation of the GPR frequency spectra in asphalt pavement assessment. *Constr. Build. Mater.* **2015**, *96*, 181–188. [CrossRef]

43. Salvi, R.; Ramdasi, A.; Kolekar, Y.A.; Bhandarkar, L.V. Use of Ground-Penetrating Radar (GPR) as an Effective Tool in Assessing Pavements—A Review. In *Geotechnics for Transportation Infrastructure*; Springer: Singapore, 2019; pp. 85–95.

44. Plati, C.; Loizos, A.; Gkyrtis, K. Assessment of Modern Roadways Using Non-destructive Geophysical Surveying Techniques. *Surv. Geophys.* **2020**, *41*, 295–430. [CrossRef]

45. Evans, R.D.; Frost, M.; Stonecliffe-Jones, M.; Dixon, N. *A Review of Pavement Assessment Using Ground Penetrating Radar (GPR) in 12th International Conference on Ground Penetrating Radar, 16–19 June 2008*; Universty of Birmingham and EuroGPR: Birmingham, UK, 2008.

46. Saarenketo, T.; Scullion, T. Road evaluation with ground penetrating radar. *J. Appl. Geophys.* **2000**, *43*, 119–138. [CrossRef]

47. Daniels, D.J. *Ground Penetrating Radar, Ser. IEE Radar, Sonar, Navigation and Avionics Series*; The Institution of Electrical Engineers: London, UK, 2004.

48. Al-Qadi, I.L.; Lahouar, S.; Loulizi, A. In situ measurements of hot-mix asphalt dielectric properties. *NDT E Int.* **2001**, *34*, 427–434. [CrossRef]

49. Evans, R.; Frost, M.; Stonecliffe-Jones, M.; Dixon, N. Assessment of in situ dielectric constant of pavement materials. *Transp. Res. Rec.* **2007**, *2037*, 128–135. [CrossRef]

50. ASTM. *D4748-06 Standard Test Method for Determining the Thickness of Bound Pavement Layers Using Short-Pulse Radar*; American Society for Testing and Materials: West Conshohocken, PA, USA, 2006.

51. Loizos, A.; Plati, C. Accuracy of pavement thicknesses estimation using different ground penetrating radar analysis approaches. *NDT E Int.* **2007**, *40*, 147–157. [CrossRef]

52. Varela-González, M.; Solla, M.; Martínez-Sánchez, J.; Arias, P. A semi-automatic processing and visualisation tool for ground-penetrating radar pavement thickness data. *Autom. Constr.* **2014**, *45*, 42–49. [CrossRef]

53. Marecos, V.; Fontul, S.; Solla, M.; de Lurdes Antunes, M. Evaluation of the feasibility of Common Mid-Point approach for air-coupled GPR applied to road pavement assessment. *Measurement* **2018**, *128*, 295–305. [CrossRef]

54. Wang, S.; Zhao, S.; Al-Qadi, I.L. Continuous real-time monitoring of flexible pavement layer density and thickness using ground penetrating radar. *NDT E Int.* **2018**, *100*, 48–54. [CrossRef]

55. Fernandes, F.M.; Pais, J.C. Laboratory observation of cracks in road pavements with GPR. *Constr. Build. Mater.* **2017**, *154*, 1130–1138. [CrossRef]

56. Saarenketo, T. *Recommendations for Guidelines for the Use of GPR in Asphalt Air Voids Content Measurement. Interreg IVA Nord Program, Mara Nord Proyect*; Päällystetutkaajat Oy: Rovaniemi, Finland, 2012; pp. 5–22.

57. Shang, J.Q.; Umana, J.A.; Bartlett, F.M.; Rossiter, J.R. Measurement of complex permittivity of asphalt pavement materials. *J. Transp. Eng.* **1999**, *125*, 347–356. [CrossRef]

58. Jaselskis, E.J.; Grigas, J.; Brilingas, A. Dielectric properties of asphalt pavement. *J. Mater. Civ. Eng.* **2003**, *15*, 427–434. [CrossRef]

59. Zhang, J.; Yang, X.; Li, W.; Zhang, S.; Jia, Y. Automatic detection of moisture damages in asphalt pavements from GPR data with deep CNN and IRS method. *Autom. Constr.* **2020**, *113*, 103119. [CrossRef]

60. Zhang, J.; Zhang, C.; Lu, Y.; Zheng, T.; Dong, Z.; Tian, Y.; Jia, Y. In-situ recognition of moisture damage in bridge deck asphalt pavement with time-frequency features of GPR signal. *Constr. Build. Mater.* **2020**, *244*, 118295. [CrossRef]

61. Benedetto, A. Water content evaluation in unsaturated soil using GPR signal analysis in the frequency domain. *J. Appl. Geophys.* **2010**, *71*, 26–35. [CrossRef]

62. Lai, W.L.; Kind, T.; Wiggenhauser, H. A study of concrete hydration and dielectric relaxation mechanism using ground penetrating radar and short-time Fourier transform. *EURASIP J. Adv. Signal Process.* **2010**, *2010*, 12. [CrossRef]

63. Dérobert, X.; Villain, G.; Cortas, R.; Chazelas, J.L. EM characterization of hydraulic concretes in the GPR frequency band using a quadratic experimental design. Presented at the NDT Conference on Civil Engineering, Nantes, France, 30 June–3 July 2009; pp. 177–182.

64. Bano, M. Modelling of GPR waves for lossy media obeying a complex power law of frequency for dielectric permittivity. *Geophys. Prospect.* **2004**, *52*, 11–26. [CrossRef]

65. Pérez-Gracia, V.; González-Drigo, R.; Di Capua, D. Horizontal resolution in a non-destructive shallow GPR survey: An experimental evaluation. *NDT E Int.* **2008**, *41*, 611–620. [CrossRef]

66. Laurens, S.; Balayssac, J.P.; Rhazi, J.; Klysz, G.; Arliguie, G. Non-destructive evaluation of concrete moisture by GPR: Experimental study and direct modeling. *Mater. Struct.* **2005**, *38*, 827–832. [CrossRef]

67. Santos-Assunçao, S.; Krishna, S.; Perez-Gracia, V. GPR analysis of water content in concrete specimen-laboratory test. In *Near Surface Geoscience 2016—22nd European Meeting of Environmental and Engineering Geophysics*; European Association of Geoscientists & Engineers: Houten, The Netherlands, 2016; Volume 2016, p. cp-495.

68. Sbartaï, Z.M.; Laurens, S.; Breysse, D. Concrete moisture assessment using radar NDT technique: Comparison between time and frequency domain analysis. Presented at the Non-Destructive Testing in Civil Engineering (NDTCE '09), Nantes, France, 30 June–3 July 2009.

69. Santos-Assunçao, S.; Perez-Gracia, V.; González-Drigo, R.; Salinas, V.; Caselles, O. Ground Penetrating Radar Applications in Seismic Microzonation. In *Near Surface Geoscience 2015—21st European Meeting of Environmental and Engineering Geophysics*; European Association of Geoscientists & Engineers: Houten, The Netherlands, 2015; Volume 2015, pp. 1–5.

70. Cui, F.; Wu, Z.Y.; Wang, L.; Wu, Y.B. Application of the Ground Penetrating Radar ARMA power spectrum estimation method to detect moisture content and compactness values in sandy loam. *J. Appl. Geophys.* **2015**, *120*, 26–35. [CrossRef]
71. Tosti, F.; Patriarca, C.; Slob, E.; Benedetto, A.; Lambot, S. Clay content evaluation in soils through GPR signal processing. *J. Appl. Geophys.* **2013**, *97*, 69–80. [CrossRef]
72. Benedetto, F.; Tosti, F. GPR spectral analysis for clay content evaluation by the frequency shift method. *J. Appl. Geophys.* **2013**, *97*, 89–96. [CrossRef]
73. Benedetto, A.; Tosti, F.; Ortuani, B.; Giudici, M.; Mele, M. Mapping the spatial variation of soil moisture at the large scale using GPR for pavement applications. *Near Surf. Geophys.* **2015**, *13*, 269–278. [CrossRef]
74. Tosti, F.; Benedetto, A.; Ciampoli, L.B.; Lambot, S.; Patriarca, C.; Slob, E.C. GPR analysis of clayey soil behaviour in unsaturated conditions for pavement engineering and geoscience applications. *Near Surf. Geophys.* **2016**, *14*, 127–144. [CrossRef]
75. Dirección General de Carreteras del Ministerio de Fomento. *Orden Circular 10/2002 Sobre Secciones de Firme y Capas Estructurales de Firmes*; Dirección General de Carreteras del Ministerio de Fomento: Madrid, Spain, 2002.
76. AASHTO. *American Association of State Highway and Transportation Officials Guide for Design of Pavement Structures*; AASHTO: Washington, DC, USA, 1986.
77. Prandi, E. The Lacroix-LCPC Deflectograph. In *Intl Conf Struct Design Asphalt Pvmts*; The National Academies of Sciences, Engineering, and Medicine: Washington, DC, USA, 1967.
78. Santos.Assunçao, S. Ground Penetrating Radar Applications in Seismic Zonation: Assessment and Evaluation. Ph.D. Thesis, Universitat Politècnica de Catalunya (UPC), Barcelona, Spain, 2014.
79. Solla, M.; González-Jorge, H.; Lorenzo, H.; Arias, P. Uncertainty evaluation of the 1 GHz GPR antenna for the estimation of concrete asphalt thickness. *Measurement* **2013**, *46*, 3032–3040. [CrossRef]
80. Lorenzo Cimadevilla, H. *Prospección Geofísica de Alta Resolución Mediante Geo-Radar*; Aplicación a Obras Civiles; Laboratorio de Geotecnia del CEDEX: Madrid, Spain, 1996; p. 196.
81. Sudyka, J.; Krysiński, L. Radar Technique Application in Structural Analysis and Identification of Interlayer Bonding. *Int. J. Pavement Res. Technol.* **2011**, *4*, 176–184.

remote sensing

Article

Spatial Representation of GPR Data—Accuracy of Asphalt Layers Thickness Mapping

Šime Bezina, Ivica Stančerić *, Josipa Domitrović and Tatjana Rukavina

Department of Transportation Engineering, Faculty of Civil Engineering, University of Zagreb, Fra Andrije Kačića-Miošića 26, 10 000 Zagreb, Croatia; sime.bezina@grad.unizg.hr (Š.B.); josipa.domitrovic@grad.unizg.hr (J.D.); tatjana.rukavina@grad.unizg.hr (T.R.)
* Correspondence: ivica.stanceric@grad.unizg.hr; Tel.: +385-1-4639-310

Abstract: Information on pavement layer thickness is very important for determining bearing capacity, estimating remaining life and strengthening planning. Ground-penetrating radar (GPR) is a nondestructive testing (NDT) method used for determining the continuous pavement layer thickness in the travel direction. The data obtained with GPR in one survey line is suitable for the needs of repair and rehabilitation planning of roads and highways, but not for wider traffic areas such as airfield pavements. Spatial representation of pavement thickness is more useful for airfield pavements but requires a 3D model. In the absence of 3D GPR, a 3D model of pavement thickness can be created by additional processing of GPR data obtained from multiple survey lines. Five 3D models of asphalt pavements were created to determine how different numbers of survey lines affect their accuracy. The distance between survey lines ranges from 1 to 5 m. The accuracy of the 3D models is determined by comparing the asphalt layer thickness on the model with the values measured on 22 cores. The results, as expected, show that the highest accuracy is achieved for the 3D model created with a distance of 1 m between survey lines, with an average relative error of up to 1.5%. The lowest accuracy was obtained for the 3D model created with a distance of 4 m between the survey lines, with an average relative error of 7.4%.

Keywords: ground-penetrating radar (GPR); asphalt layer thickness; nondestructive testing (NDT); 3D modelling; spatial representation; airfield pavement; apron

Citation: Bezina, Š.; Stančerić, I.; Domitrović, J.; Rukavina, T. Spatial Representation of GPR Data—Accuracy of Asphalt Layers Thickness Mapping. *Remote Sens.* **2021**, *13*, 864. https://doi.org/10.3390/rs13050864

Academic Editor: Mercedes Solla

Received: 29 January 2021
Accepted: 23 February 2021
Published: 25 February 2021

Publisher's Note: MDPI stays neutral with regard to jurisdictional claims in published maps and institutional affiliations.

1. Introduction

Pavement maintenance procedures and measures are important to ensure safe and unobstructed traffic flow and to maintain pavement condition-prescribed engineering and operational values. Maintenance procedures and measures should be based on data on the actual condition of the pavement and its physical properties. These data are traditionally collected through digging test-pits and by extracting cores [1]. These methods cannot provide a complete picture of pavement conditions because the data are related to a specific location. In addition, these methods are destructive because they require disruption of traffic and repair of a pavement section. Therefore, data about the pavement condition are very often obtained by nondestructive testing (NDT) methods, such as pavement deflection testing [2], laser scanning [3], infrared thermography (IRT) [4] and ground-penetrating radar (GPR) [5,6]. According to [7], the data collected with GPR are essential for the pavement rehabilitation project.

The GPR method is based on the emission of low power electromagnetic waves to obtain images of the subsurface layers [6]. The reflection and scattering of wide-band electromagnetic waves transmitted by radar occur as a result of discontinuities in the electrical and magnetic properties of the studied structure. The echoes detected in the examined structures or subsurface layers are then converted into images using signal processing and imaging techniques.

The application of GPR as an effective tool for subsurface inspection of transportation infrastructure is constantly evolving. It is used to determine the location of reinforcements [8], the condition of pipes [9], the degree of compaction of the asphalt layer [10], and delamination between asphalt layers [11], as well as to detect moisture damage in the asphalt pavement [12]. However, GPR has been the most widely used method for determining the pavement layer thickness, which was its primary function [5,13]. Pavement layer thickness provides very important information for determining bearing capacity, estimating remaining pavement life, strengthening planning of existing pavement and quality control during and after construction. From the beginning of its application, GPR has shown a high degree of accuracy in estimating asphalt layer thickness.

In the following sections, studies on the accuracy of determining asphalt layer thickness using GPR with air-coupled antennas are presented. For newly constructed pavements with asphalt layer thicknesses of 100 to 250 mm, [14] showed a thickness error of 2.9%. In the 2000s, the accuracy of GPR measurements was systematically researched in the USA [15–17]. According to a test conducted in Virginia [15], an HMA layer thickness error of about 3% was found when the individual layers were resolved in the reflected GPR signal. The error increased to 12% when the entire HMA layer was considered without resolving the thin layers. In accordance with [16], conducted on heavily trafficked highways in Virginia, the error in determining asphalt layer thickness ranged from 3.7% to 8.4%, with a mean of 5.7%. The GPR error in determining asphalt layer thickness ranged from 3.7% to 11.8%, with a mean of 8.0%, as reported by [17]. Analysis of the GPR data collected from different sites showed that thickness error increases with pavement age—4.4% error for 0- to 5-year-old pavements and 5.8% error for pavements older than 20 years having surfaces older than 10 years [14]. In Croatia, studies have been conducted on motorways, regional and county roads and on runways [18,19]. The error for new pavements of motorways was mostly less than 10% and varied from 0.16% to 12.32% [18]. For regional and county roads that have been in service for years, the error ranges from 6.70% to 14.83% [18]. The thickness of asphalt overlays on runways was found to have a relative error ranging from 1.7% to 10.3% [19].

The repeatability of GPR measurements has also been investigated. In [20], where three sets of data were collected at the same locations and at different time periods, thickness errors ranging from 5.9% to 12% were found. The discrepancy in the results can be explained by changes in the value of the dielectric constant due to the different moisture content of the test location, and same day measurements showed good repeatability. The speed of the survey had no significant effect on the performance of the GPR [17]. It was found that the thickness error was 6.7% for steady state measurements, 7.9% for low speed and 8.3% for high speed measurements [17]. Thus, GPR is capable of acquiring data at speeds up to 100 km/h [21].

Although GPR applications on airfields are similar to those on road pavements, less research has been conducted on airfields. GPR has been successfully applied to determine the thicknesses of all runway pavement layers [19]. According to [22], GPR and infrared thermography are two nondestructive testing (NDT) methods that are efficient in airfield inspection. Infrared thermography was used to locate cracks and other anomalies and GPR was used to determine the depth and thickness of these defects. Testing GPR along with other NDT methods to detect voids under airfield pavements showed that the advantage of GPR is that it can quickly process a large amount of data [23]. GPR has been successfully used at airfields in Poland to determine the direction of cracks and structural voids and to detect dowels and anchors in concrete slabs [24].

3D GPR is a useful tool for airfields, as it allows fast and economical surveying of the large areas of runways and taxiways [25]. As stated by [26], 3D GPR mapping and imaging is an efficient tool for airfield inspection and construction planning.

The data obtained by a GPR measurement only provides information about the thickness of asphalt layers in the longitudinal survey lines compared to the 3D GPR. GPR data are sufficient for the rehabilitation planning of road pavements, but not for airfield pavements. Airfield pavements are much wider than roads and have many sections with different construction histories and thus different pavement structures. To deal with this problem, a method to create a spatial representation (3D model) based on the GPR data was presented [27]. The main disadvantage of this method is that it is time consuming, requiring a large number of survey lines and additional processing of the GPR data. This disadvantage is especially a problem at airports where measurements must be made with traffic at short intervals. However, the value of the collected and interpreted data overcomes the above disadvantage.

The aim of this study is to investigate the influence of different distances between survey lines on the accuracy of 3D models of asphalt layer thickness.

2. Materials and Methods

The research was conducted on the apron of Pula airport in Croatia. The apron area is approximately 60,000 m^2. The structure of the apron pavement consists of asphalt layers placed on an unbound base course. Some sections of the apron have asphalt layers that are more than 30 years old—i.e., well beyond their expected service life (Figure 1). The pavement had many cracks propagating in various directions, both longitudinally and transversely, as well as radially in all directions.

Figure 1. Pula Airport apron with pavement distresses.

Research procedure consisted of the following steps:

- GPR data collection;
- interpreting GPR data;
- creating a 3D model with a spatial representation;
- core extraction.

2.1. GPR Data Collection

The GPR system shown in Figure 2 was used to collect the data on the total asphalt layer thickness of the apron pavement. The system consists of two air-coupled antennas (1.0 and 2.0 GHz), a central unit for connecting the system components (SIR 20), a computer for processing and storing the data and a Distance Measuring Instrument (DMI). The system is supplemented by a high resolution digital camera. Antennas with 1.0 and 2.0 GHz central frequencies offer a very good compromise between the possible depth and the recording resolution for determining the pavement thickness; the transmitters and receivers are located in each antenna. They are separated by a duplexer that allows each antenna to transmit and receive electromagnetic waves (monostatic type of radar).

Figure 2. Measurement vehicle at the Pula Airport with Ground-penetrating radar (GPR) antennas.

Before starting data collection, certain parameters and filters need to be set (Table 1). Position correction is the parameter that controls the length of the time that the system will acquire data. Range gain controls the time-variable gain. Gain is signal amplification used to compensate for the natural effects of signal attenuation. As the transmitted signal passes through a material, it will attenuate as the material absorbs some signal. Gain amplifies that signal after it is received to compensate for signal losses and make weaker reflectors easier to see. A Finite impulse response (FIR) filter and Infinite impulse response (IIR) filter are useful for reducing high and low frequency noise in the data.

Table 1. Parameters and filters that were used during the data collection.

Parameters and Filters	Variable	1 GHz Antenna	2 GHz Antenna
Position/Range	Range [ns]	20	15
	Position [ns]	96	96
Range gain	Point	1	1
	Number of points	1	1
	Value	13	17
FIR filter	Low pass [MHz]	5000	6000
	High pass [MHz]	300	300
	Filter type	boxcar	boxcar
IIR filter	Horizontal low pass [scans]	0	0
	Horizontal high pass [scans]	0	0
	Vertical low pass [MHz]	0	0
	Vertical high pass [MHz]	0	0

Unfortunately, the data acquired with the 2.0 GHz antenna showed a high noise level. From previous experience, a possible reason for this noise could be the GPR signal interfering with the signals from an air traffic control tower. During the measurement, all flight control systems were on. Equipment and installations of a military base for unmanned aerial vehicles (UAVs), which were in the vicinity, could also have had a negative influence. Only the data acquired with 1.0 GHz antenna were used in this research.

For detailed determination of asphalt layer thickness, data collection was performed on 247 lines, of which 160 survey lines had lengths of 150 m and 87 survey lines had lengths of 409 m (Figure 1). The mutual distance between lines was 1 m. The data were collected in the west–east direction. Data were collected every 10 cm along the lines. Near physical obstacles (traffic lights, stop barriers), data could not be collected along the entire survey line.

Two metal plates with widths of 10 cm and lengths of 150 cm were placed at the beginning and at the end of each survey line so that the exact position of the beginning and the end of the measurements could be clearly seen during data interpretation, since the electromagnetic (EM) wave reflection from the metal plates is complete in the radargram (Figure 3).

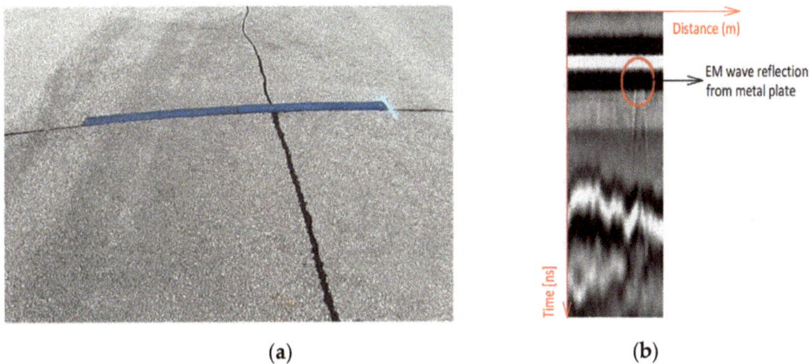

Figure 3. (**a**) Metal plates; (**b**) electromagnetic (EM) wave reflection on radargram.

To ensure straight-line movement of the vehicle on the test site, guidance lines were delineated. For each guidance line, the surveyor set the start and end points. Cones with offset rods were placed at intervals of approximately 15 m between them. To prevent the driver from "wandering", a wooden visor was mounted on the vehicle to facilitate driving in the direction of the guidance line. A low driving speed of (20 km/h) also proved to be crucial. After collecting data on a line, the cones with offset rods were moved laterally by 1 m. Prior to the measurement, this approach was tested on a polygon. The deviation between the guidance line and the GPR survey line was not more than 5 cm.

2.2. Interpreting GPR Data

By interpreting the data collected with the GPR, the values of the total asphalt layer thickness for each survey line were determined. Pavement cracks did not pose a problem in interpreting the radargram. The thickness of the asphalt layer was determined based on the reflection method. The principle of using GPR reflections to calculate the layer thickness and dielectric constant was explained in [28].

According to [28], the velocity of the electromagnetic (EM) wave through a given medium (air or pavement layer) is affected by the dielectric constant of a single layer. Equation (1) is used to determine the thickness of each layer (h):

$$h = v \cdot \frac{\Delta t}{2} \tag{1}$$

where (v) is the velocity of the EM wave through the layer and (Δt) is the propagation time of the wave reflected at the layer's base. The velocity of the EM wave v through the layer is determined by means of Equation (2):

$$v = \frac{c}{\sqrt{\varepsilon_r}} \tag{2}$$

where (c) is the velocity of the EM wave through the vacuum (approximately 3×10^8 m/s) and (ε_r) is the dielectric constant of a layer. If the dielectric constant ε_{r1} of a top layer is known, the thickness of that layer h_1 can be calculated by Equation (3):

$$h_1 = \frac{c \cdot \Delta t_1}{2 \cdot \sqrt{\varepsilon_{r1}}} \qquad (3)$$

EM waves have larger amplitudes at the boundaries between layers (Figure 4), and this amplitude depends on the dielectric constant of each layer. The greater the difference in dielectric constant between layers, the greater the amplitude of the reflected EM wave. The dielectric constant ranges from 1 for air (vacuum) to 81 for water. For road construction materials, the dielectric constant ranges from 2 to 30 [29]. For air-coupled antennas, the surface reflection method is used to calculate the dielectric constant of the layer. In this method, a metal plate is used because the metal plate completely reflects the EM waves so that the amplitude of reflection is maximum [28,30]. Since the dielectric constant of air is known, the dielectric constant of the top layer can be calculated according to Equation (4):

$$\varepsilon_{r1} = \left(\frac{A_m + A_1}{A_m - A_1} \right)^2 \qquad (4)$$

where (A_m) is the amplitude of the metal plate reflection and (A_1) is the top layer amplitude. Similarly, it is possible to calculate the dielectric constant of the next pavement layer according to Equation (5):

$$\varepsilon_{r2} = \varepsilon_{r1} \cdot \left[\frac{1 - \left(\frac{A_1}{A_m} \right)^2 + \left(\frac{A_2}{A_m} \right)}{1 - \left(\frac{A_1}{A_m} \right) + \left(\frac{A_2}{A_m} \right)} \right]^2 \qquad (5)$$

where (ε_{r2}) is the dielectric constant of the middle layer and (A_2) is the middle layer reflection amplitude.

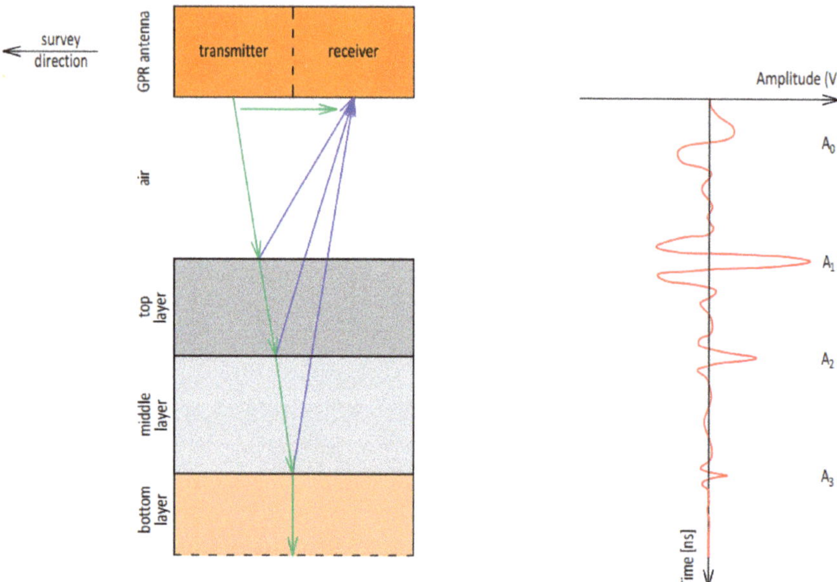

Figure 4. Scheme of electromagnetic wave transmission to the pavement.

The processing and interpretation of the collected GPR data were performed in RADAN 6.6 software in accordance with [31]. In the processing stage, the raw GPR data were combined with the calibration data collected over a metal plate to obtain processed data. This step was necessary to compensate for the antennas bouncing during the data collection, adjust the time-zero correction and calculate the velocity of the GPR signal. The reflections of the EM waves at the different layer boundaries within the pavement were visually identified in the radargram (Figure 5a). Discontinuities and larger dielectric contrasts between media imply more prominent reflection. As shown in Figure 5a, a continuous strong reflection defines an obvious interface between air and asphalt layers and the asphalt layer and unbound base layer. The layer depth was determined by manually controlled semiautomatic interpretation based on finding the nearest peak. The results are shown in the depth and distance diagrams for each measurement line (Figure 5b).

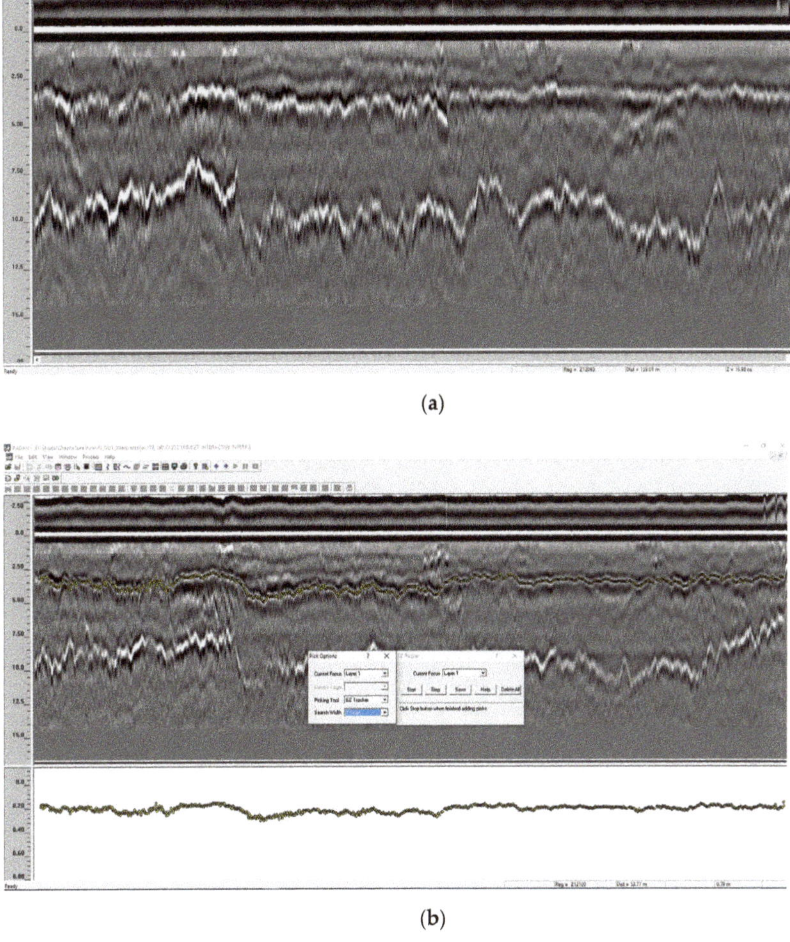

(a)

(b)

Figure 5. Interpretation of the collected GPR data by 1 GHz antenna: (**a**) radargram; (**b**) layer depth determination.

2.3. Creating a 3D Model with a Spatial Representation

The profile changes of asphalt layer thickness in different longitudinal lines are not sufficient for a useful representation because it is difficult to see how the thickness of asphalt layers changes over the observed apron. Therefore, a spatial representation is required. Spatial representation can be achieved on a 3D model. In the absence of 3D GPR, a 3D model of asphalt layer thickness can be created by additional processing of GPR data. The procedure for creating a 3D model is explained in detail in [27]; in brief, it consists of three steps:

1. Determining the spatial coordinates (x, y, and z) for all data points acquired by the GPR measurement. Point coordinate x is the distance from the start of the measurement. Point coordinate y is the lateral displacement between the start line and the adjacent survey line. Point coordinate z is the thickness of the asphalt layers and is determined after processing the GPR measurement data.
2. Import all data points with spatial coordinates into the software to create a 3D model (Figure 6a).
3. Create a 3D model of asphalt layer thickness by Delaunay triangulation of all data points (Figure 6b). The developed 3D model of asphalt layer usually consists of triangles connecting all points in the form of a regular square grid. Based on a 3D model of the surface, it is possible to create contours (lines representing points with the same thickness of the asphalt layers) (Figure 6c) or bands (areas with the same thickness range of the asphalt layers).

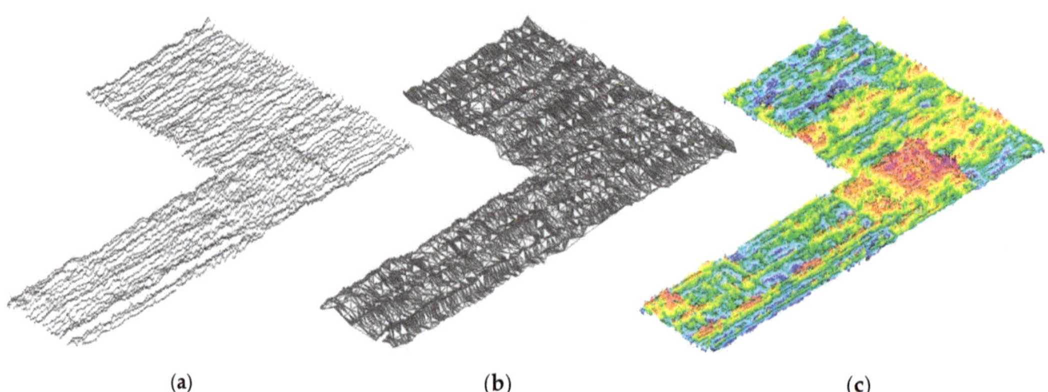

(a)   (b)   (c)

Figure 6. The procedure for creating 3D model of an asphalt layer: (**a**) 3D points; (**b**) 3D model with triangulated regular square grid; (**c**) 3D model with contours.

2.4. Cores Extraction

Twenty-two cores were extracted to determine the thickness of the asphalt layers, seven of which were 300 mm in diameter and 15 of which were 100 mm in diameter. The positions of the cores were recorded by GPS and are shown in Figure 7.

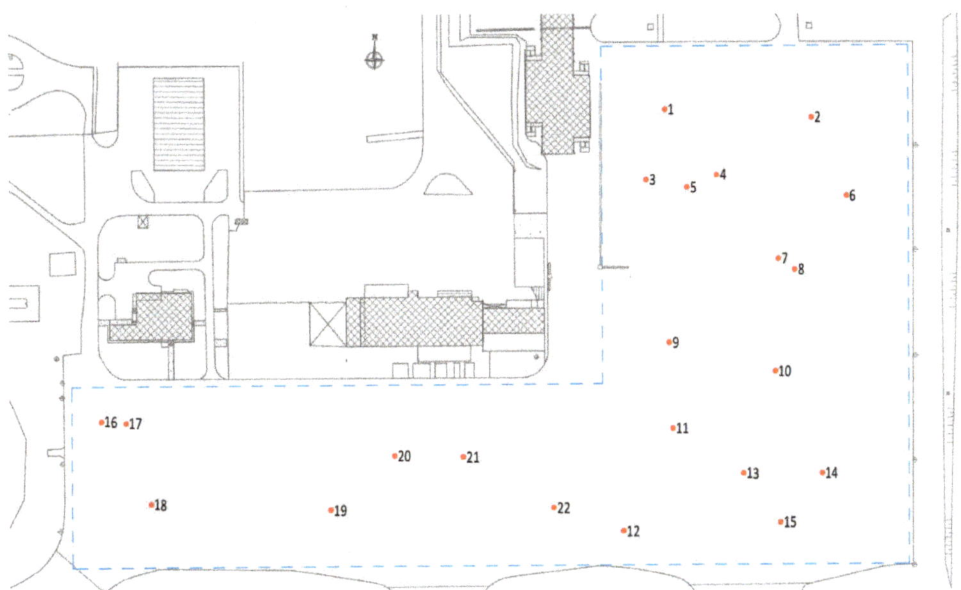

Figure 7. Location of core samples on the apron.

3. Results

Determining asphalt layer thickness for wide traffic areas, such as aprons, is very time consuming, especially if the measurement is made for a large number of survey lines. The larger the number of survey lines, the larger the amount of collected data that need to be processed. In this paper, the possibility of optimizing the whole process of acquiring, processing and displaying asphalt layer thickness measurement data and determining its accuracy is researched.

Based on the analysis of GPR data, the total asphalt layer thickness was determined at 597,348 measurement points. The total asphalt layer thickness ranged from a minimum of 55 mm to a maximum of 371 mm. The collected asphalt layer thickness data were additionally processed and x, y, and z coordinates were assigned for all measurement points according to the instructions in Chapter 2. A 3D model of the asphalt layers (MC1) was created from the measurement points by Delaunay triangulation using AutoCAD Civil 3D software. The MC1 model is graphically represented on a contour map—i.e., the points with the same asphalt layer thickness are connected by contour lines (Figure 8). The contour lines have an equidistance value of 5 mm, which is considered accurate enough for the rehabilitation projects, although other equidistance values can be used if necessary.

In addition to the MC1 model, four other models were created—MC2, MC3, MC4, and MC5 (Figures 9–12). All models were created based on identical data collected by the measurement described above, but they differ in the number of points considered in their creation. By omitting a specific longitudinal series of points, a simulation to determine the total asphalt layer thickness was performed with a smaller number of survey lines at a greater mutual spacing. The number of survey lines and their mutual spacing, as well as the number of points used to create a particular model, are presented in Table 2, which shows the effect of the number of survey points on the representation of asphalt layer thickness and the accuracy of a particular model.

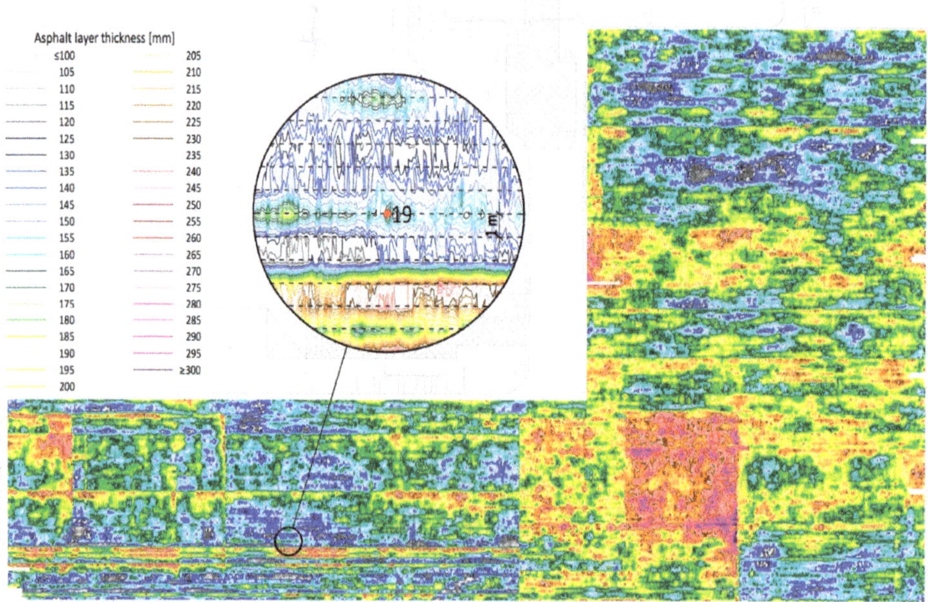

Figure 8. MC1 model represented with contours.

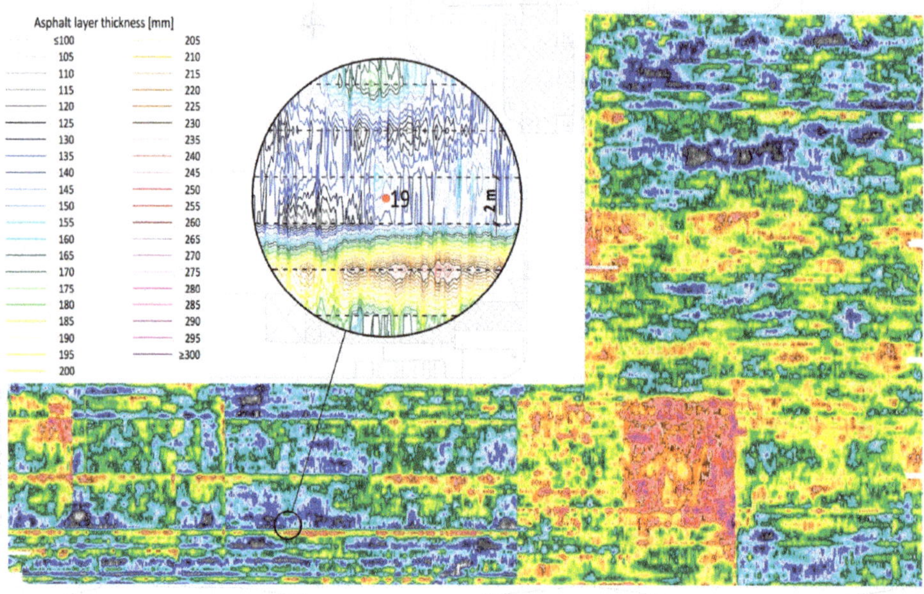

Figure 9. MC2 model represented with contours.

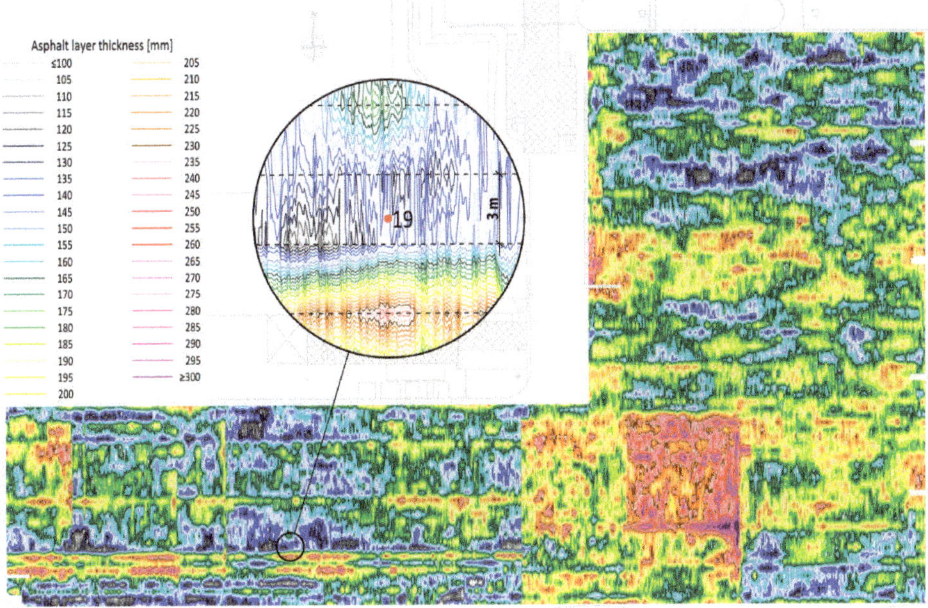

Figure 10. MC3 model represented with contours.

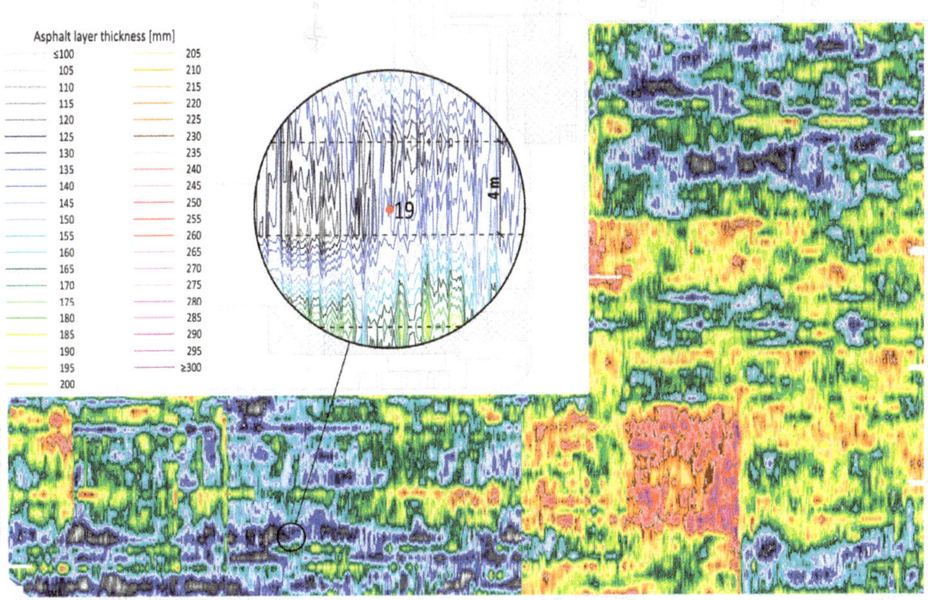

Figure 11. MC4 model represented with contours.

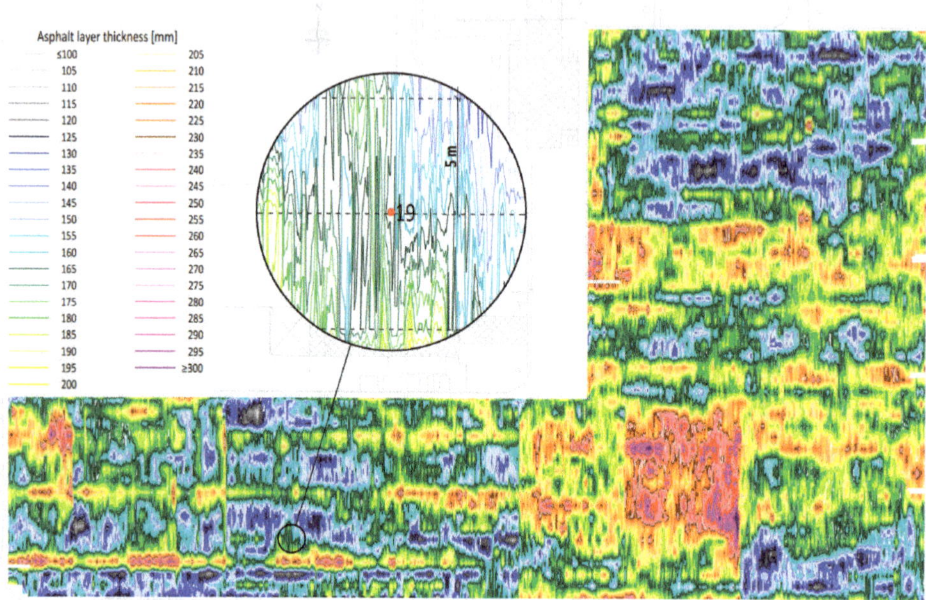

Figure 12. MC5 model represented with contours.

Table 2. Settings of 3D models with contours.

Model Settings	3D Models				
	MC1	MC2	MC3	MC4	MC5
Distance between survey line [m]	1	2	3	4	5
Number of survey lines	247	125	84	63	51
Number of points	597,348	303,562	203,343	153,788	125,560
Contour equidistance [mm]	5	5	5	5	5

To determine the accuracy of the model, asphalt layer thickness values obtained by coring were compared to the values shown on the contour map. The asphalt layer thickness values on the map were determined by the nearest contour line at the core location. The relative error of the thickness values presented on the map was calculated as the ratio between the absolute error (the difference between the thickness on the map and that of the core) and the core thickness.

The MC1 model with a distance of 1 m between survey lines resulted in relative errors ranging from 0.0% to 5.6%, with a mean error of 1.5% (Figure 13). This shows the high accuracy of the created 3D model.

The MC2 model with a distance of 2 m between survey lines resulted in relative errors ranging from 0.0% to 18.5%, with a mean error of 6.6% (Figure 14).

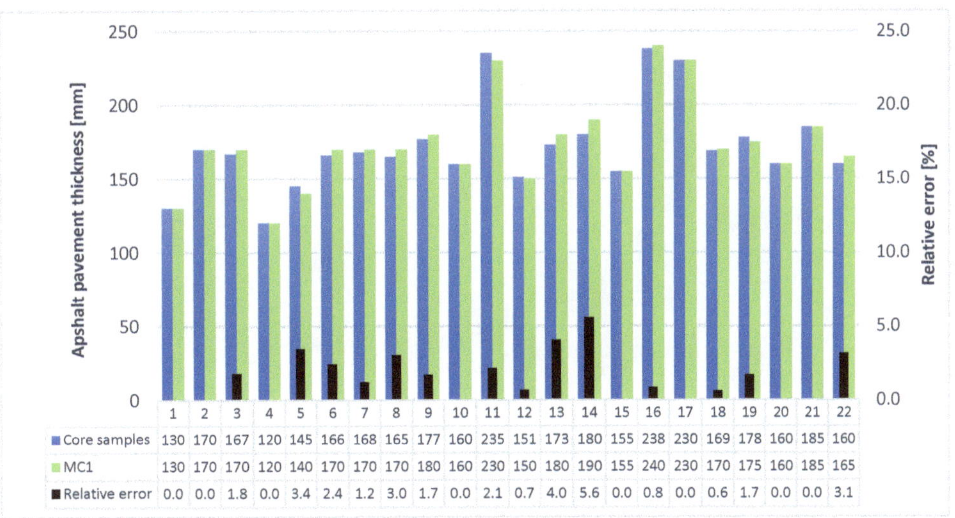

Figure 13. Model MC1—asphalt layer thickness and relative error in the core locations.

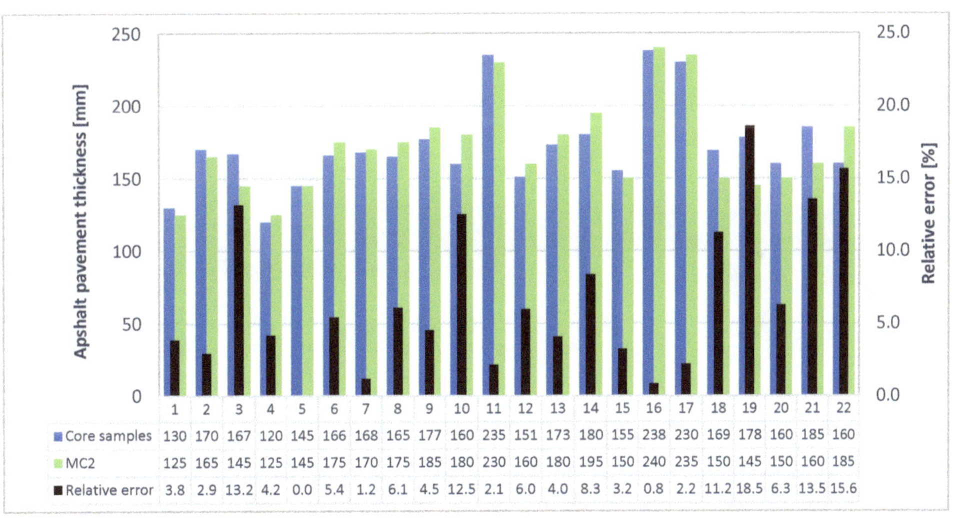

Figure 14. Model MC2—asphalt layer thickness and relative error in the core locations.

The MC3 model with a distance of 3 m between survey lines resulted in relative errors ranging from 0.0% to 24.2%, with a mean error of 4.6% (Figure 15).

The MC4 model with a distance of 4 m between survey lines resulted in relative errors ranging from 0.0% to 24.2%, with a mean error of 7.4% (Figure 16).

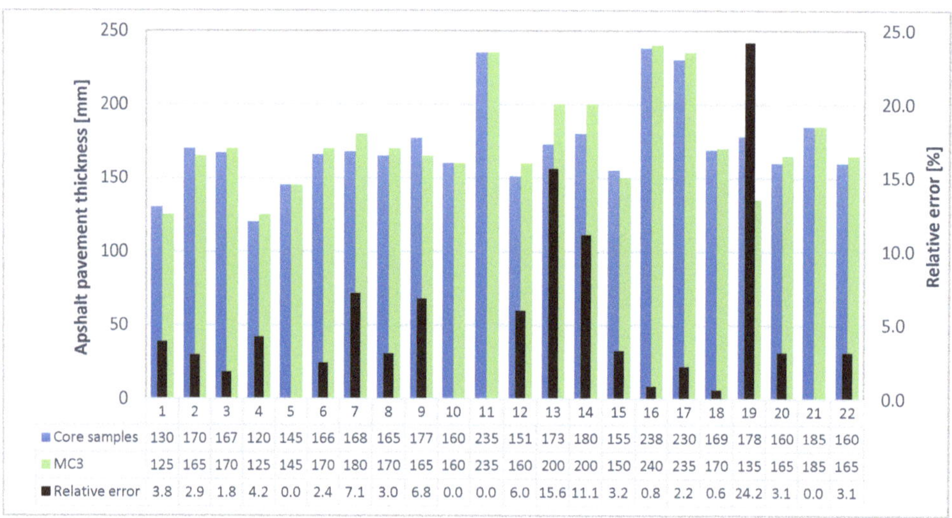

Figure 15. Model MC3—asphalt layer thickness and relative error in the core locations.

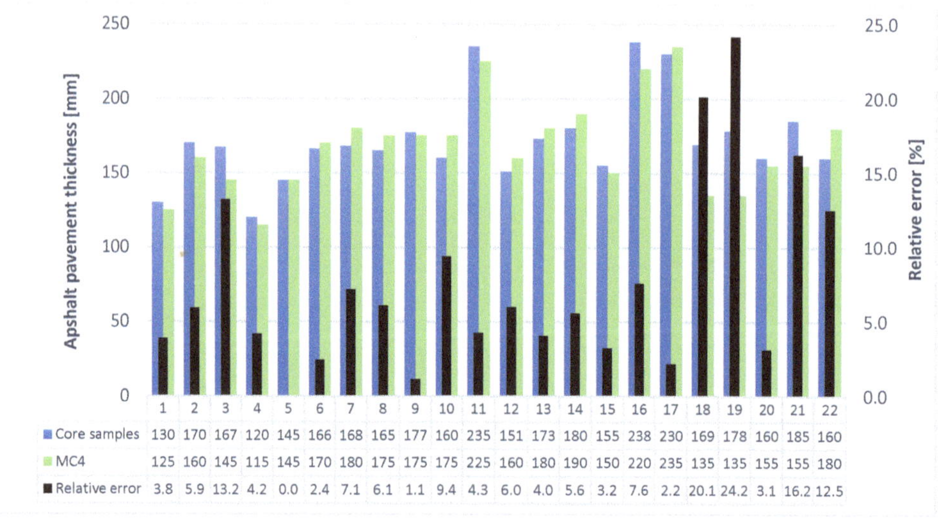

Figure 16. Model MC4—asphalt layer thickness and relative error in the core locations.

The MC5 model with a distance of 5 m between survey lines resulted in relative errors ranging from 0.0% to 14.5%, with a mean error of 6.5% (Figure 17).

The accuracy of the GPR measurement was determined by comparing the thicknesses measured on the cores with those from the GPR data. Only three cores (3, 15, and 22) were placed on survey lines, and the relative errors of the GPR measurement there ranged from 0.0% to 3.1% (Figure 13, Table 3).

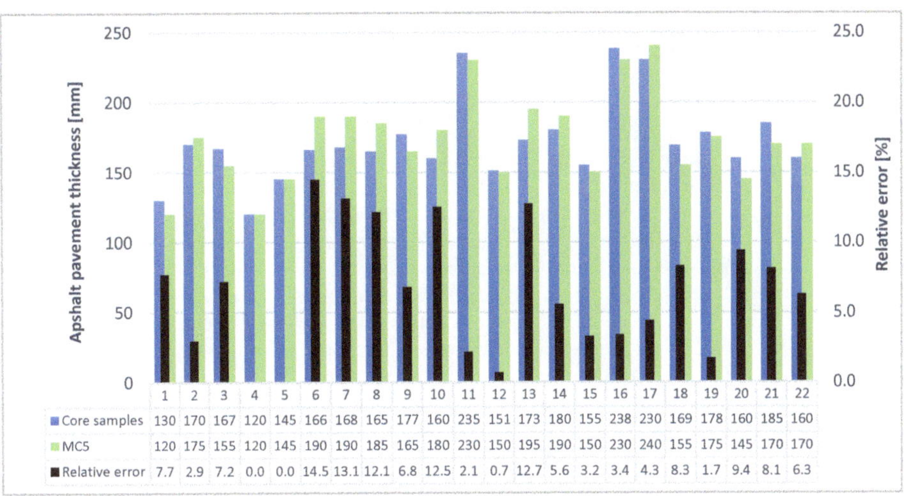

Figure 17. Model MC5—asphalt layer thickness and relative error in the core locations.

Table 3. Distance between core samples and survey line in different models and relative error.

Core Sample	Distance between Core and Survey Line [m]				
	MC1	**MC2**	**MC3**	**MC4**	**MC5**
1	0.43	0.43	0.43	0.43	1.43
2	0.37	0.63	0.37	1.36	1.63
3	0.00	1.00	0.00	1.00	2.00
4	0.40	0.40	0.40	1.60	0.60
5	0.40	0.40	0.40	0.40	0.40
6	0.33	0.33	0.67	0.33	1.33
7	0.26	0.26	0.74	1.74	1.26
8	0.09	0.91	0.09	1.09	1.09
9	0.24	0.24	0.76	1.76	0.76
10	0.21	0.79	0.21	0.79	2.21
11	0.25	0.75	0.75	1.25	0.25
12	0.26	0.26	1.26	0.26	1.26
13	0.06	0.06	0.94	0.06	0.94
14	0.50	0.50	0.50	0.50	0.50
15	0.00	1.00	0.00	1.00	1.00
16	0.25	0.25	0.25	1.75	1.25
17	0.50	0.50	0.50	1.50	0.50
18	0.03	0.97	0.03	0.97	2.03
19	0.09	0.91	1.09	1.09	0.09
20	0.14	0.86	0.14	1.14	0.86
21	0.23	0.77	0.23	1.23	0.77
22	0.00	1.00	0.00	1.00	2.00
Average	0.23	0.60	0.44	1.01	1.10

To get a better insight into the reason why the errors in the models are larger than the GPR measurement errors, the distances between the cores and the nearest survey line were measured and then compared to the relative errors of the model. The result was that the average distance between cores closest to the survey lines was 0.23 m for the MC1 model, while the average distance between cores farthest from the survey lines was 1.10 m for the MC5 model (Table 3). The smallest average relative errors found in the MC1 and MC3 models coincide with the smallest average distances between cores and survey lines.

As expected, the comparison of the measured distances and the relative errors of the model showed that the relative error of the model increases with increasing distance between the core and the survey line (Table 3). However, the largest relative error, 24.2%, was not registered at the farthest core, but at core 19, which was 1.09 m away from the survey line in the MC3 and MC4 models. The most distant core 10 (model MC5) showed a relative error of 12.5%. Since the largest error of the model was not recorded at the furthest point, other influencing factors that may have affected the error occurring at core 19 were considered. It was found that within a 3.0 m radius of Core 19, there is a significant variation in asphalt layer thickness from 120 to 225 mm (Figures 8–12).

Contour maps provide a detailed representation of asphalt layer thickness and allow determination of model accuracy but are not suitable for defining homogeneous areas used in selection of rehabilitation technology. Maps with bands are more suitable for defining homogeneous areas because the bands represent the pavement area with a specific range of asphalt layer thickness. Six bands (Table 4) were defined to determine the effect of the number of survey lines on the size of the homogeneous areas on each of the five 3D models (Figures 18–22). Band B1 includes all asphalt layer thickness values less than 100 mm. Bands B2–B5 include thickness values from 100 to 300 mm, with each band being 50 mm. Band B6 includes all asphalt layer thickness values greater than 300 mm.

Table 4. Band surface areas.

Band Label	Asphalt Layers Thickness [mm]	Area [m^2]				
		MC1	MC2	MC3	MC4	MC5
B1	<100	24.58	8.49	20.89	17.10	1.44
B2	100–150	13,144.01	12,179.82	12,119.81	13,406.81	11,154.58
B3	150–200	36,568.32	37,835.70	38,322.33	36,576.57	38,676.12
B4	200–250	8561.50	8354.39	7863.45	8330.34	8486.47
B5	250–300	691.45	618.36	655.45	667.93	672.51
B6	≥300	13.62	6.72	21.55	4.73	12.36

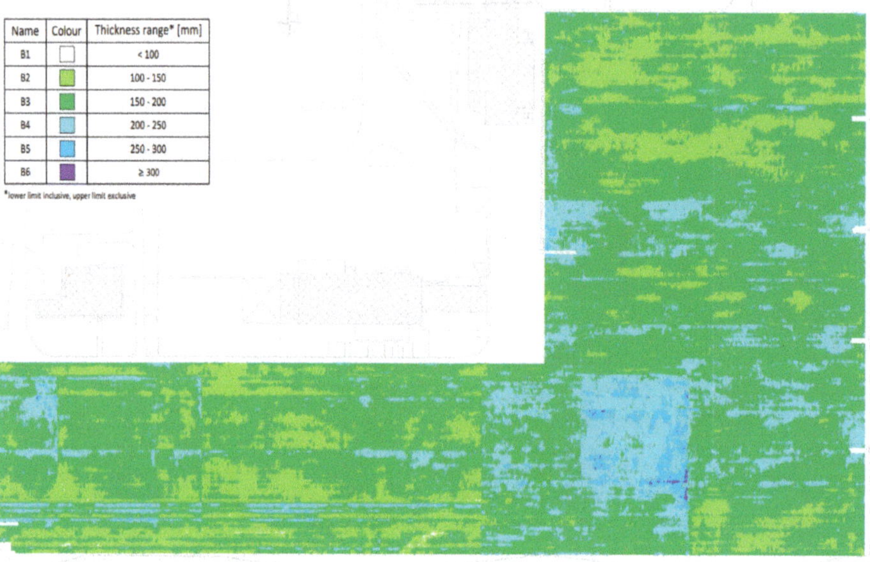

Figure 18. MC1 model represented with bands.

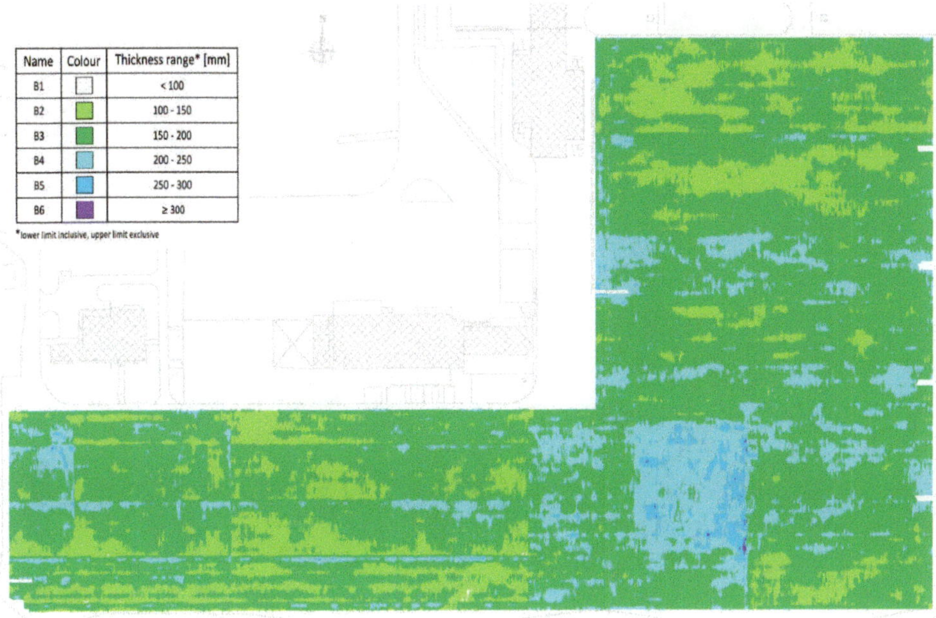

Figure 19. MC2 model represented with bands.

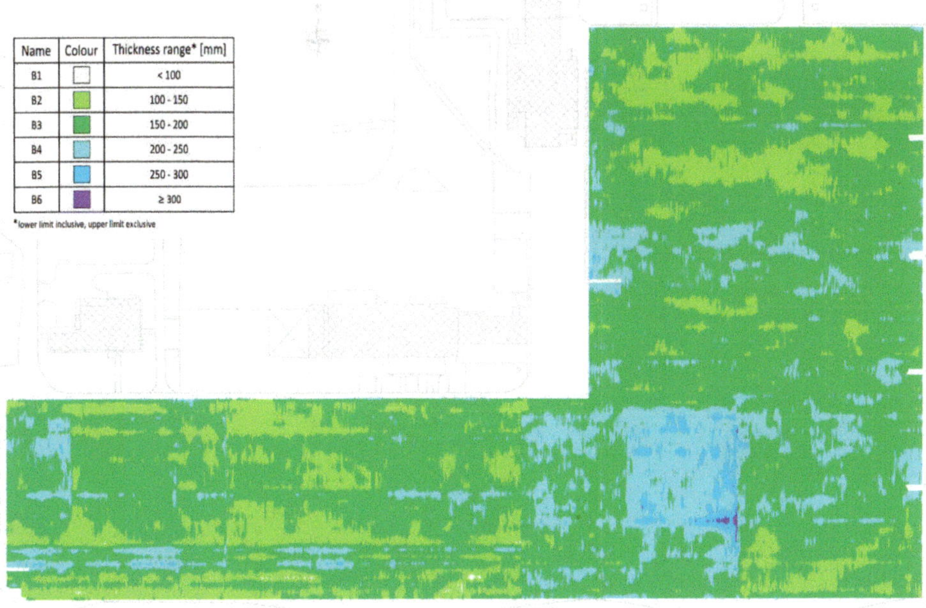

Figure 20. MC3 model represented with bands.

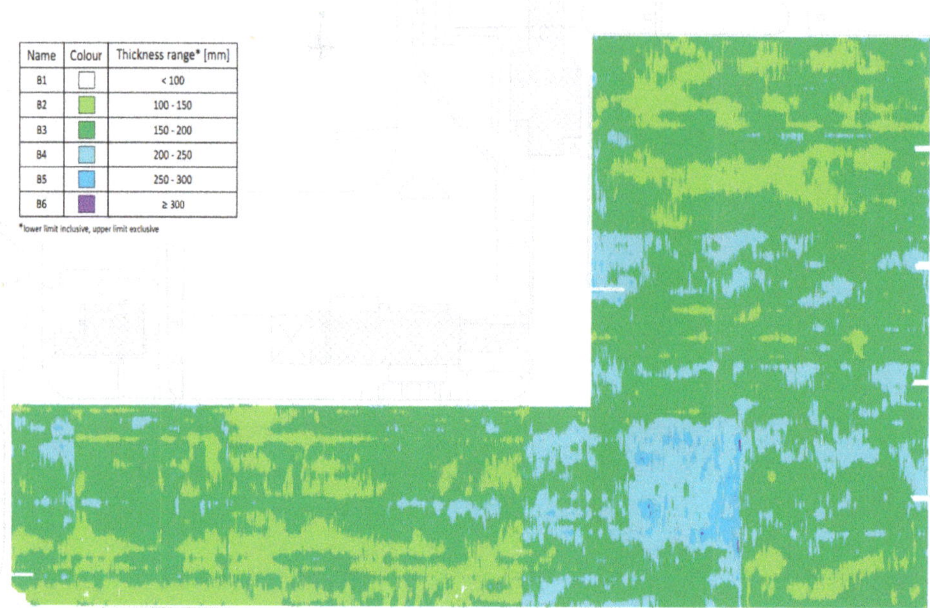

Figure 21. MC4 model represented with bands.

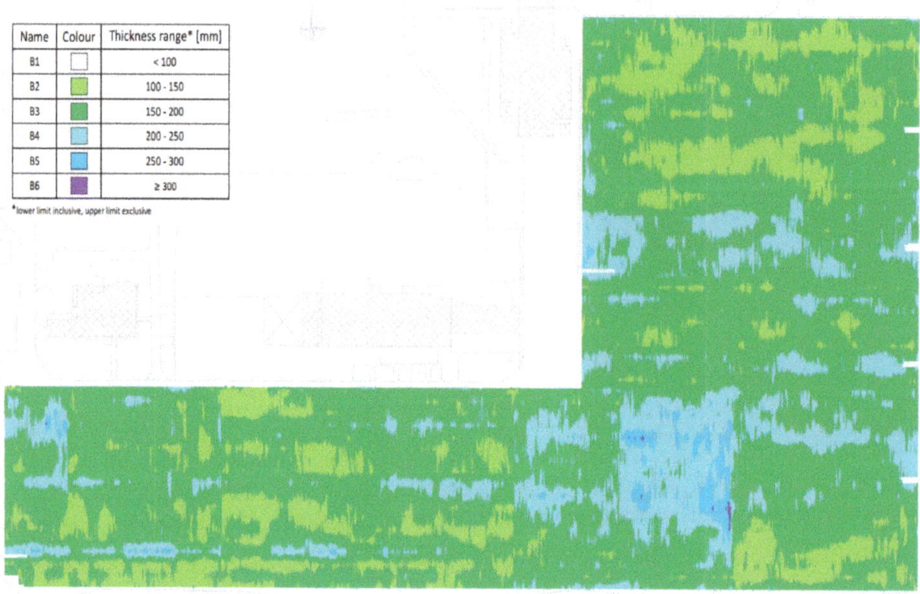

Figure 22. MC5 model represented with bands.

The results in Tables 4 and 5 show that the asphalt layer thicknesses on the apron are mostly between 150 and 200 mm, while their area percentage varies between 61.98%

and 65.55% depending on the 3D model. Asphalt layer thicknesses below 100 mm and above 300 mm have the lowest percentages of total apron area ($\leq$0.04%). The differences in surface area between the same bands in different models are not significant. The largest difference in surface area between the same bands was found in the MC1 and MC5 models. The differences in surface area between the two models in band B2 and band B3 are 3.37% and 3.57%, respectively.

Table 5. Band surface share in the total apron area.

Band Label	Asphalt Layers Thickness [mm]	Area [m^2]				
		MC1	MC2	MC3	MC4	MC5
B1	<100	0.04	0.01	0.04	0.03	0.00
B2	100–150	22.28	20.64	20.54	22.72	18.91
B3	150–200	61.98	64.12	64.95	61.99	65.55
B4	200–250	14.51	14.16	13.33	14.12	14.38
B5	250–300	1.17	1.05	1.11	1.13	1.14
B6	$\geq$300	0.02	0.01	0.04	0.01	0.02

4. Discussion

The classical two-dimensional representation of asphalt layer thickness is not satisfactory for the rehabilitation plans of wide surfaces, such as airfield pavements, because it is necessary to know the thickness profile over the entire surface. Therefore, a spatial representation of asphalt layer thickness is more useful on such surfaces. The spatial representation is usually achieved based on data collected by 3D GPR. However, it is also possible based on GPR data, but with additional activities. Additional activities are related to GPR data collection on more parallel survey lines that must cover the whole area under consideration, definition of the y-coordinate (a lateral displacement between the starting and adjacent survey lines) on all measurement points and creation of the 3D model. The main disadvantage of such a method is that it is time consuming; it requires a large number of survey lines, which can be a problem at airports where measurements have to be made at short intervals between regular aircraft operations without airport closure.

3D models of asphalt layer thickness can be displayed on contour maps or maps with bands. The different ways of representing asphalt layer thickness allow researchers and designers to better determine homogeneous areas, which is important for the proper selection of pavement rehabilitation technology. Specialized software allows the representation of 3D models with contours with different equidistance values and bands with different ranges of layer thickness. In this research, the division of the bands into ranges of 50 mm was chosen. This division of the bands allows the analysis of the existing pavement structure with sufficient accuracy and allows the possibility to choose the optimal rehabilitation solution. It should be emphasized that such a division of the bands is particularly important when there is no possibility to change the vertical alignment of the surface to be rehabilitated. Two cases should be distinguished. The first case concerns pavements with a total thickness of asphalt layers greater than the thickness of the newly planned layers. In this case, when the portion of existing layers are milled off, a certain minimum thickness of them must remain (50 mm). If the bands are created with a range of 50 mm, it is possible to estimate the area with a thickness of asphalt layers smaller than the specified 50 mm. On such areas it will be necessary to completely remove the asphalt layers and rehabilitate the subgrade. The second case relates to the asphalt overlay on the concrete pavement, which is usually made of the asphalt concrete AC 16 surf type with a minimum technological thickness of 42 mm. In practice, the thickness of this layer is usually 50 mm. If the thickness of the existing overlay is less than 50 mm, bands with a range of 50 mm provide information about the need to remove part of the concrete pavement.

Of course, it is possible to make divisions in bands with a smaller range, but such a choice is suitable for smaller areas where there are significant variations in the thickness of asphalt layers. Finer division, e.g., the range of 10 mm, leads to many homogeneous areas with smaller dimensions (smaller share in the total area of the apron, large scatter).

The number of survey lines affects model accuracy, but is not the only factor, as the model with the lowest number of survey lines is not the least accurate. Accuracy is affected by the distance between the core and the survey line, while local variations in layer thickness can also have a significant impact. The accuracy of the model is also affected by the triangulation procedure. Because of the regular square grid of points in the longitudinal and transverse directions, triangulation forms triangles simply by connecting points on adjacent lines. The elevation of any point on the triangulated surface is determined by interpolating the elevations of the vertices of the triangles. If the variations in the asphalt layer thickness are small, i.e., the section is homogeneous, then the influence of triangulation on the accuracy of the 3D model is smaller.

The large spacing between survey lines reduces the time required to collect and process the GPR data. If the distance between the survey lines is increased, there is a risk that significant local variations in thickness will remain undetected. Similarly, when the survey line spacing is large, if the location of the local variation in layer thickness coincides with the location of the survey line, the effect of the local variation will be visible over a larger area due to the triangulation process described previously.

The high accuracy of the 3D model suggests that such an approach to data analysis, processing and representation is satisfactory and can compensate the 3D GPR.

5. Conclusions

3D models of asphalt layers provide a clear insight into the variations of asphalt layer thickness in longitudinal and transverse directions. This research analyses the accuracy of different 3D models of asphalt layers using Airport Pula apron as an example. A total of five 3D models were created based on the different number of survey lines in which GPR data were collected and subsequently presented on contour maps or maps with bands. The different numbers of survey lines simulated the smaller or larger number of survey points that were considered in the creation of the 3D model.

Contour maps and cores were used to analyze the accuracy of the 3D models. As expected, it was found that the relative error of the 3D model was lowest at the narrowest survey line distance of 1 m and highest at the 4.0 m distance. It was found that the accuracy of a 3D model depends primarily on the accuracy of the GPR measurement, the distance of the core from the survey line—i.e., triangulation process—but also on the local thickness variations.

Maps with bands provided a representation of homogeneous surfaces with respect to the range of asphalt layer thickness. It was found that the size of certain homogeneous surfaces did not vary much considering the number of survey lines used in the creation of the 3D model.

The results of the analysis showed that the previously described methodology of GPR data collection, processing and spatial representation can be applied to airfield asphalt pavements with high accuracy. Furthermore, if the GPR measurement, data processing and interpretation have to be performed in the shortest possible time, a smaller number of survey lines can be selected, resulting in a slightly lower accuracy of the 3D model. The research has shown that GPR data can be initially collected on survey lines spaced more than 1.0 m apart, assuming that the observed section does not have significant variations in asphalt layer thickness as a result of different construction histories. The number of survey lines, and therefore the accuracy of the model, should be adjusted to suit the purpose of the research.

Author Contributions: Conceptualization, Š.B., I.S. and J.D.; methodology, Š.B. and I.S.; software, Š.B., I.S. and J.D.; validation, Š.B., I.S., J.D. and T.R.; formal analysis, Š.B., I.S. and J.D.; investigation, Š.B., I.S., J.D. and T.R.; resources, T.R.; data curation, Š.B., I.S. and J.D.; writing—original draft preparation, Š.B., and I.S.; writing—review and editing, Š.B. and I.S.; visualization, Š.B. and I.S.; supervision, J.D. and T.R.; project administration, T.R.; funding acquisition, T.R. All authors have read and agreed to the published version of the manuscript.

Funding: This research was funded by UNIVERSITY OF ZAGREB, CROATIA based on the decision on the distribution of advances for the basic financing of scientific and artistic activities, referenced as 402-08/19-03/17.

Data Availability Statement: Data available from the authors on request.

Conflicts of Interest: The authors declare no conflict of interest.

References

1. Loizos, A.; Plati, C. Accuracy of pavement thicknesses estimation using different ground penetrating radar analysis approaches. *NDT E Int.* **2007**, *40*, 147–157. [CrossRef]
2. Singh, A.; Sharma, A.; Chopra, T. Analysis of The Flexible Pavement Using Falling Weight Deflectometer for Indian National Highway Road Network. *Transp. Res. Procedia* **2020**, *48*, 969–3979. [CrossRef]
3. Zhang, D.; Zou, Q.; Lin, H.; Xu, X.; He, L.; Gui, R.; Li, Q. Automatic pavement defect detection using 3D laser profiling technology. *Autom. Constr.* **2018**, *96*, 350–365. [CrossRef]
4. Solla, M.; Lagüela, S.; González-Jorge, H.; Arias, P. Approach to identify cracking in asphalt pavement using GPR and infrared thermographic methods: Preliminary findings. *NDT E Int.* **2014**, *62*, 55–65. [CrossRef]
5. Solla, M.; Lorenzo, H.; Martínez-Sánchez, J.; Pérez-Gracia, V. Applications of GPR in association with other non-destructive testing methods in surveying of transport. In *Civil Engineering Applications of Ground Penetrating Radar*, 1st ed.; Benedetto, A., Pajewski, L., Eds.; Springer International Publishing: Bern, Switzerland, 2015; pp. 327–342.
6. Pajewski, L.; Fontul, S.; Solla, M. Ground-penetrating radar for the evaluation and monitoring of transport infrastructures. In *Innovation in Near-Surface Geophysics: Instrumentation, Application, and Data Processing Methods*, 1st ed.; Persico, R., Piro, S., Linford, N., Eds.; Elsevier: Amsterdam, The Netherlands, 2018; pp. 341–398.
7. Hu, J.; Pavana, K.R.V.; White, D.J.; Beresnev, I. Pavement thickness and stabilised foundation layer assessment using ground-coupled GPR. *Nondestruct. Test. Eval.* **2015**, *31*, 1–21. [CrossRef]
8. Agred, K.; Klysz, G.; Balayssac, J.-P. Location of reinforcement and moisture assessment in reinforced concrete with a double receiver GPR antenna. *Constr. Build. Mater.* **2018**, *188*, 1119–1127. [CrossRef]
9. Zhao, S.; Al-Qadi, I. Pavement drainage pipe condition assessment by GPR image reconstruction using FDTD modeling. *Constr. Build. Mater.* **2017**, *154*, 1283–1293. [CrossRef]
10. Wang, S.; Al-Qadi, I.L.; Cao, Q. Factors Impacting Monitoring Asphalt Pavement Density by Ground Penetrating Radar. *NDT E Int.* **2020**, *115*, 102296. [CrossRef]
11. Liu, H.; Deng, Z.; Han, F.; Xia, Y.; Liu, Q.H.; Sato, M. Time-frequency analysis of air-coupled GPR data for identification of delamination between pavement layers. *Constr. Build. Mater.* **2017**, *154*, 1207–1215. [CrossRef]
12. Zhang, J.; Zhang, C.; Lu, Y.; Zheng, T.; Dong, Z.; Tian, Y.; Jia, Y. In-situ recognition of moisture damage in bridge deck asphalt pavement with time-frequency features of GPR signal. *Constr. Build. Mater.* **2020**, *244*, 118295. [CrossRef]
13. Pedret Rodes, J.; Martínez-Sánchez, J.; Pérez-Gracia, V. GPR Spectra for Monitoring Asphalt Pavements. *Remote Sens.* **2020**, *12*, 1749. [CrossRef]
14. Al-Qadi, I.L.; Lahouar, S.; Jiang, K.; McGhee, K.K.; Mokarem, D. Accuracy of Ground-Penetrating Radar for Estimating Rigid and Flexible Pavement Layer Thicknesses. *Transp. Res. Rec. J. Transp. Res. Board* **2005**, *1940*, 69–78. [CrossRef]
15. Al-Qadi, I.L.; Lahouar, S.; Loulizi, A. *Ground-Penetrating Radar Calibration at the Virginia Smart Road and Signal Analysis to Improve Prediction of Flexible Pavement Layer Thicknesses*; Technical Report No. FHWA/VTRC 05-CR7; Virginia Tech Transportation Institute: Charlottesville, Virginia, 2005.
16. Flintsch, G.W.; Al-Qadi, I.L.; Loulizi, A.; Lahouar, S.; McGhee, K.; Clark, T. *Field Investigation of High Performance Pavements in Virginia*; Technical Report, Report No. VTRC 05-CR9; Virginia Tech Transportation Institute: Charlottesville, Virginia, 2005.
17. Holzschuher, C.; Lee, H.S.; Greene, J. *Accuracy and Repeatability of Ground Penetrating Radar for Surface Layer Thickness Estimation of Florida Roadways*; Technical Report No. FL/DOT/SMO/07-505; Florida Department of Transportation, State Material Office: Gainesville, FL, USA, 2007.
18. Ožbolt, M.; Rukavina, T.; Domitrović, J. Comparison of the Pavement Layers Thickness Measured by Georadar and Conventional Methods—Examples from Croatia. *Balt. J. Road Bridge. Eng.* **2012**, *7*, 30–35. [CrossRef]
19. Domitrović, J.; Rukavina, T.; Bezina, Š.; Stančerić, I. Mapping of Runway Pavement Layers Thickness by GPR. In Proceedings of the International Conference on Sustainable Materials, Systems and Structures (SMSS2019)—Challenges in Design and Management of Structures, Rovinj, Croatia, 20–22 March 2019; Ivanković, A.M., Kušter Marić, M., Strauss, A., Kišiček, T., Eds.; RILEM Publications, S.A.R.L.: Paris, France, 2019; pp. 208–215.

20. Al-Qadi, I.L.; Lahouar, S.; Loulizi, A. Ground Penetrating Radar Evaluation for Flexible Pavement Thickness Estimation. In Proceedings of the Pavement Evaluation Conference, Roanoke, VA, USA, 21–25 October 2002.
21. Plati, C.; Loizos, A.; Gkyrtis, K. Integration of non-destructive testing methods to assess asphalt pavement thickness. *NDT E Int.* **2020**, *115*, 102292. [CrossRef]
22. Moropoulou, A.; Avdelidis, N.P.; Koui, M.; Aggelopoulos, A.; Karmis, P. Infrared thermography and ground penetrating radar for airport pavements assessment. *Nondestruct. Test. Eval.* **2002**, *18*, 37–42. [CrossRef]
23. Malvar, L.J. Detecting Voids under Pavements: Update on Approach of U.S. Department of Defense. *Transp. Res. Rec. J. Transp. Res. Board* **2010**, *2170*, 28–35. [CrossRef]
24. Graczyk, M.; Krysiński, L.; Topczewski, Ł.; Sudyka, J. The Use of Three-dimensional Analysis of GPR Data in Evaluation of Operational Safety of Airfield Pavements. *Transp. Res. Procedia* **2016**, *14*, 3704–3712. [CrossRef]
25. Saarenketo, T. NDT transportation. In *Ground Penetrating Radar: Theory and Applications*, 1st ed.; Jol, H.M., Ed.; Elsevier: Amsterdam, The Netherlands, 2009; pp. 395–444.
26. Eide, E.; Sandnes, P.A.; Nilssen, B.; Tjora, S. Airfield Runway Inspection Using 3 Dimensional GPR. In *Proceedings of 3rd International Workshop on Advanced Ground Penetrating Radar IWAGPR 2005*; Delft, The Netherland, 2–3 May 2005, Lambot, S., Gorriti, A.G., Eds.; Institute of Electrical and Electronics Engineers: Delft, The Netherlands, 2005; pp. 87–91.
27. Domitrović, J.; Bezina, Š.; Stančerić, I.; Rukavina, T. 3D modelling of asphalt overlays based on GPR data. *Int. J. Pavement Eng.* submitted for publication.
28. Maser, K.R.; Scullion, T. Automated Detection of Pavement Layer Thicknesses and Subsurface Moisture Using Ground Penetrating Radar. *Transp. Res. Rec.* **1991**, *1344*, 22–30.
29. Daniels, D.J. *Ground Penetrating Radar*, 2nd ed.; The Institution of Electrical Engineering: London, UK, 2004; p. 90.
30. Saarenketo, T.; Scullion, T. Road evaluation with ground penetrating radar. *J. Appl. Geophys.* **2000**, *43*, 119–138. [CrossRef]
31. Geophysical Survey Systems, Inc. Available online: https://www.geophysical.com/wp-content/uploads/2017/11/GSSI-RADAN-6_6-Manual.pdf (accessed on 18 February 2021).

Article

Frequency–Wavenumber Analysis of Deep Learning-based Super Resolution 3D GPR Images

Man-Sung Kang and Yun-Kyu An *

Department of Architectural Engineering, Sejong University, Seoul 05006, Korea; kms102353@sju.ac.kr
* Correspondence: yunkyuan@sejong.ac.kr; Tel.: +82-2-3408-4331

Received: 19 August 2020; Accepted: 17 September 2020; Published: 18 September 2020

Abstract: This paper proposes a frequency–wavenumber (f–k) analysis technique through deep learning-based super resolution (SR) ground penetrating radar (GPR) image enhancement. GPR is one of the most popular underground investigation tools owing to its nondestructive and high-speed survey capabilities. However, arbitrary underground medium inhomogeneity and undesired measurement noises often disturb GPR data interpretation. Although the f–k analysis can be a promising technique for GPR data interpretation, the lack of GPR image resolution caused by the fast or coarse spatial scanning mechanism in reality often leads to analysis distortion. To address the technical issue, we propose the f–k analysis technique by a deep learning network in this study. The proposed f–k analysis technique incorporated with the SR GPR images generated by a deep learning network makes it possible to significantly reduce the arbitrary underground medium inhomogeneity and undesired measurement noises. Moreover, the GPR-induced electromagnetic wavefields can be decomposed for directivity analysis of wave propagation that is reflected from a certain underground object. The effectiveness of the proposed technique is numerically validated through 3D GPR simulation and experimentally demonstrated using in-situ 3D GPR data collected from urban roads in Seoul, Korea.

Keywords: ground penetrating radar (GPR); frequency–wavenumber (f–k) analysis; super resolution (SR) image; deep learning; noise reduction; directivity analysis

---

1. Introduction

In the past few decades, sinkhole accidents of urban roads have posed a serious hazard to buildings, infrastructures and especially inhabitants of the area [1,2]. Although vision-based road surface inspection techniques have been widely proposed for road degradation evaluation [3,4], early detection of sinkholes, which are typically invisible from the road surface, is still challenging. To effectively detect underground cavities which are most likely extended to sinkholes, various nondestructive testing (NDT) techniques have taken the limelight. Ground penetrating radar (GPR) is one of the widely accepted NDT tools thanks to its high sensitivity to underground media change and rapid inspection capability for broad target areas [5–7]. However, the physical interpretation of field GPR data for underground object detection and classification is still challenging in some cases, because electromagnetic waves, which are reflected from a target underground object, are often weaker than underground media's inhomogeneity and undesired measurement noises [8,9]. In general, most of the dominant signals reflected from the road surface often hinder the precise data interpretation of relatively weak signals coming from underground media under air-coupled GPR data acquisition conditions [8,10,11].

To enhance the GPR data interpretability, a number of signal and image processing techniques, such as time-varying gain [10,11], subtraction [8], migration [12], deconvolution [13], basis-pursuit [9], compressive sensing [14], velocity analysis [15], radon transform [16], discrete wavelet transform [17] and empirical mode decomposition [18], have been proposed. Although these techniques have tried to make GPR data interpretation easier, their results still highly depend on experts' experiences and are often susceptible to undesired noises. Thus, a number of researchers have proposed artificial neural networks to automate the GPR data interpretation [19–21]. Recently, deep learning networks have been actively applied to the GPR data interpretation for minimizing users' intervention. For instance, Kim et al. [22] proposed a convolutional neural network (CNN) combined with a statistical thresholding technique to classify underground objects using GPR B-scan images. Then, more advanced deep learning networks based on the combination of B- and C-scan GPR images [23,24], as well as triplanar GPR images [25], were also developed to improve the data classification performance.

However, the undesired noise problems are inevitable in reality and still disturb the proper data interpretation. In particular, incoherent GPR data caused by inhomogeneity of arbitrary underground medium, measurement noises and systematic error are often misinterpreted [26]. A frequency–wavenumber (f–k) analysis has been developed to address the similar noise issue in ultrasonic NDT fields [27–29]. The f–k analysis, which transforms time–space (t–s) signals to the corresponding f–k space, is able to effectively filter out the noise components by removing undesired wave patterns in the f–k domain. The filtered f–k domain signals are then restored in the t–s domain signals without unwanted noise patterns, making it possible to highlight real wave components reflected from a target underground object. In addition, the wave propagation direction in the t–s domain can be precisely decomposed, which is useful to recognize the wave scatter size and location as well as to classify the object type. In spite of the f–k analysis's benefits, 3D GPR data are not often suitable for the f–k analysis. High resolution GPR images, which are composed of dense spatial GPR data considering the minimum target underground object size, are necessary for the proper f–k analysis. However, lack of GPR image resolution determined by the number of GPR channels and spatial scanning speed, which leads to analysis distortion, is often caused by the fast or coarse spatial scanning mechanism in reality [30–32].

To tackle the image resolution issue, a number of image resolution enhancement techniques have been proposed in computer vision fields. For example, super resolution (SR) images have been artificially generated by various image processing methods such as an image prediction model [33], an image statistical method [34] and a patch-based method [35]. However, the high frequency regions, such as textures and edge components in the target image, are not properly generated by the conventional image processing methods. Recently, numerous deep learning-based SR image generation techniques have been proposed. Dong et al. proposed an SR image enhancement network using CNN as the first SR network [36]. Then, more advanced SR networks such as very deep SR [37], generative adversarial network-based SR network [38] and residual channel attention networks [39] have been developed. More recently, a residual learning-based deep CNN has been intensively studied for improving its training efficiency [40,41]. As for the GPR application, Kang et al. proposed a deep learning-based SR GPR image generation network for enhancing underground cavity detectability [42].

In this paper, the f–k analysis incorporated with a deep learning-based SR network is proposed for unwanted noise reduction and electromagnetic wavefield decomposition. First, a deep learning-based SR GPR image enhancement network is described in Section 2. The f–k analysis corresponding to the SR images is then proposed in Section 3. The effectiveness of the proposed technique is numerically validated using 3D GPR simulation data and experimentally demonstrated using in-situ 3D GPR data obtained from complex urban roads at Seoul, Korea.

2. Deep Learning-based SR GPR Image Enhancement

When multi-channel GPR scans along the region of interest, B- and C-scan images can be typically constructed by collecting multiple A-scan data along the scanning direction. The B-scan image at each GPR channel includes the parabola features that are reflected from underground objects, and the C-scan image at a certain depth is able to display the circular features corresponding to the parabola features. These parabola and circular features in the B- and C-scan images have been widely used as one of the main distinctive features for underground object identification and classification. However, lack of image resolution often hinders the feature recognition. To obtain high resolution B- and C-scan images, slow scanning speed and dense GPR antenna arrangement are necessary. Unfortunately, the resolution issues are, however, a trade-off with respect to time and cost in reality.

To effectively tackle the lack of resolution issue in existing 3D GPR data without data acquisition condition change, the deep learning-based SR GPR image enhancement network based on a deep residual channel attention network [39] is developed as shown in Figure 1. The deep residual channel attention network is one of the SR image enhancement networks based on CNN, which consists of 500 layers and 1.6 M parameters. This network increases the LR image resolution four times, comprises the four main steps: (1) shallow feature extraction, (2) deep feature extraction, (3) upscaling and (4) reconstruction. First, the shallow feature extraction step, which consists of a single convolution layer with 64 kernels of 3×3 size and stride of 1, extracts shallow features from the input low resolution (LR) image.

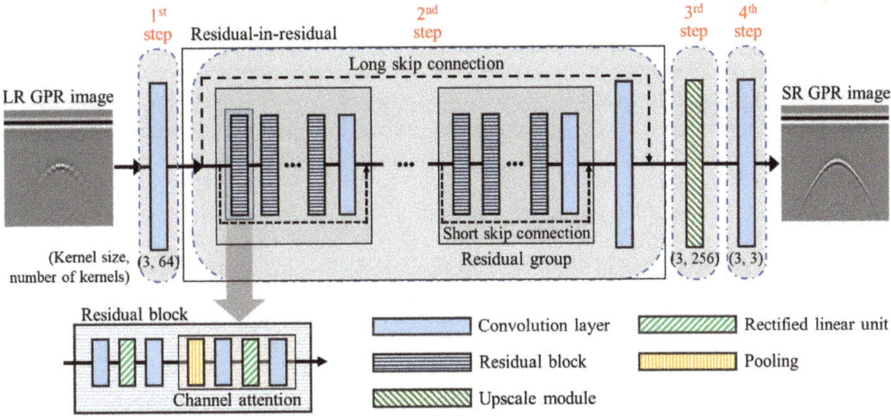

Figure 1. Super resolution (SR) ground penetrating radar (GPR) image enhancement network. The 1st step: shallow feature extraction; 2nd step: deep feature extraction; 3rd step: upscaling; 4th step: reconstruction (LR: low resolution, SR: super resolution).

Subsequently, deep features are extracted through the residual-in-residual structure in the second step of Figure 1. The main purpose of the residual-in-residual structure learning is to allow very deep networks to achieve easy and powerful performance in the training process. Here, the deep feature means high frequency information composed of lines or edges, which is the biggest difference between the SR and LR images. The residual-in-residual structure is the very deep network which consists of 10 residual groups, and each residual group includes 20 residual blocks and 1 convolution layer as shown in Figure 1. To effectively train the deep network, the skip connections are embedded in the residual-in-residual structure in terms of the short and long skip connections. Each residual group is connected by the long skip connection, and the residual blocks inside the residual group are connected by the short skip connection as shown in Figure 1. The multiple skip connections allow enough shallow features to be bypassed, enabling the main network to focus on learning deep features. The skip connections make it possible to stabilize the training performance of the very deep network with residual learning, as the main superiority of the residual-in-residual structure. Each convolution layer inside the residual-in-residual structure similarly works as the convolution layer of the first step. To more efficiently train the high frequency regions of the image, the channel attention mechanism is performed inside each residual block through global average pooling, convolution, rectified linear unit (ReLU) layer and sigmoid function. The channel attention mechanism improves the discriminative learning capabilities by focusing on more useful channels based on the average value extracted from each channel. Each average value is obtained via the global average pooling layer. Then, the convolution, ReLU layer and sigmoid function provide non-linearity between the channels and make multiple channel-wise features to be emphasized for a non-mutually exclusive relationship. Next, the upscaling step, as the third step, extends the feature data of the image to the SR resolution size. It consists of the deconvolution layer with 256 kernels of 3×3 size and the stride of 1, which increases the size of each pixel by four times in this network. Finally, the SR image is generated through a single convolution layer comprised of three kernels of 3×3 size and stride of 1 in the reconstruction step, as the fourth step, of Figure 1.

3. SR GPR Image-Based *f–k* Analysis

Once the SR GPR images are generated, the SR C-scan images can be obtained in the *t–s* domain for the subsequent *f–k* analysis, as shown in Figure 2. Note that the proposed *f–k* analysis can be easily extended to B- or D-scan images, although the sequential C-scan images along the depth direction are used in this study. Even though the SR C-scan images contain meaningful wave signals mixed with noise components, it is difficult to remove the undesired noises caused by arbitrary underground medium inhomogeneity and data measurement procedure in the *t–s* domain. On the other hand, the noise components can be effectively eliminated in the *f–k* domain. Moreover, wavefield decomposition in the *f–k* domain is able to precisely analyze wave propagation directivity by reconverting the decomposed wavefield data to the *t–s* domain data as shown in Figure 2. First, the SR C-scan data in the *t–s* domain are transformed into the *f–k* domain through 3D Fourier transform which is given by:

$$U(k_x, k_y, w) = \iiint_{-\infty}^{\infty} E(x, y, t) e^{-i(k_x x + k_y y + wt)} dx dy dt \tag{1}$$

where E and U denote the electromagnetic wavefields of the SR C-scan data in the *t–s* and *f–k* domains, respectively. k, w and t are the wavenumber, angular frequency and time, respectively. x and y are the spatial cartesian coordinates.

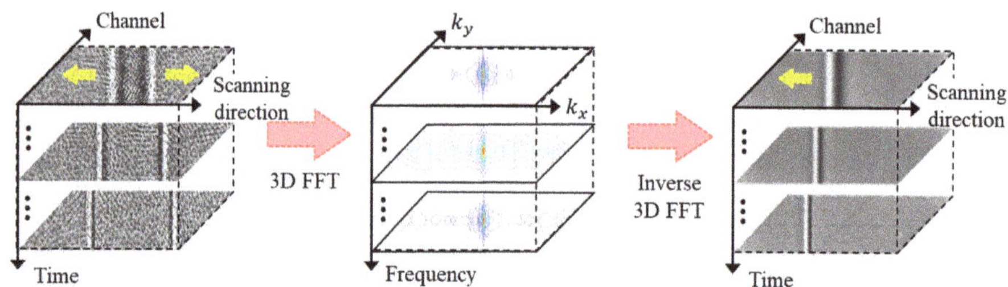

Figure 2. Frequency–wavenumber (*f–k*) analysis using the SR C-scan images: SR C-scan images in the *t–s* domain are transformed into the *f–k* domain through 3D Fourier transform. The undesired incoherent noises and electromagnetic wave propagation directivity are then filtered out and analyzed, respectively, in the *f–k* domain. Finally, the filtered data in the *f–k* domain are reconstructed to the *t–s* domain through the inverse 3D Fourier transform.

Subsequently, the tailored *f–k* filter is designed. First, a lowpass filter is applied in the *f* domain so that the measurement noises can be eliminated outside the excitation frequency range as shown in Figure 3. The filtering frequency bandwidth can be determined by considering the excitation frequency range. The *k* domain filter is then developed. Since the electromagnetic waves propagating along arbitrary underground media are partially and randomly reflected from the media's inhomogeneity and numerous small porosities, these reflection signals randomly appear and look like non-propagating wave components in the C-scan images of the *t–s* domain. It physically means that no spatially propagating wave can be observed if there is no certain object inside the underground media. Even if there are meaningful signals reflected from a certain object, such physical phenomenon can be clearly observed, as shown in Figure 3a. The wave energy is highly concentrated near zero k_x value, which are undesired noise components to be eliminated. One more interesting thing to see here is that these randomly reflected signals in the *f–k* domain may have dominant energy due to their high repetition rate, although each reflection signal intrinsically has small amplitude. Based on the physical observation, the *k* domain filter is established using a Laplacian of Gaussian window (Φ_k):

$$\Phi_k = \left[\frac{k_x{}^2 + k_y{}^2 - 2\sigma^2}{\sigma^4}\right] e^{-\frac{k_x{}^2 + k_y{}^2}{2\sigma^2}} \quad \forall\, w \tag{2}$$

where σ is the standard deviation.

Once the *f–k* filter is designed, the filtered SR C-scan data (U_f) can be obtained in the *f–k* domain. Figure 3b shows that the measurement noises, as well as non-propagating components, are clearly filtered out, and meaningful wave components remain.

$$U_f\left(k_x,\, k_y, w\right) = U\left(k_x,\, k_y, w\right) \cdot \Phi_k \tag{3}$$

Furthermore, the electromagnetic wavefields can be decomposed in the *f–k* domain so that the wave propagation directivity can be precisely analyzed, which is useful to recognize underground objects' size and location as well as to classify the object type. For instance, a $\pm x$ directional window filter ($\Phi_{\pm k_x}$) can be designed for decomposing U_f to the $+x$ or $-x$ directional wavefield ($U_{\pm k_x}$) in the *f–k* domain, which is given by:

$$U_{\pm k_x}\left(k_x,\, k_y, w\right) = U_f\left(k_x,\, k_y, w\right) \cdot \Phi_{\pm k_x}$$

$$\Phi_{+k_x} = \begin{cases} 0 & k_x \leq 0 \\ 1 & k_x > 0 \end{cases} \quad \Phi_{-k_x} = \begin{cases} 1 & k_x < 0 \\ 0 & k_x \geq 0 \end{cases} \tag{4}$$

In a similar fashion, it can be readily extended to the $\pm y$ directional filter.

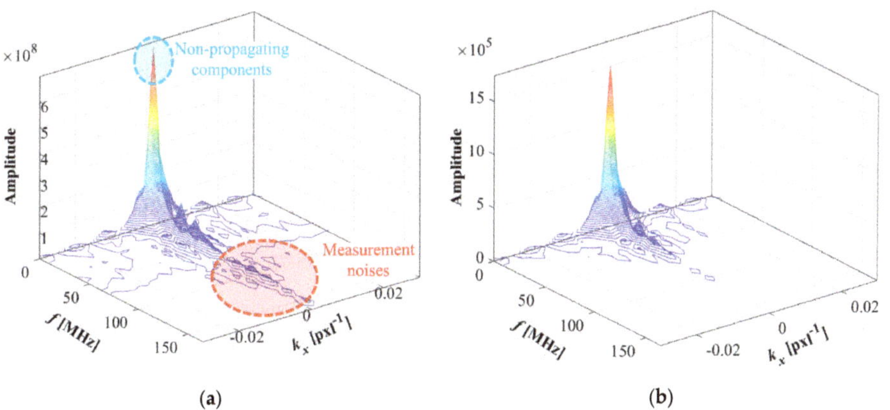

Figure 3. The representative *f–k_x* plots at the center of k_y (a) before and (b) after filtering.

Next, the resultant C-scan data ($E_{\pm k_x}$) in the *t–s* domain can be reconstructed using the following inverse 3D Fourier transform:

$$E_{\pm k_x}(x, y, t) = \frac{1}{2\pi} \iiint_{-\infty}^{\infty} U_{\pm k_x}\left(k_x,\, k_y, w\right) e^{i(k_x x + k_y y + wt)} dk_x dk_y dw \tag{5}$$

As one of the representative examples, only the $-x$ directional wavefield (E_{-k_x}) remains, and it reveals much higher signal-to-noise ratio (SNR) than E without pixel information loss and distortion, as shown in Figure 2. Note that the wavefield decomposition process is optional in the algorithm, thus $U_{\pm k_x}$ can be replaced by U_f in Equation (5), resulting in the filtered SR C-scan data (E_f) in the t–s domain.

4. Numerical and Experimental Validations

The proposed f–k analysis technique is numerically and experimentally validated through 3D GPR simulation using gprMax [43] and in-situ 3D GPR data obtained from urban roads in Seoul, Korea.

4.1. Numerical Validation

The target 3D model is comprised of $8 \times 2.975 \times 2.75$ m³ soil layer, $8 \times 0.525 \times 2.75$ m³ air layer and steel pipe with a diameter of 500 mm, as depicted in Figure 4. It was modelled so that the pipe was buried perpendicular to the GPR scanning direction inside the soil layer. Note that the pipe was intentionally selected in this study, because it is one of the representative wave scatters which can be clearly reflected in all GPR channels constituting the C-scan images. Here, the relative permittivity values of air, soil and pipe were set to 1, 5 and infinity, respectively. The transmitter (Tx) was 50 mm apart from the receiver (Rx), and the GPR data reflected from the pipe were acquired by moving the Tx and Rx antennas along the soil layer surface, as shown in Figure 4. A finite difference time domain method [44] was used to simulate electromagnetic wave propagation. To simulate the similar conditions with the real-world GPR scanning, the spatial discretization was set to 20 mm, which is equivalent to 20 km/h scanning speed with 20 GPR channels in the real-world application. Here, the 20 GPR channels are able to cover a road width of 1.5 m. The excitation electromagnetic wave was normalized by the second derivative of a Gaussian waveform with a center frequency of 1.8 GHz. In addition, Gaussian random noises, which are equivalent to 25% of magnitude of the maximum value of the GPR signal, were artificially added to simulate the arbitrary underground medium inhomogeneity and undesired measurement noises.

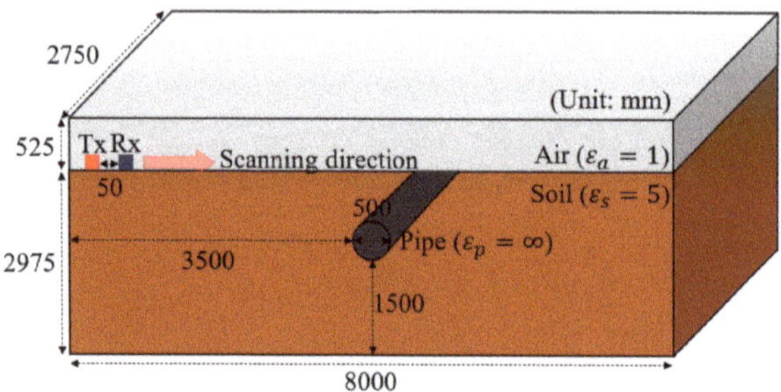

Figure 4. Three-dimensional GPR simulation setup: Tx and Rx are the transmitter and receiver, respectively. ε means the permittivity.

Figure 5 shows the representative GPR B- and C-scan images obtained from the simulation model. The original LR B- and C- scan images are clearly shown in Figure 5a, although the relatively slow scanning of 20 km/h with dense GPR channel arrangement was modelled in this simulation. On the other hand, the SR B- and C-scan images show that the edges of informative parabola and line features are well enhanced without pixel information loss and distortion, as displayed in Figure 5b. Although the image resolution is successfully enhanced, the arbitrary underground medium inhomogeneity and undesired measurement noises still remain in Figure 5b.

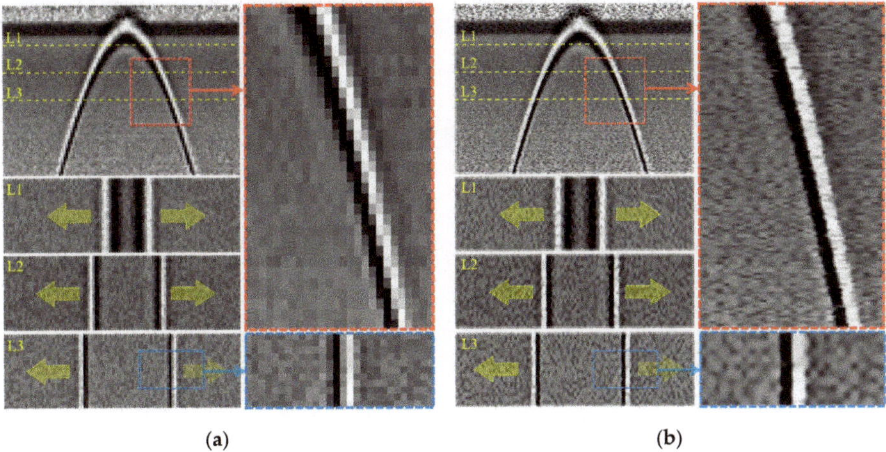

(a)   (b)

Figure 5. Representative image enhancement results of the simulation data: (**a**) original LR B- and C-scan images, (**b**) enhanced SR B- and C-scan images.

Figure 6a shows the representative k_x–k_y plots at 300 MHz. As expected, the non-propagating components caused by incoherent noise components are highly concentrated on the origin of the k_x–k_y plane. To remove the non-propagating components, the f–k filter is applied to U shown in Figure 6a using Equation (2). The lowpass filter was designed by fitting an exponential function with a rate parameter of 0.05. As for the k domain filter, σ was set to 1 considering k to all excitation frequency ranges. After applying the f–k filter, the non-propagating components are remarkably reduced in U_f, while meaningful wave components reflected from the pipe remain, as shown in Figure 6b. Subsequently, U_{-k_x} and U_{+k_x} are obtained by applying $\Phi_{\pm k_x}$ using Equation (4) as shown in Figure 6c,d, respectively.

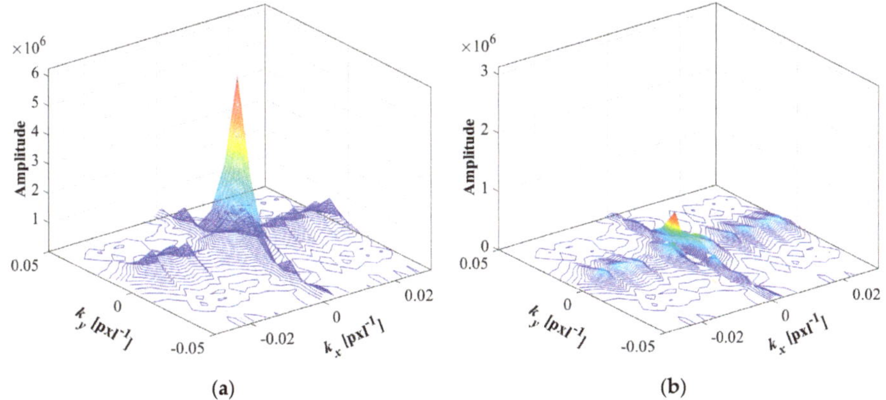

(a)   (b)

Figure 6. *Cont.*

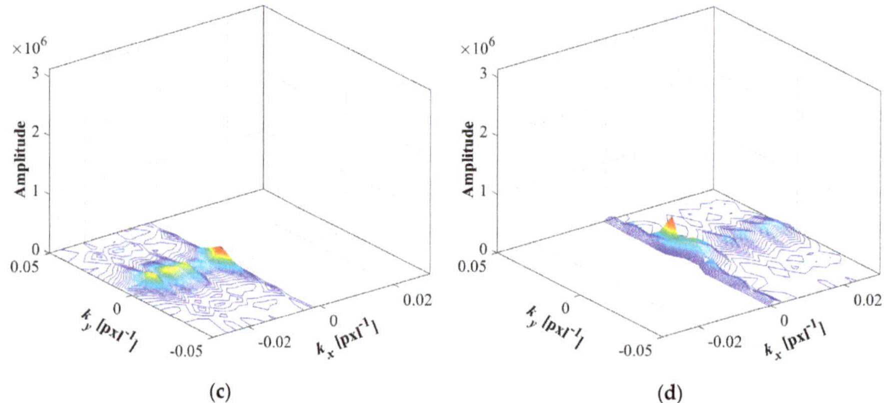

Figure 6. Representative simulation k_x–k_y plots at 300 MHz in the f–k domain: (**a**) U, (**b**) U_f, (**c**) U_{-k_x} and (**d**) U_{+k_x}.

Figure 7 shows the resultant t–s domain images corresponding to Figure 6, which are reconstructed using Equation (5). Compared to Figure 7a, it is clearly observed that the incoherent and random noises are significantly eliminated in Figure 7b. To quantitatively estimate the results, the SNR values of the representative A-scan signals along the vertical white dash-dotted lines in Figure 7a,b were compared. Figure 8 shows the A-scan signals with the reference signals obtained by smoothing spline curve fitting. Figure 8a reveals that the A-scan signal of E is quite different from the reference signal, resulting in SNR of 19.2 dB as summarized in Table 1. Once the f–k filter is applied, Figure 8b shows that the A-scan signal of E_f is well matched with the reference signal, which has 54.1 dB SNR as shown in Table 1. It can be confirmed that the proposed f–k filter is very effective in removing incoherent and random noise components. In addition, Figure 7c,d, respectively, show that E_{-k_x} and E_{+k_x} are successfully decomposed along the $-x$ and $+x$ directions. Again, the wavefield decomposition is very powerful in identifying the underground object boundary and classifying the object type.

Table 1. SNR comparison of the simulation A-scan signals between E and E_f.

	A-Scan of E	A-Scan of E_f
SNR (dB)	19.2	54.1

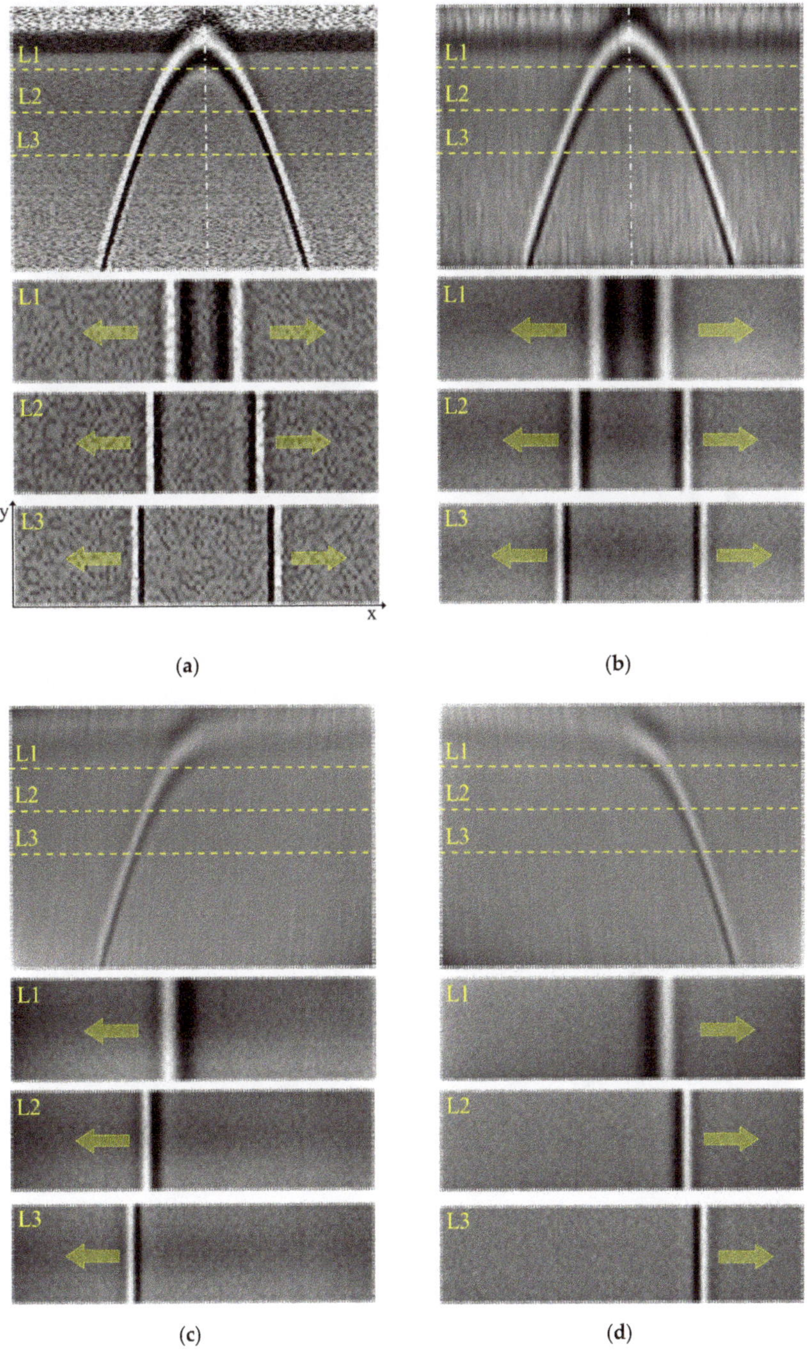

Figure 7. Representative simulation resultant images in the *t–s* domain: (**a**) E, (**b**) E_f, (**c**) E_{-k_x} and (**d**) E_{+k_x}.

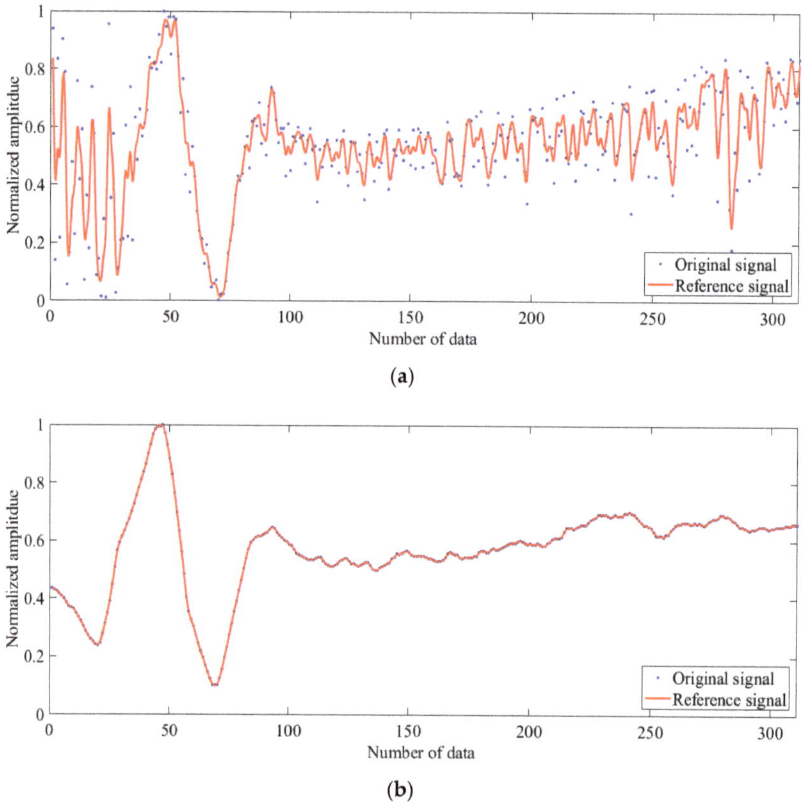

Figure 8. Representative GPR A-scan signals with the reference signals obtained from (**a**) E and (**b**) E_f.

4.2. Experimental Validation Using In-Situ 3D GPR Data

The proposed f–k analysis technique was also experimentally validated using 3D GPR data collected from complex urban roads in Seoul, Korea. Figure 9a shows the 3D GPR-mounted van for the field tests. The 3D GPR consisted of bow-tie monopole 20 Tx and Rx antennas generate a step frequency with wide frequency bandwidth ranging from 100 MHz to 3 GHz. Here, the 20-channeled GPR device has 1.5 m scanning width, which can typically cover a single road lane as shown in Figure 9a,b. The average scanning speed was approximately 20 km/h for avoiding traffic congestion in urban roads. The GEOSCOPE MK IV data acquisition system shown in Figure 9c, which has 3 GHz sampling rate and 250 ns time range, was used in the field tests.

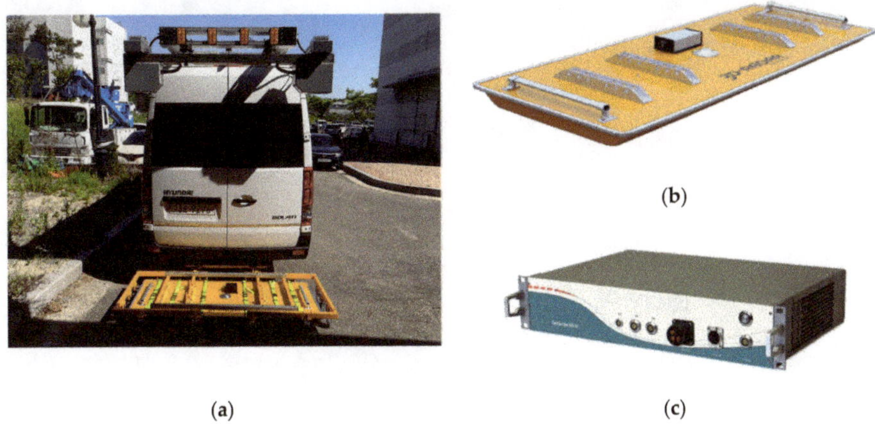

(a)

(b)

(c)

Figure 9. Experimental setup: (**a**) 3D GPR-mounted van with (**b**) 3D GPR device and (**c**) data acquisition system.

Figures 10a and 11a show the representative experimental results including B- and C-scan images obtained from two different underground pipes that are defined as pipe cases 1 and 2. Compared to the simulation results of Figure 5a, the experimental LR images show lower resolution and more noises which are most likely caused by arbitrary underground medium inhomogeneity and undesired measurement noises. In particular, the B- and C-scan images of Figures 10a and 11a do not have sufficient pixel resolution for the f–k analysis. On the other hand, the SR B- and C-scan images reveal that informative edges of the parabola and line features are well reconstructed without pixel information loss and distortion, as displayed in Figures 10b and 11b.

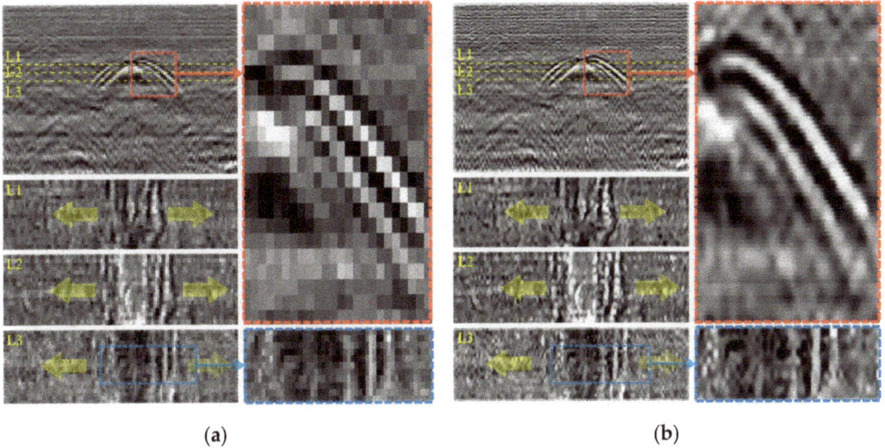

(a)

(b)

Figure 10. Representative image enhancement results of the experimental pipe case 1: (**a**) original LR B- and C-scan images, (**b**) enhanced SR B- and C-scan images.

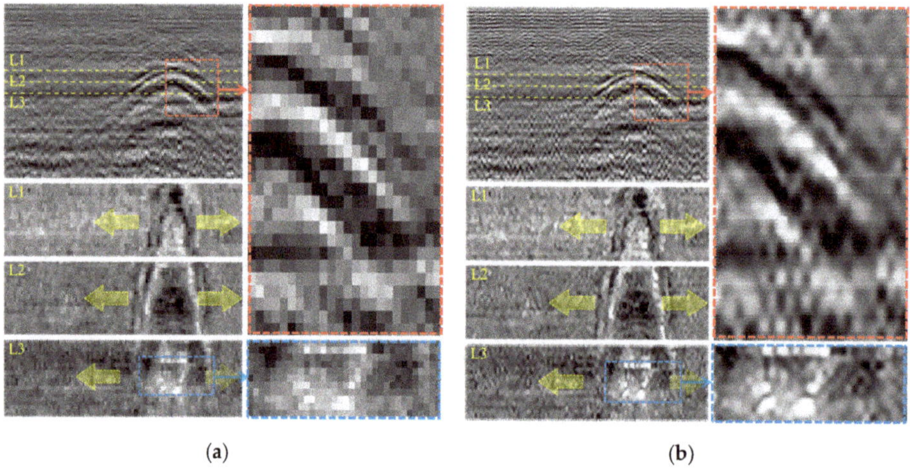

(a)   (b)

Figure 11. Representative image enhancement results of the experimental pipe case 2: (**a**) original LR B- and C-scan images, (**b**) enhanced SR B- and C-scan images.

Figures 12 and 13 show the representative k_x–k_y plots at 500 MHz of the pipe cases 1 and 2. Similarly, the non-propagation components caused by the incoherent noises are concentrated on the origin of the k_x–k_y plane in U, as shown in Figures 12a and 13a. Then, U_f's of Figures 12b and 13b show that the undesired non-propagating components are remarkably reduced by applying the filtering parameters, i.e., rate parameter of 0.05 and σ of 1 in Equation (2). To decompose U_f into U_{-k_x} and U_{+k_x}, the $\Phi_{\pm k_x}$ window filter is similarly applied using Equation (4), as shown in Figure 12c,d and Figure 13c,d.

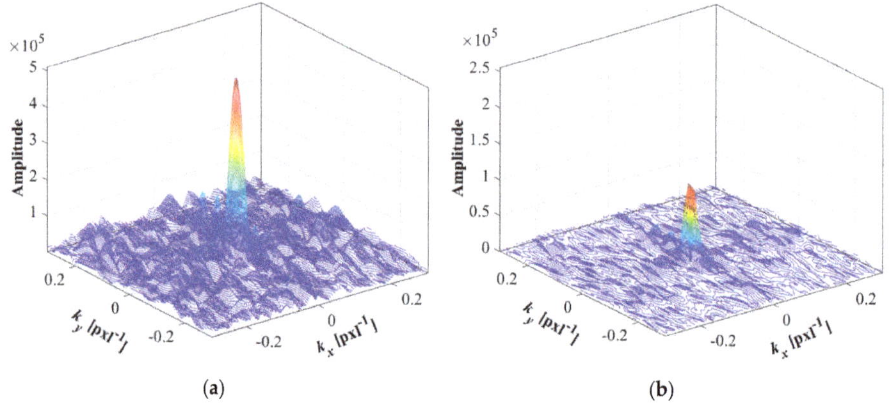

(a)   (b)

Figure 12. *Cont.*

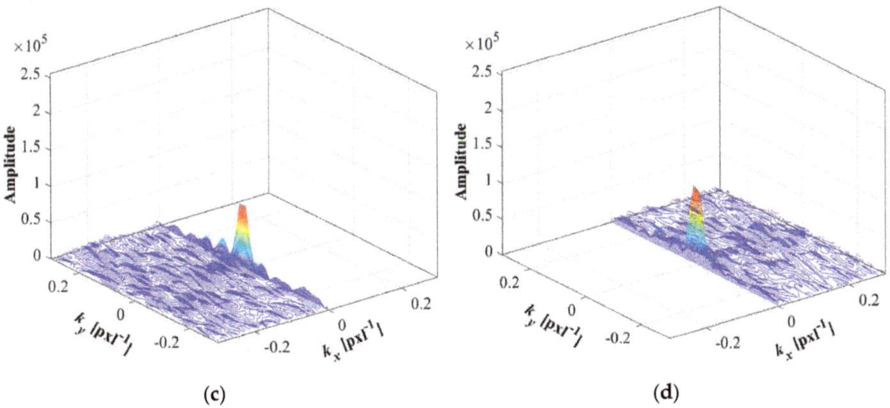

(c)

(d)

Figure 12. Representative k_x–k_y plots at 500 MHz of the experimental pipe case 1: (**a**) U, (**b**) U_f, (**c**) U_{-k_x} and (**d**) U_{+k_x}.

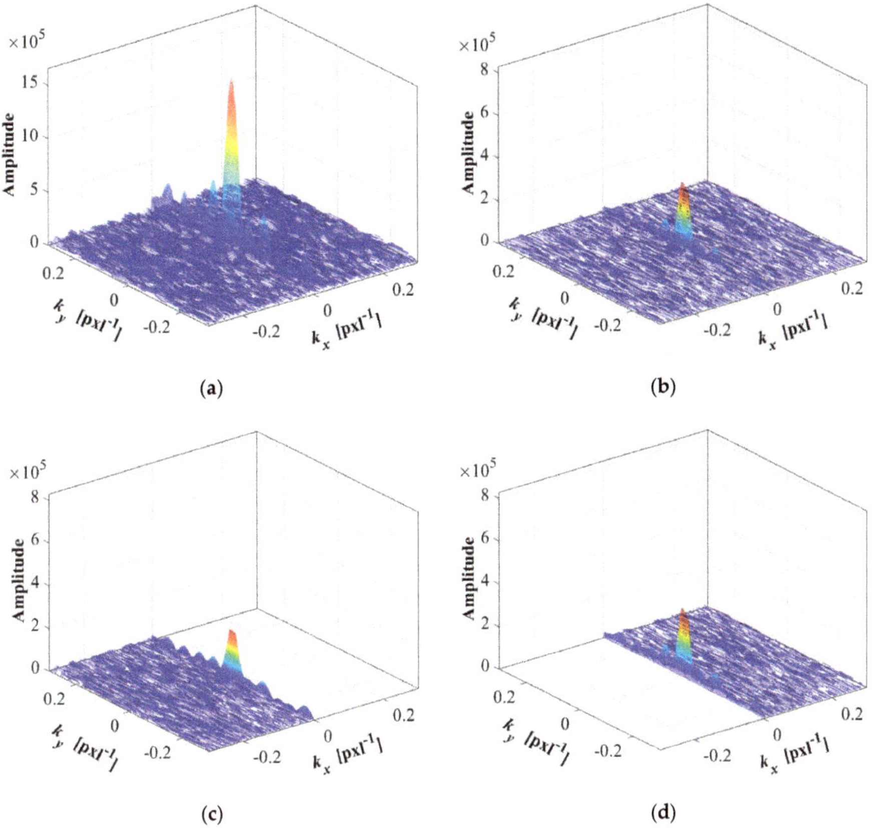

(a)

(b)

(c)

(d)

Figure 13. Representative k_x–k_y plots at 500 MHz of the experimental pipe case 1: (**a**) U, (**b**) U_f, (**c**) U_{-k_x} and (**d**) U_{+k_x}.

Figures 14 and 15 show the resultant t–s domain images of the pipe cases 1 and 2 corresponding to Figures 12 and 13. The incoherent and random noises are remarkably removed in Figures 14b and 15b compared to Figures 14a and 15a. E_{-k_x} and E_{+k_x} are also successfully decomposed using Equation (5), as shown in Figure 14c,d and Figure 15c,d, respectively.

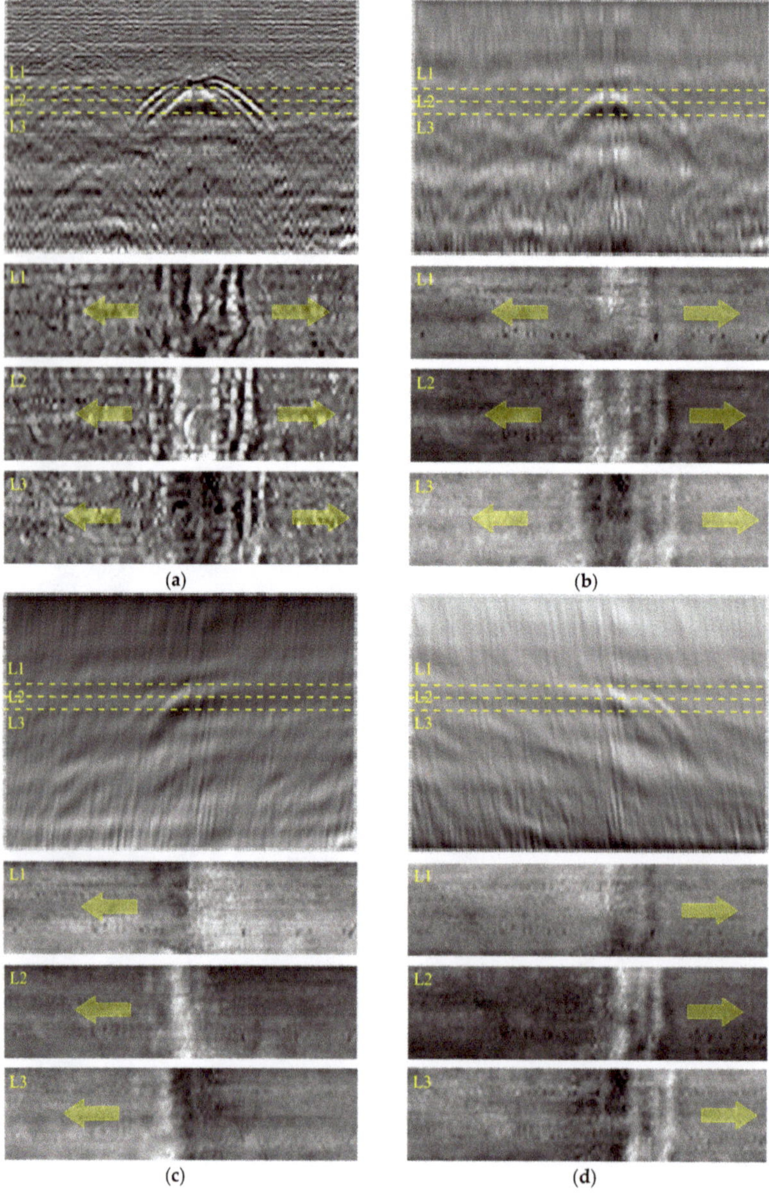

Figure 14. Representative resultant images in the t–s domain of the experimental pipe case 1: (**a**) E, (**b**) E_f, (**c**) E_{-k_x} and (**d**) E_{+k_x}.

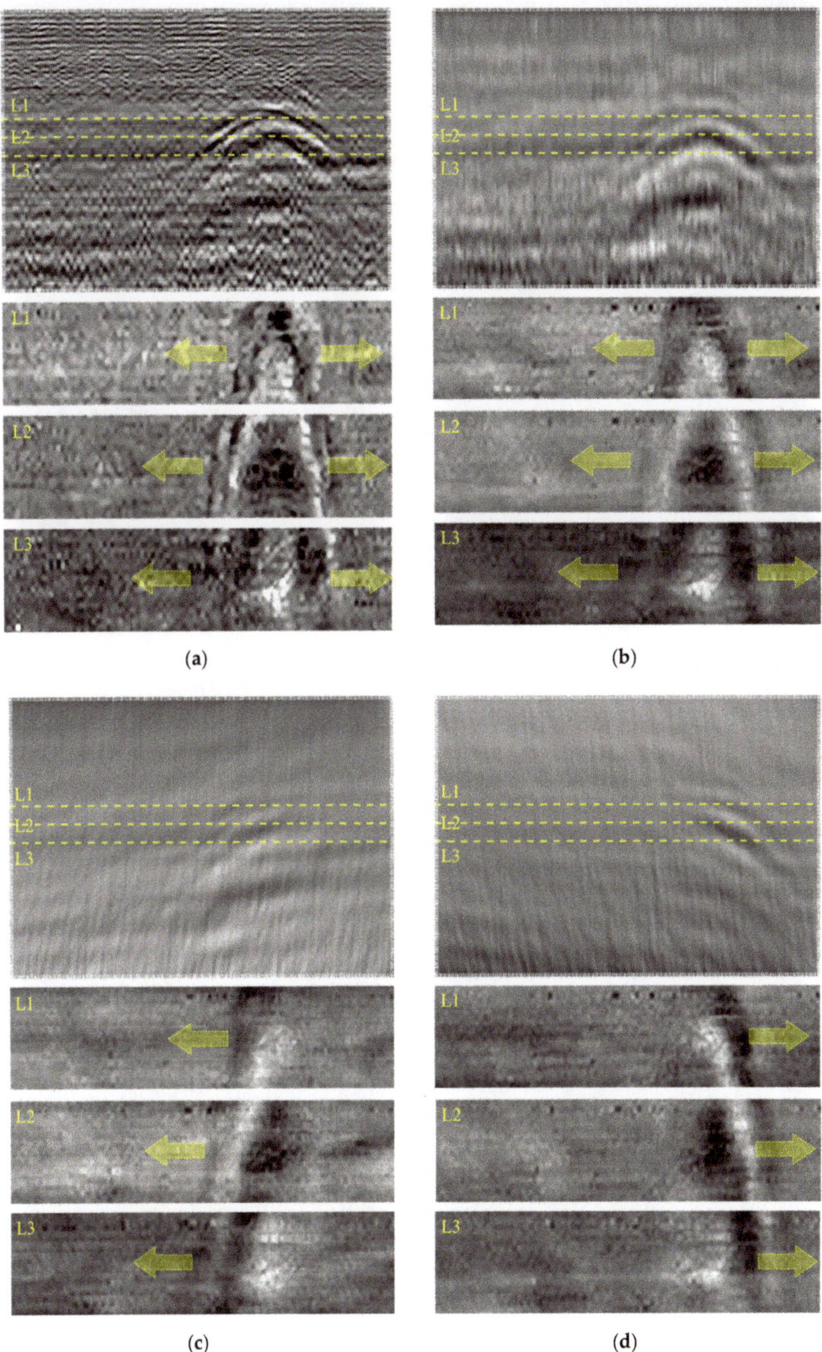

Figure 15. Representative resultant images in the *t–s* domain of the experimental pipe case 2: (**a**) E, (**b**) E_f, (**c**) E_{-k_x} and (**d**) E_{+k_x}.

Similar to the simulation one, the quantitative comparison results using SNR are summarized in Table 2. Both the pipe cases 1 and 2 show about 75% improvement after applying the proposed technique, which is consistent with the simulation results.

Table 2. SNR comparison of the experimental A-scan signals between E and E_f.

		A-Scan of E	A-Scan of E_f
SNR (dB)	Pipe 1	29	50
	Pipe 2	28	50.3

5. Discussion

The proposed f–k analysis technique based on SR GPR images was well validated via the numerical simulation and field tests. Note that the SNR improvement rate of simulation is higher than the experimental ones, because the incoherent noises were simply assumed using ideal Gaussian random noises in the simulation. One more interesting thing is that the pipe cases 1 and 2 show the similar SNR improvement rates, which means that the proposed f–k analysis technique is robust against the test environmental variation. In other words, the performance of the proposed technique would be consistent regardless of underground site conditions. Although the pipe cases, which is one of the most dominant features for clearer validation, are shown in the paper, the proposed technique can be easily extended to other types of underground objects. In addition, the cylindrical coordinate can be employed for the directivity analysis depending on the target objects' shapes, rather than the Cartesian coordinate.

6. Conclusions

This paper proposes a frequency–wavenumber (f–k) technique of 3D ground penetrating radar (GPR) data, which enables one to effectively eliminate incoherent noises and precisely analyze the electromagnetic wave propagation directivity. This technique is newly proposed and validated using super resolution (SR) artificially generated by the deep learning network. Three-dimensional GPR data collected using the existing GPR devices typically suffer from the lack of resolution problem, making it difficult to be analyzed in the f–k domain. To avoid the f–k analysis distortion, a deep learning-based SR GPR image enhancement network incorporated with the f–k analysis is developed. The proposed technique is able to effectively eliminate incoherent noises caused by arbitrary underground medium inhomogeneity and undesired measurement noises, which is one of the biggest technical conundrums in real-world GPR data interpretation. In addition, electromagnetic wave propagation directivity can be precisely analyzed through wavefield decomposition, which is another strong benefit of the f–k analysis. The proposed f–k analysis technique is successfully validated via 3D GPR simulation, as well as field tests, revealing the pretty consistent and outstanding performances. The proposed f–k analysis would be a promising tool for 3D GPR data interpretation especially obtained from complex urban roads.

As the follow-up work, it is warranted that wavefield decomposition-based underground object characterization is thoroughly studied using more GPR data obtained from various in-site roads. Moreover, the proposed f–k analysis can be combined with deep learning-based automated data classification, making it possible to outperform the existing deep learning networks. It is envisioned that this novel f–k analysis can be helpful for not only underground object identification but also concrete structure inspection using GPR.

Author Contributions: Y.-K.A. designed the study. M.-S.K. acquired simulation and experimental data. Y.-K.A. and M.-S.K. analyzed the simulation and experimental data. M.-S.K. and Y.-K.A. wrote the manuscript. Y.-K.A. made critical revisions to the manuscript. Y.-K.A. and M.-S.K. approved the submission of this manuscript. All authors have read and agreed to the published version of the manuscript.

Funding: This research was funded by faculty research fund of Sejong University in 2020.

Acknowledgments: This work was supported by the National Research Foundation of Korea (NRF) grant funded by the Korea Government (MSIT) (2018R1A1A1A05078493).

Conflicts of Interest: The authors declare no conflict of interest.

References

1. Brinkmann, R.; Parise, M.; Dye, D. Sinkhole distribution in a rapidly developing urban environment: Hillsborough Country, Tampa Bay area, Florida. *Eng. Geol.* **2008**, *99*, 169–184. [CrossRef]
2. Strzalkowski, P. Sinkhole formation hazard assessment. *Earth Sci.* **2018**, *78*, 9. [CrossRef]
3. Huang, J.; Liu, W.; Sun, X. A pavement crack detection method combining 2D with 3D information based on dempster-shafer theory. *Comput. Aided Civ. Infrastruct. Eng.* **2014**, *29*, 299–313. [CrossRef]
4. Guan, H.; Li, J.; Yu, Y.; Chapman, M.; Wang, H.; Wang, C.; Zhai, R. Iterative tensor voting for pavement crack extraction using mobile laser scanning data. *IEEE Trans. Geosci. Remote Sens.* **2014**, *53*, 1527–1537. [CrossRef]
5. Toksoz, D.; Yilmaz, I.; Seren, A.; Mataraci, I. A study on the performance of GPR for detection of different types of buried objects. *Procedia Eng.* **2016**, *161*, 399–406. [CrossRef]
6. Sun, H.; Pashoutani, S.; Zhu, J. Nondestructive evaluation of concrete bridge decks with automated acoustic scanning system and ground penetrating radar. *Sensors* **2018**, *18*, 1955. [CrossRef]
7. Ukaegbu, I.K.; Gamage, K.A.A.; Aspinall, M.D. Integration of ground-penetrating radar and gamma-ray detectors for nonintrusive characterization of buried radioactive objects. *Sensors* **2019**, *19*, 2743. [CrossRef]
8. Sharma, P.; Kumar, B.; Singh, D.; Gaba, S.P. Critical analysis of background subtraction techniques on real GPR data. *Def. Sci. J.* **2017**, *67*, 559–571. [CrossRef]
9. Park, B.J.; Kim, J.G.; Lee, J.S.; Kang, M.-S.; An, Y.-K. Underground object classification for urban roads using instantaneous phase analysis of GPR data. *Remote Sens.* **2018**, *10*, 1417. [CrossRef]
10. Daniels, D.J. *Ground Penetrating Radar*, 2nd ed.; The Institution of Electrical Engineers: London, UK, 2004.
11. Ciampoli, L.B.; Tosti, F.; Economou, N.; Benedetto, F. Signal processing of GPR data for road surveys. *Geosciences* **2019**, *9*, 96. [CrossRef]
12. Feng, X.; Yu, Y.; Liu, C.; Fehler, M. Combination of H-alpha decomposition and migration for enhancing subsurface target classification of GPR. *IEEE Trans. Geosci. Remote Sens.* **2015**, *53*, 4852–4863. [CrossRef]
13. Economou, N.; Vafidis, A. GPR data time varying deconvolution by kurtosis maximization. *J. Appl. Geophys.* **2012**, *81*, 117–121. [CrossRef]
14. Gurbuz, A.C.; McClellan, J.H.; Scott, W.R. Compressive sensing for subsurface imaging using ground penetrating radar. *Signal. Process.* **2009**, *89*, 1959–1972. [CrossRef]
15. Pue, J.D.; Meirvenne, M.V.; Cornelis, W.M. Accounting for surface refraction in velocity semblance analysis with air-coupled GPR. *IEEE J. Top. Appl. Earth Obs. Remote Sens.* **2015**, *9*, 60–73. [CrossRef]
16. Nuzzo, L. Coherent noise attenuation in GPR data by linear and parabolic radon transform techniques. *Ann. Geophys.* **2003**, *46*, 533–547.
17. Baili, J.; Lahouar, S.; Hergli, M.; Al-Qadi, I.L.; Besbes, K. GPR signal de-noising by discrete wavelet transform. *NDT E Int.* **2009**, *42*, 696–703. [CrossRef]
18. Ostoori, R.; Goudarzi, A.; Oskooi, B. GPR random noise reduction using BPD and EMD. *J. Geophys. Eng.* **2018**, *15*, 347–353. [CrossRef]
19. Nunez-Nieto, X.; Solla, M.; Gomez-Perez, P.; Lorenzo, H. GPR signal characterization for automated landmine and UXO detection based on machine learning techniques. *Remote Sens.* **2014**, *6*, 9729–9748. [CrossRef]
20. Klesk, P.; Godziuk, A.; Kapruziak, M.; Olech, B. Fast analysis of C-scans from ground penetrating radar via 3-D Haar-Like features with application to landmine detection. *IEEE Trans. Geosci. Remote Sens.* **2015**, *53*, 3996–4009. [CrossRef]
21. Mazurkiewicz, E.; Tadeusiewicz, R.; Tomecka-Suchon, S. Application of neural network enhanced ground penetrating radar to localization of burial sites. *Appl. Artif. Intell.* **2016**, *30*, 844–860. [CrossRef]
22. Kim, N.G.; Kim, K.D.; An, Y.-K.; Lee, H.J.; Lee, J.J. Deep learning-based underground object detection for urban road pavement. *Int. J. Pavement Eng.* **2018**, 1–13. [CrossRef]
23. Kang, M.-S.; Kim, N.G.; Lee, J.J.; An, Y.-K. Deep learning-based automated underground cavity detection using three-dimensional ground penetrating radar. *Struct. Health Monit.* **2019**, *19*, 173–185. [CrossRef]
24. Kim, N.G.; Kim, S.H.; An, Y.-K.; Lee, J.J. A novel 3D GPR image arrangement for deep learning—based underground object classification. *Int. J. Pavement Eng.* **2019**, 1–12. [CrossRef]

25. Kim, N.G.; Kim, S.H.; An, Y.-K.; Lee, J.J. Triplanar imaging of 3-D GPR data for deep-learning-based underground object detection. *IEEE J. Sel. Top. Appl. Earth Obs. Remote Sens.* **2019**, *12*, 4446–4456. [CrossRef]

26. Benedetto, F.; Tosti, F. A signal processing methodology for assessing the performance of ASTM standard test methods for GPR systems. *Signal. Process.* **2017**, *132*, 327–337. [CrossRef]

27. Ruzzene, M. Frequency-wavenumber domain filtering for improved damage visualization. *Smart Mater. Struct.* **2007**, *16*, 2116–2129. [CrossRef]

28. An, Y.-K.; Park, B.J.; Sohn, H. Complete noncontact laser ultrasonic imaging for automated crack visualization in a plate. *Smart Mater. Struct.* **2013**, *22*, 1–10. [CrossRef]

29. An, Y.-K.; Kwon, Y.S.; Sohn, H. Noncontact laser ultrasonic crack detection for plates with additional structural complexities. *Struct. Health Monit.* **2013**, *12*, 522–538. [CrossRef]

30. Miwa, T.; Arai, I. Super-resolution imaging for point reflectors near transmitting and receiving array. *IEEE Trans. Antennas Propag.* **2004**, *52*, 220–229. [CrossRef]

31. Yamaguchi, T.; Mizutani, T.; Tarumi, M.; Su, D. Sensitive damage detection of reinforced concrete bridge slab by "Time-variant deconvolution" of SHF-band radar signal. *IEEE Trans. Geosci. Remote Sens.* **2019**, *57*, 1478–1488. [CrossRef]

32. Chang, P.; Flatau, A.; Liu, S. Review paper: Health monitoring of civil infrastructure. *Struct. Health Monit.* **2003**, *2*, 257–267. [CrossRef]

33. Irani, M.; Peleg, S. Improving resolution by imaging registration. *CVGIP Graph. Models Image Process.* **1991**, *53*, 231–239. [CrossRef]

34. Kim, K.I.; Kwon, Y. Single-image super-resolution using sparse regression and natural image prior. *IEEE Trans. Pattern Anal.* **2010**, *32*, 1127–1133.

35. Yang, J.; Wright, J.; Huang, T.S.; Ma, Y. Image super-resolution via sparse representation. *IEEE Trans. Image Process.* **2010**, *19*, 2861–2873. [CrossRef]

36. Dong, C.; Loy, C.C.; He, K.; Tang, X. Image super-resolution using deep convolutional networks. *IEEE Trans. Pattern Anal.* **2015**, *38*, 295–307. [CrossRef] [PubMed]

37. Kim, J.; Lee, J.K.; Lee, K.M. *Accurate Image Super-Resolution Using Very Deep Convolutional Networks*; IEEE Computer Vision and Pattern Recognition: Las Vegas, NV, USA, 2016.

38. Ledig, C.; Theis, L.; Huszar, F.; Caballero, J.; Cunningham, A.; Acosta, A.; Aitken, A.; Tejani, A.; Totz, J.; Wang, Z.; et al. *Photo-Realistic Single Image Super-Resolution Using a Generative Adversarial Networks*; IEEE Computer Vision and Pattern Recognition: Honolulu, HI, USA, 2017.

39. Zhang, Y.; Li, K.; Li, K.; Wang, L.; Zhong, B.; Fu, Y. Image Super-Resolution Using Very Deep Residual Channel Attention Networks. In Proceedings of the 15th European Conference on Computer Vision, Munich, Germany, 8–14 September 2018.

40. Tai, Y.; Yang, J.; Liu, X. *Image Super-Resolution via Deep Recursive Residual Network*; IEEE Computer Vision and Pattern Recognition: Honolulu, HI, USA, 2017.

41. Bae, H.; Jang, K.; An, Y.-K. Deep super resolution crack network (SrcNet) for improving computer vision-based automated crack detectability in in situ bridges. *Struct. Health Monit.* **2020**. [CrossRef]

42. Kang, M.-S.; Kim, N.G.; Im, S.B.; Lee, J.J.; An, Y.-K. 3D GPR image—Based UcNet for enhancing underground cavity detectability. *Remote Sens.* **2019**, *11*, 2545. [CrossRef]

43. Warren, C.; Giannopoulos, A.; Giannakis, I. gprMax: Open source software to simulate electromagnetic wave propagation for ground penetrating radar. *Comput. Phys. Commun.* **2016**, *209*, 163–170. [CrossRef]

44. Yee, K. Numerical solution of initial boundary value problems involving Maxwell's equations in isotropic media. *IEEE Trans. Antennas Propag.* **1966**, *14*, 302–307.

Article

Application of Combining YOLO Models and 3D GPR Images in Road Detection and Maintenance

Zhen Liu [1], Wenxiu Wu [2], Xingyu Gu [1,*], Shuwei Li [1], Lutai Wang [1] and Tianjie Zhang [3]

[1] Department of Roadway Engineering, School of Transportation, Southeast University, Nanjing 211189, China; 230208344@seu.edu.cn (Z.L.); 220182999@seu.edu.cn (S.L.); 213172518@seu.edu.cn (L.W.)
[2] Highway and Transportation Management Center, Jinhua 321000, China; 220173126@seu.edu.cn
[3] Zhejiang Scientific Research Institute of Transport, Hangzhou 310023, China; 220162634@seu.edu.cn
* Correspondence: guxingyu1976@seu.edu.cn; Tel.: +86-025-86342563

Abstract: Improving the detection efficiency and maintenance benefits is one of the greatest challenges in road testing and maintenance. To address this problem, this paper presents a method for combining the you only look once (YOLO) series with 3D ground-penetrating radar (GPR) images to recognize the internal defects in asphalt pavement and compares the effectiveness of traditional detection and GPR detection by evaluating the maintenance benefits. First, traditional detection is conducted to survey and summarize the surface conditions of tested roads, which are missing the internal information. Therefore, GPR detection is implemented to acquire the images of concealed defects. Then, the YOLOv5 model with the most even performance of the six selected models is applied to achieve the rapid identification of road defects. Finally, the benefits evaluation of maintenance programs based on these two detection methods is conducted from economic and environmental perspectives. The results demonstrate that the economic scores are improved and the maintenance cost is reduced by \$49,398/km based on GPR detection; the energy consumption and carbon emissions are reduced by 792,106 MJ/km (16.94%) and 56,289 kg/km (16.91%), respectively, all of which indicates the effectiveness of 3D GPR in pavement detection and maintenance.

Keywords: ground-penetrating radar; road defect detection; YOLOv5 models; road defects image recognition; road maintenance benefit; road maintenance effectiveness

Citation: Liu, Z.; Wu, W.; Gu, X.; Li, S.; Wang, L.; Zhang, T. Application of Combining YOLO Models and 3D GPR Images in Road Detection and Maintenance. *Remote Sens.* **2021**, *13*, 1081. https://doi.org/10.3390/rs13061081

Academic Editor: Mercedes Solla

Received: 10 February 2021
Accepted: 10 March 2021
Published: 12 March 2021

Publisher's Note: MDPI stays neutral with regard to jurisdictional claims in published maps and institutional affiliations.

1. Introduction

The quality parameters for structural layers of pavement are obtained through reasonable setpoint, drilled core on-site and laboratory testing in core sample detection. However, the inspection results cannot reflect the true conditions of the road at the scene because the setpoint is random and incidental [1,2]. In addition, the defect conditions of the road surface are acquired by manual-based patrol and judgement, which cannot detect the internal defects. These methods have the characteristics of low efficiency, poor presentation, and destructiveness that have led to a considerable increase in the cost of road maintenance. Thus, the traditional testing methods fail to meet the growing demands of road maintenance.

With the development of science and technology, new nondestructive testing (NDT) devices, such as ground-penetrating radar (GPR), the nuclear-free densitometer, laser detector, and ultrasonic depth finder have been used in fast nondestructive and precise testing. GPR is already well recognized for its role in improving the efficiency, security, and anti-interference [3,4]. Radar-collected data can provide the basis for recognizing hidden defects and be used to conduct the later maintenance and management of roads [5]. The development of 3D GPR further reinforces these effects [6]. Nevertheless, this technology has limitations, such as tedious data post-processing and a lack of evaluation criteria, which have resulted in a failure to provide automatic detection and quantitative evaluation of road testing and maintenance [7].

Recently, several efforts have been made in terms of the data processing of GPR inspection that includes signal processing and image recognition. Zhao et al. [8] proposed a nonlinear optimization method based on gradient descent to analyze the collected GPR signals in the thickness detection of asphalt pavement, which needs a prior knowledge of road structure. Liu et al. [9] used the frequency domain focusing technology of synthetic aperture radar (SAR) to aggregate scattered GPR signals for acquiring testing images. The noise of primordial signals was removed through the designed low-pass filter, and the profiles of detecting objects were extracted via the edge detection technique using the background information. Moreover, Mezgeen et al. [10] presented a formula relating the hidden crack width with the relative amplitude measured in the vertex of the hyperbola. However, a major drawback is that this research only considered regular single cracks.

As for image recognition of GPR detection, many researchers have tried to apply the complex manual processes to automatically inspect internal defects in a road, but this goal is difficult to realize [11,12]. It was not until the appearance of deep learning (DL), the real, efficient, automatic detection of concealed defects became possible in asphalt pavement [13,14]. As a result, the combination of deep convolutional neural network (CNN) models and GPR images has become a mainstream research direction. Tong et al. [15–17] used a CNN algorithm to achieve the automatic localization of internal cracking based on GPR testing images, which used the GPR signals as an input value to import into the CNNs.

However, although the region proposal types of CNN series models have the advantage of high accuracy, the limitation of detection speed loss has been reported. This limitation has promoted the development of more advanced DL models. Another regression method (also known as the one-stage method) substantially enhances the speed of defect detection by streamlining the workflow. This method primarily includes YOLO [18–20], RetinaNet [21], the single shot multibox detector (SSD) [22,23], and CenterNet [24]. Above all, YOLO version 3 (YOLOv3) is a mainstream method, and it has been widely used in remote sensing [25,26], agriculture [27], and energy [28]. It has also been successfully applied in transportation infrastructure, e.g., for the detection of pavement potholes and cracking [28–30]. Currently, the latest YOLO version 4 (YOLOv4) [31] and YOLO version 5 (YOLOv5) [32] have become more effective for object detection by integrating the most advanced methods.

On the other hand, researchers have performed many studies of the standardization of road testing and maintenance [33,34]. The group criteria of technical guidelines for ground-penetrating radar detection of the internal condition of highway asphalt pavement has been published by the China Highway and Transportation Society (CHTS) [35], which has provided a scientific reference for future exploration. However, these attempts are far from numerous. Therefore, in the present study, we developed a method for evaluating the maintenance benefits by comparing the traditional detection and GPR detection in asphalt pavement.

This work proposes a method for combining the YOLO series with GPR images to recognize the internal defects in asphalt pavement and compares the effectiveness of traditional detection and GPR detection by evaluating the maintenance benefits. The technical roadmap is shown in Figure 1. An introduction to the tested roads and traditional detection method are given in Section 2. In Section 3, the GPR detection process, which includes testing equipment, a testing scheme, data processing, and testing results, is elaborated. Moreover, the YOLOv3 and YOLOv5 models are applied to defect detection for better accuracy and efficiency. The fourth section discusses the maintenance programs and maintenance benefits based on two types of detection methods. Section 5 concludes the research.

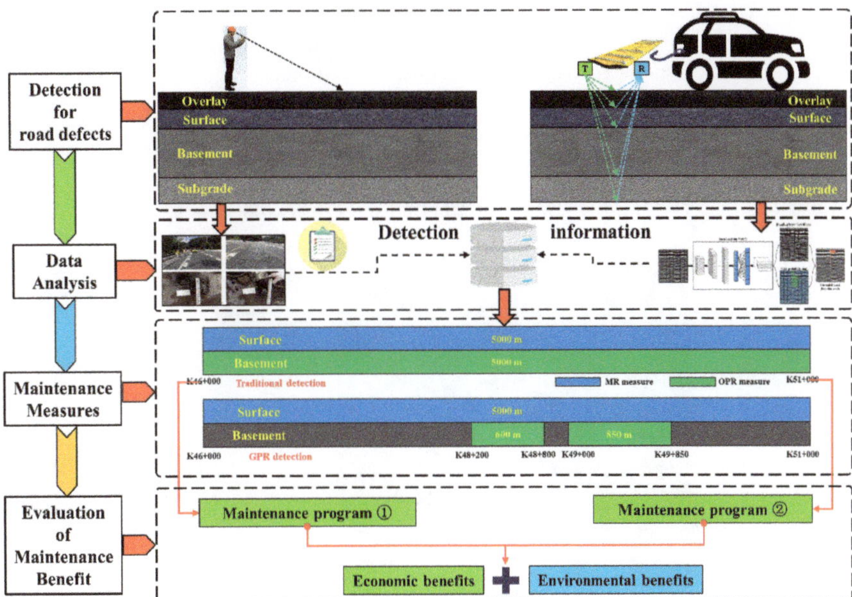

Figure 1. Technical roadmap of the detection, data analysis, maintenance measures, and benefits analysis based on traditional detection and ground-penetrating radar (GPR) detection in this research.

2. Traditional Pavement Detection

2.1. Tested Road Sections

Figure 2 shows the tested provincial road sections, which are called the Tonglu-Yiwu (TY) line (S210) and are located in Zhejiang province, China. Traditional and GPR inspection were implemented on this asphalt pavement from K46+000 to K51+000 and had a total length of 5 km. The structure layers, materials, and position of the tested road are indicated below.

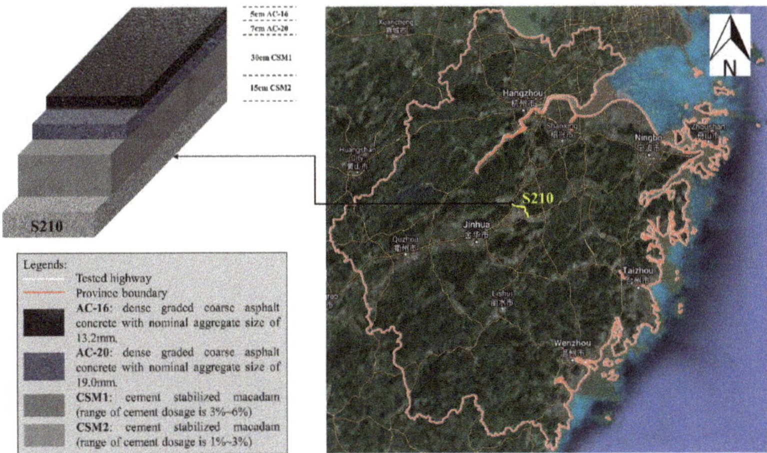

Figure 2. View of the tested road network, construction scheme and materials. The map used is the free version of Google Earth 2020.

2.2. Testing Process and Results

As shown in Figure 3, visual surveying and measurement were adopted by testing personnel to determine the damage condition of the pavement, and an inspection van was used to survey the surface roughness and skidding resistance of the pavement. Moreover, a coring survey was taken to obtain an accurate thickness of the asphalt pavement according to the highway performance assessment standards (JTG 5210-2018) and the specifications for maintenance design of highway asphalt pavement (JTG 5421-2019), which was enacted by the Ministry of Transport of the People's Republic of China.

Figure 3. (**a**) Pavement testing by inspection van, (**b**) pavement damage conditions, (**c**) the core sample of left-pavement, and (**d**) the core sample of right-pavement.

After these detections and observations, the results of the pavement defects investigation are shown in Table 1.

Table 1. The detection results of pavement defects investigation.

Disease Structural Layer			Cracking				Settlement			
	N^l	L^l	D^l	N^r	L^r	D^r	N^l	A^l	N^r	A^r
Asphalt surface	126	337.6	1.4	84	174.1	0.7	1	4.5	2	7.3
Base	–	–	–	–	–	–	–	–	–	–
Subbase	–	–	–	–	–	–	–	–	–	–

[l] Left side, [r] right side (N-number, L-length (m), D-density (N/m), A-area (m²)).

3. Nondestructive Testing of Pavement Based on GPR

3D GPR is a new type of nondestructive testing equipment, and its testing work will not damage the pavement. 3D GPR emits penetrating high-frequency electromagnetic waves to the pavement structure through the fixed distance transmitting antenna and receives the directional reflection signals by the paired receiving antenna. Then, through data processing and analysis of the radar host, the 3D detection information of the pavement structure is reconstructed in the computer.

3.1. Testing Equipment

The 3D GPR system (3d-Radar Company, Trondheim, Norway) was used to inspect the internal damage of the road, which substantially reduced the misjudgment rate of interior conditions due to 2D imaging. The radar host of GeoScope^{TM} MKIV (Figure 4a), multi-channel DXG^{TM} 1820 ground-coupled antenna arrays (Figure 4b), Examiner^{TM} 3 data analysis software, and GPS-RTK equipment (Figure 4c) was included in the 3D GPR system. GeoScope^{TM} MKIV enables high-density, high-speed data acquisition while combining deeper detection capabilities with high resolution. By optimizing the signal bandwidth and the best possible resolution, high-speed surveying and a large scan width can be realized without losing the image details for the study of different depth layers underground. The multi-channel DXG^{TM} 1820 ground-coupled antenna arrays have the advantage of high resolution that can collect 3D GPR data from up to 41 survey lines in a single pass in a continuous frequency range of 200 MHz to 3 GHz. In addition, the road conditions are detected from the surface of the road to a depth of 3 m by this DXG^{TM} antenna, which is well-suited for the detection requirements of highway subgrade and pavement.

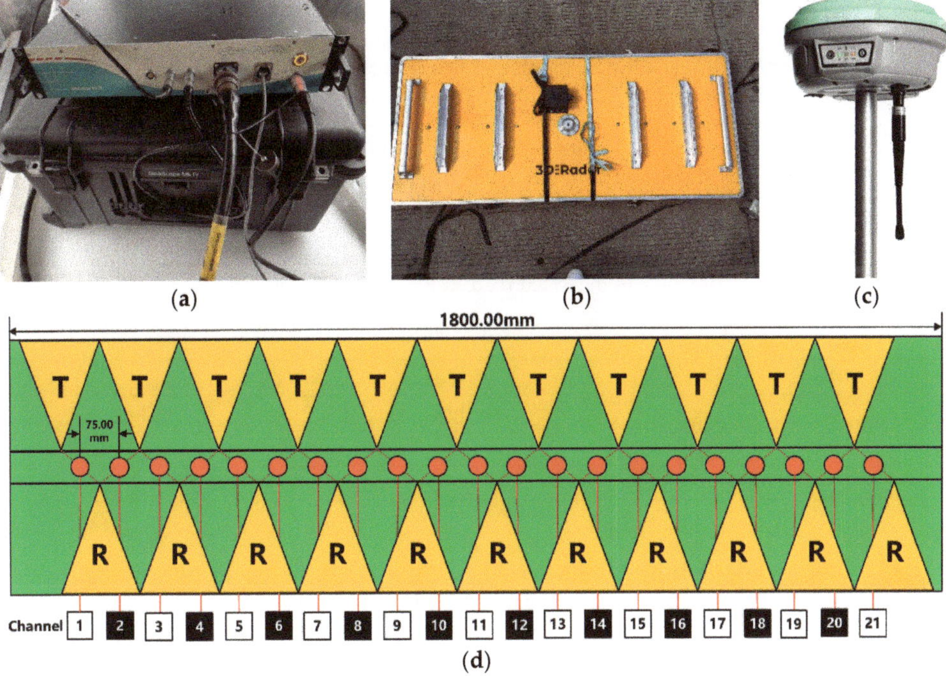

Figure 4. (**a**) GeoScope^{TM} MKIV, (**b**) DXGTM 1820 ground-coupled antenna arrays, (**c**) GPS-RTK, and (**d**) transmit and receive arrays of Radar.

Combined with the unique ability of the stepped-frequency radar host GeoScope^{TM} MKIV and VX series antennas to collect 3D radar data with a certain scan line density, the real 3D radar data processing is realized. As shown in Figure 4d, these antenna arrays combine different transmitting/receiving antenna pairs, allowing the user to collect multiple channels of data at once. By setting up, the user can collect data in a 7.5 cm × 7.5 cm grid (cover 1.5 m scanning) to obtain a true 3D image. The remaining technical parameters are shown in Table 2.

Table 2. The technical specifications of GeoScope^{TM} MKIV and DXG^{TM} 1820.

GeoScope^{TM} MKIV Data Acquisition		DXG^{TM} 1820 Ground-Coupled Antenna Array	
Indicators	Parameters	Indicators	Parameters
Antennas	Compatible with all 3D-RADAR DX and DXG antenna array models	Width	1.8 m
Number of channels	0~21	Frequency range	200–3000 MHz
Scan pattern	Liner scan, multi-offset, and common mid-point	Number of channels	21
Frequency bandwidth	2.9 GHz (100–300 MHz)	Channels spacing (Crpss-line)	75 mm
Resolution (time)	$\geq$0.34 ns	Effective scan width	1.5 m
Time range	$\leq$250 ns	Direct wave suppression	>50 dB
Scan rate	13,000 A-scans per second	Polarization (in-line direction)	Linear

3.2. Testing Scheme

3D GPR was adopted in this research to realize a full scan covering the road cross-section of the TY line (S210). According to the stake number, horizon, area, volume and width of the characteristic signal of internal road defects, and the details of some typical defects were detected, including subsidence of internal road structure (position, the maximum height difference, and area), bad interlayer bonding (position and area), general transverse cracking (position and length), general longitudinal cracking (position and length), penetrating cracking (position and length), water-rich zones (position and area), void zones (position and volume), and relaxing zones (position and degree). The information of the tested road section is shown in Table 3.

Table 3. Tested road section.

Location	Stake Number of Starting Point	Stake Number of Ending Point	Breadth of Road	Length/m	Number of Lines	Testing Content	Number of Repeated Scans	Testing Mileage/m
S210	K46 + 000	K51 + 000	Full width	5000	2	disease	1	10,000

As shown in Figure 5, the 3D GPR detection was conducted lane by lane and covered all the lanes. Some vehicles were arranged to follow the inspection van during the detection process by the proprietor, which assured the security of detection personnel and equipment. Under suitable conditions, the detection process should be closed to traffic based on the Safety Work Rules for Highway Maintenance, JTG H30-2015 (Ministry of Transport of the People's Republic of China).

3.3. Data Processing

The construction of a deep, learning-based road internal defect identification model requires a 3D GPR image dataset to provide the training set, verification, and testing required for model construction. This process was conducted by taking the steps in Figure 6.

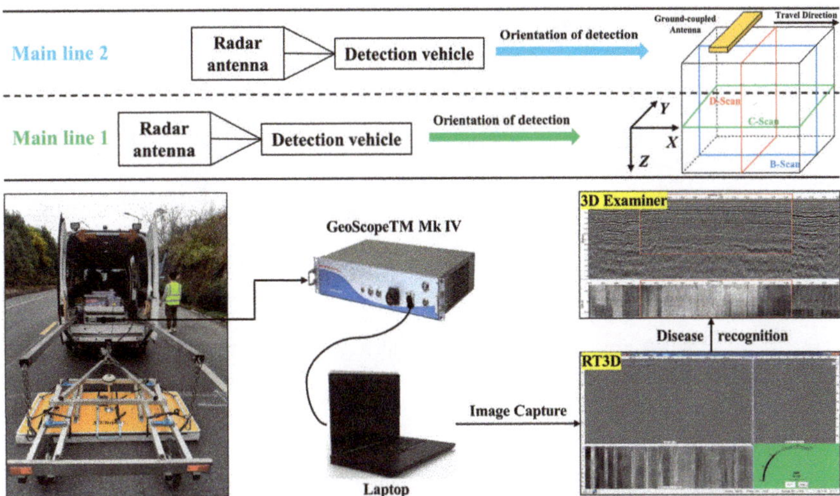

Figure 5. The full field-testing workflow of GPR in Asphalt pavement (B-scan indicates the longitudinal vertical section, C-scan indicates the horizontal section, and D-scan indicates the lateral vertical section).

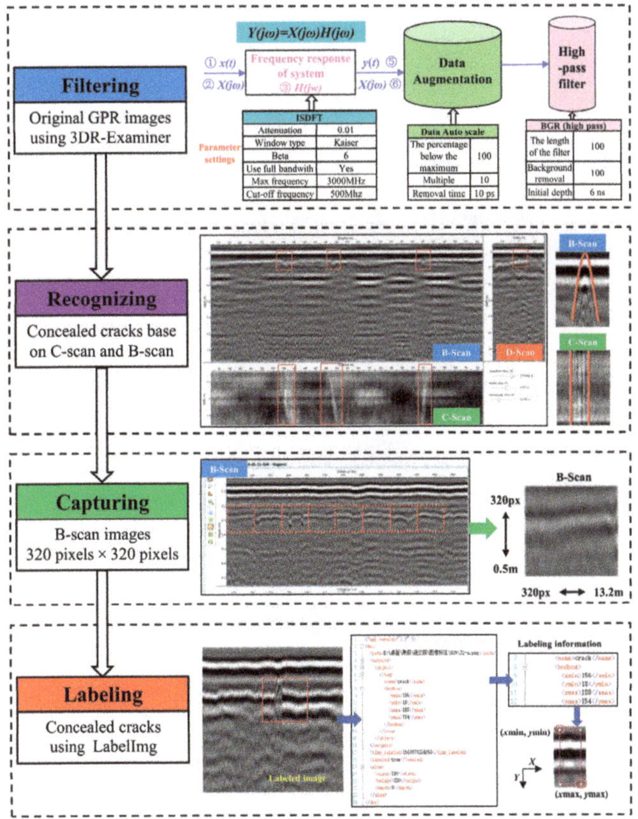

Figure 6. The workflow of dataset construction for GPR images [36].

3.3.1. Filtering for GPR Data

After the GPR data acquisition, the augmentation and filtering of these images were to be performed. Based on the GPR system data processing software (Examiner^{TM} 3), the inverse discrete Fourier transform (ISDFT), data autoscale, and background removal (BGR) (high pass) were used for data processing. The specific settings of the filtering parameters are shown in Figure 6.

3.3.2. Recognizing for GPR Data

Cracking, void, and settlement are the three main defects to be classified and identified in this research. However, the settlement defect was not included in the identification model because the scale of settlement is much larger than the other two defects, and its characteristics are distinctive. Therefore, according to Technical Guideline for Ground Penetrating Radar Detection for Internal Conditions of Highway Asphalt Pavement promulgated by China Highway and Transportation Society, the basis for judgement of cracking and void was determined by summarizing the features of the B-scan and C-scan of these two defects in Table 4.

Table 4. The classification criteria of defects.

Typical Images of Abnormal GPR Signals				Judgement	Excavation for Verifying
B-Scan	**Description**	**C-Scan**	**Description**		
	Both sides of the waveform in-phase axial near horizontal distribution accompanied interruption or dislocation		Similar to the shape pf cracking (long strip)	Cracking	
	Reflected waves of in-phase axial clearly protrude toward the top		Irregular bright-spots	Void	

3.3.3. Capturing for GPR Data

The B-scan images were chosen as the input images because they could reflect the most basic features of internal defects and the exact location through GPR. In addition, the images have a high identifiability degree, which is easier to recognize. The resolution of the captured images was 320×320 pixels, and the real size for the B-scan was 0.5 m $\times$ 13.2 m.

3.3.4. Labeling for GPR Data

LabelImg labeling software [37] was used to mark hidden cracking in the captured images. Based on the identification method of 3.3.2 (the void defect was manually identified because the number of samples was too small), the hyperbolic reflection wave in B-scan and the long strip in C-scan were used to mark the hidden cracking with rectangular boxes.

Then, the corresponding annotation information for the box was stored in an XML-formatted file, as shown in the bottom of Figure 6. The marking information includes the coordinates of two points on the diagonal line of the rectangular box, which can reflect the location and size of the selected cracking.

Next, according to the number of captured images in our early research, 350 sample images were labeled, and the total number of concealed cracks was 1400. Afterwards, these samples were assigned to three groups randomly in a certain ratio as follows: the training model's dataset (263 images and 1134 cracks), the verifying model's validation set (44 images and 135 cracks), and the evaluating model's test set (43 images and 131 cracks).

3.4. Testing Results

The workflow for the detection method of the YOLO models is shown in Figure 7. YOLOv3 is well known for having the most advanced one-stage detection networks. Although the updated version YOLOv5 uses new peculiarities to increase the detection efficiency, YOLOv5 and YOLOv3 still have a similar detection principle and network architecture. In brief, the latest technology has been used in YOLOv5 to update YOLOv3 in terms of Backbone and Neck. In parallel, skills are also added. Detailed information is shown in Table 5.

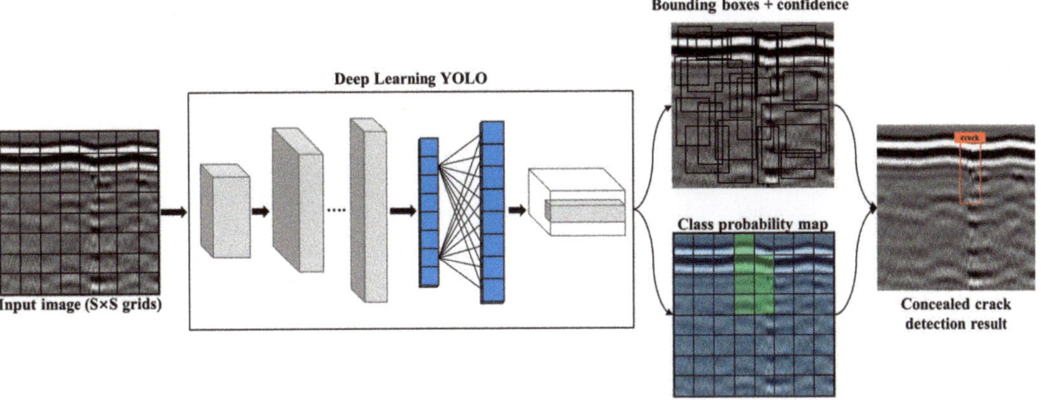

Figure 7. The workflow for detection method of the you only look once (YOLO) models.

Table 5. Architecture and improvement of you only look once version 3 (YOLOv3) and YOLOv5.

Model	Backbone	Neck	Head	Main Improvement
YOLOv3	Darknet	Feature Pyramid Network	YOLOv3	–
YOLOv5s	Cross-stage Partial Darknet	Path Aggregation Network		Mosaic (Data Augmentation) GIoU (estimating the bounding box loss) Auto-learning bounding box anchors (adjusting and optimize the choice of anchors)

Figure 7 describes the detection workflow for the YOLO models of this research. The operations of our works were based on Python 3.7, PyTorch 1.4 in Windows 10. The model trainings were performed by a computer equipped with the following features: an AMD Ryzen 5 2600X CPU and 16 GB of memory.

This study compared 6 models with 2 different kinds of versions, namely, YOLOv3, YOLOv3-tiny, YOLOv5s, YOLOv5m, YOLOv5l, and YOLOv5x [38]. YOLO-tiny was considered light YOLO to substantially increase the detection speed but brought accuracy loss. Note that the s, m, l, and x appended to YOLOv5 represent the increasing depth of the model.

YOLOv3 predicts an objectness score for each bounding box based on logistic regression. As for the loss of the bounding box regression, intersection over union (IoU) [39] is the most popular metric for calculating loss. YOLOv5 uses the same backbone of YOLOv3 and utilizes GIoU to estimate the bounding box loss. Besides, it also uses auto-learning bounding box anchors to adjust and optimize the choice of anchors.

The network has a relatively large number of parameters and a small dataset, which could result in overfitting. Therefore, transfer learning was adopted to train the models to overcome this hidden danger [40]. The COCO dataset includes over 500,000 image data points belonging to 80 different categories. Consequently, the pretrained weights by the COCO dataset were used to initialize the model to be trained. The other hyperparameters of the model were set as shown below: the initial learning rate was 0.001; the size of the batch and mini-batch were 16 and 4, respectively; the momentum and weight decay were 0.9 and 0.0005, respectively; the epoch was 300; and the other parameters were set to their default values.

As is shown in Figure 8, the loss and mAP curves of the YOLO models were compared. The value of loss represents the difference between the predicted value and true value. The smaller the value of loss, the better training effect. Moreover, the high mAP also denotes a great performance of the training models.

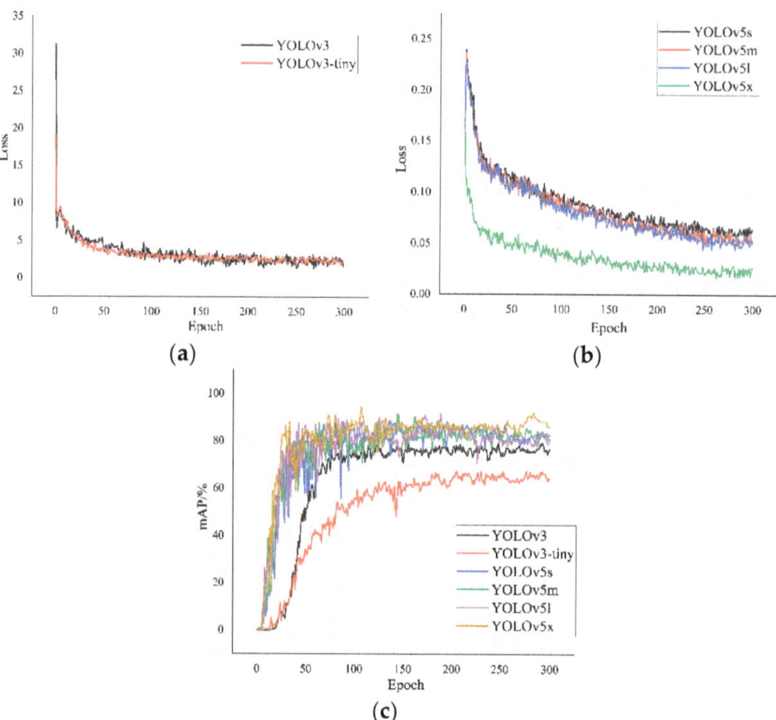

Figure 8. (**a**) The loss curve of the YOLOv3 models, (**b**) the loss curve of the YOLOv5 models, (**c**) the mAP Curve of the YOLOv3 and YOLOv5 models.

According to Figure 8a,b, the final converged loss value of YOLOv3 was approximately 2, whereas that of YOLOv5 was lower than 0.2, which suggested that the YOLOv5 model performed substantially better than YOLOv3 because the lower loss indicates better training effects. Moreover, all mAP values of YOLOv5 were higher than those of YOLOv3. Taken together, we concluded that the performance of the YOLOv5 models was superior.

The specific training results of the YOLOv3 and YOLOv5 models are summarized in Table 6. All mAP values of the YOLOv5m, YOLOv5l, and YOLOv5x models were higher than 90% (the highest value was 94.45%), which is commendable for the small training set.

Another finding may be summarized as the higher the number of weights is, the higher the model's mAP value will be, suggesting that an appropriate increase in model depth favored the enhancement of the training performance. However, with increasing weights, the frames per second (FPS) were reduced, while the inference time was prolonged. It is not difficult to find that the FPS of the YOLOv3 and YOLOv5 models were poorly differentiated when the values of the weights were similar. What needs illustration is that the model with faster inference speed was preferentially selected based on the requirements of rapid detection. Ultimately, the YOLOv5m model with the most even performance was used to detect internal defects in roads according to the integrated consideration of mAP and FPS.

Table 6. Training results of these six models of YOLO series.

Model	P	R	F1	mAP/%	FPS	Inference Time/ms	Weights/MB
YOLOv3	0.73	0.86	0.79	80.11	0.52	1920.65	235
YOLOv3-tiny	0.66	0.65	0.69	67.59	4.52	221.48	33.1
YOLOv5s	0.79	0.87	0.85	87.53	3.45	289.81	26.4
YOLOv5m	0.76	0.94	0.82	91.61	1.36	735.54	83.2
YOLOv5l	0.77	0.95	0.86	91.59	0.66	1526.37	190
YOLOv5x	0.75	0.95	0.85	94.45	0.37	2735.15	364

(P = precision, R = recall rate, F_1 score = 2*P*R/(P+R), mAP = mean average precision, FPS = frames per second).

Depending on the training results obtained via YOLOv5m, the statistical information of the defects is listed in Tables 7 and 8 (the raveling and settlement were manually recognized).

Table 7. The defects detection results of the left side of road.

Disease / Structural Layer	Cracking			Void		Raveling		Settlement	
	N [1]	L [2]	D [3]	N	A [4]	N	A	N	A
Asphalt surface	132	354.9	1.5	4	13	–	–	1	4.5
Base	103	238.5	1.0	5	16	1	3.8	2	8.2
Subbase	–	–	–	–	–	–	–	–	–

[1] N-number, [2] L-length (m), [3] D-density (N/m), [4] A-area (m^2).

Table 8. The defects detection results of the right side of road.

Disease / Structural Layer	Cracking			Void		Raveling		Settlement	
	N [1]	L [2]	D [3]	N	A [4]	N	A	N	A
Asphalt surface	92	189	0.8	–	–	–	–	2	7.3
Base	86	151.6	0.6	13	52	–	–	1	4.2
Subbase	–	–	–	–	–	–	–	–	–

[1] N-number, [2] L-length (m), [3] D-density (N/m), [4] A-area (m^2).

A schematic of the position and size of the defects is plotted in Figure 9 based on the recognition results (the detailed analysis is in the following Section 4.1).

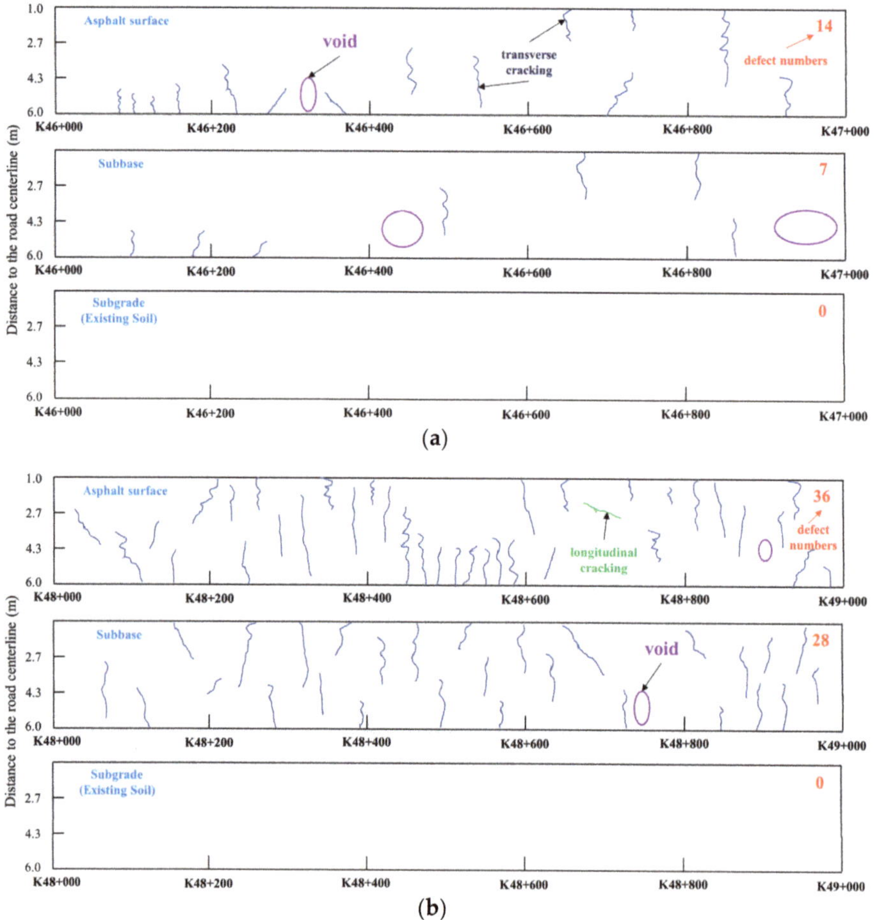

Figure 9. A schematic for detection results of partial tested road. (**a**) General defect severity and (**b**) high defect severity.

4. Discussion of Maintenance Benefits

4.1. Disease Characteristics and Analysis

4.1.1. Traditional Detection

The primary defect types of the tested road section are the dominant cracking and settlement on the road surface. The maintenance measures were conducted at the surface and basement of the tested road because of the unclear information of the internal defects.

4.1.2. GPR Detection

The main defect types of the tested road section are cracking (more than 90%), void zones, and raveling. Therefore, the characteristics of the cracking were the focus of the analysis. First, the overall cracking density of the proposed maintenance roads was low. Specifically, the cracking density of the surface was $1.5 \, \text{m/m}^2$, the cracking density of the basement was $1.0 \, \text{m/m}^2$, and the number of void defects was 9 (the total area was $27 \, \text{m}^2$) in the left side of the tested road. Moreover, the cracking density of the surface was $0.8 \, \text{m/m}^2$, the cracking density of the basement was $0.6 \, \text{m/m}^2$, and the number of void defects was 13 (the total area was $52 \, \text{m}^2$) in the right side of the tested road.

As for the development horizon of the cracking, there were three types, as shown in Figure 10.

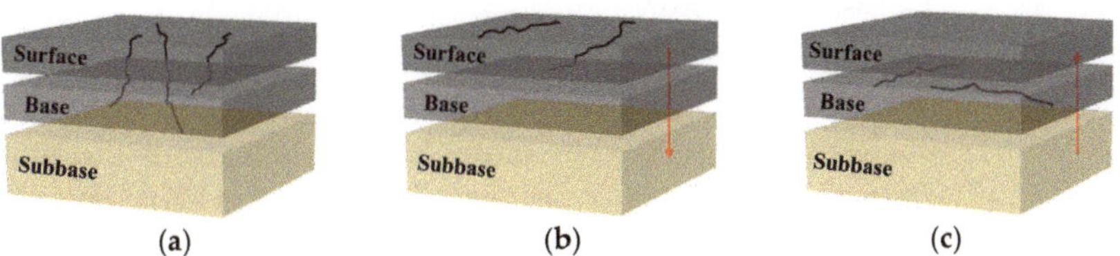

Figure 10. (**a**) The up and down cracking, (**b**) the top-down developing cracking, and (**c**) the bottom-up developing cracking.

1. The up and down cracking (the pumping defect had emerged, Figure 10a).
2. The top-down developing cracking (the cracking had emerged on the surface but not at the basement, Figure 10b).
3. The bottom-up developing cracking (the cracking had emerged on the basement but not at the surface, Figure 10c).

From the perspective of the regional distribution of defects, distinct characteristics of partial defect concentration were in the tested roads. The defects of the basement were lesser than those of other structural layers in general road sections. On the other hand, the distribution of defects was more concentrated at the surface and basement in the road sections with severe defects.

Finally, the prediction of the development of defects was conducted based on the results above. Notably, the third type of developing cracking would gradually undergo a transition to the first type with the arrival of freeze-thawing during rainy and winter seasons, which would lead to the appearance of more pumping mud.

4.2. Maintenance Program

As shown in Figure 11, the following two maintenance programs were determined according to different defect severities.

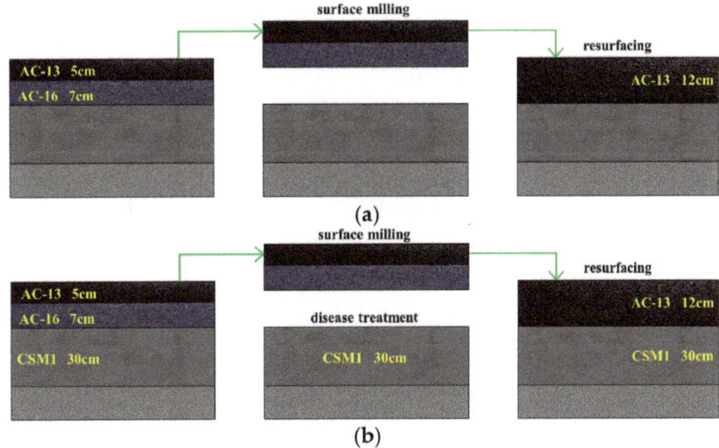

Figure 11. (**a**) The first maintenance measure and (**b**) the second maintenance measure.

First, for the general road sections with low defect severity, the milling measure for the original surface (5 cm AC-13 and 7 cm AC-16) should be performed. Then, the new surface (12 cm AC-13, a previous study demonstrated that the maintenance measure of

AC-13 has the highest comprehensive benefit [41]) is spread on the basement (milling and resurfacing, MR). In terms of the road sections with severe defects, after the surface milling measure, the treatment of defects (overlay paving for reinforcement, OPR) is conducted on the basement before the resurfacing.

Finally, according to the analysis for defect characteristics and maintenance measures based on conventional detection and GPR detection, the maintenance programs were established for these two detection methods. As presented in Figure 12, the MR measure was conducted for surface maintenance based on both detection methods. However, this was not the same case for the basement. Specifically, the OPR measure was adopted for the full range of basement with conventional detection, while only 1450 m for the basement of serious diseases with GPR detection.

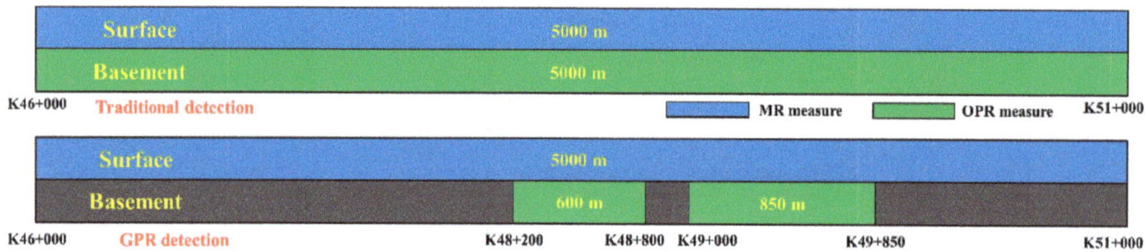

Figure 12. The selection of maintenance schemes based on traditional detection and GPR detection.

4.3. Benefits Analysis

The service life and pavement performance of maintenance measures have been used to evaluate the long-term benefits in many studies [42,43]. The present work drew on previous studies and used the economic and environmental benefits as evaluation criteria for comparison of traditional detection and GPR detection.

The fundamental assumptions for the calculation of benefits were as follows. The material haul-lengths of asphalt, gravel, and asphalt mixture are 100, 60, and 50 km, respectively. The density of hot-mix asphalt mixture is 2.45 t/m^3. The thickness of treatment is typically 4 cm. The per unit of maintenance area is calculated as 375 m^2 (100 m × 3.75 m, single lane).

In this study, the tested road was a two-way four-lane road of 5 km. According to Figure 12, the total area of the first and second maintenance measures was 75,000 m^2 based on traditional detection. As for GPR detection, the area of the first maintenance measure was 53,250 m^2 and that of the second was 21,750 m^2.

4.3.1. Economic Benefits

The average cost and economic effectiveness (the evaluation index for maintenance economic-benefits obtained by some pavement performance indexes) [44] of the MR and OPR measures are listed in Table 9. The actual thickness of treatment was 0.12 m in the road surface. Therefore, the final results in Figure 13 were obtained by multiplying by 3.

Table 9. The average cost and economic effectiveness of the two maintenance measures.

Maintenance Measures	Thickness of Treatment (m)	Average Cost ($/m^2)	Economic Effectiveness [1]	Economic Effectiveness [2]
MR	0.04	11.12	93.88	88.62
OPR	0.01	4.64	90.73	85.47

[1] Low traffic, [2] high traffic.

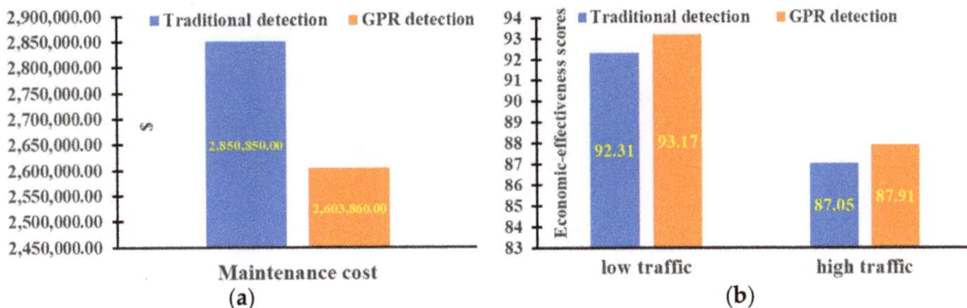

Figure 13. The contrast for maintenance cost (**a**) and economic effectiveness (**b**) of the two detecting methods.

Figure 13 shows that the maintenance cost based on GPR detection was lower than that of traditional detection. More specifically, the reducing cost is $49,398/km. In addition, the economic scores were higher based on GPR detection than traditional detection in low-traffic and high-traffic road sections.

4.3.2. Environmental Benefits

Table 10 lists the energy consumption and carbon emissions of the MR and OPR measures, including the milling, raw materials production, mixture, transport, spreading, and compaction sessions.

Table 10. The energy consumption (left) and carbon emissions (right) of the two maintenance measures [45].

Maintenance Sessions	MR	OPR	Maintenance Sessions	MR	OPR
Milling	1770.83		Milling	131.22	
Raw materials production	12,298.53	12,298.53	Raw materials production	756.05	756.05
Mixture	11,469.15	11,469.15	Mixture	925.58	925.58
Transport	2146.60	2146.60	Transport	159.06	159.06
Spreading	681.09	681.09	Spreading	50.47	50.47
Compaction	1225.96	1225.96	Compaction	90.84	90.84
Totally	29,592.17	27,821.34	Totally	2113.23	1982.01
Thickness of treatment (m)	0.04	0.04	Thickness of treatment (m)	0.04	0.04
Energy consumption (MJ/m^2)	78.91	74.19	CO_2 emissions (kg/m^2)	5.64	5.29
——	——	——	Carbon emissions (kg/m^2)	1.54	1.44

(The construction process of some sessions in the MR and OPR measures are the same)

The contrast for energy consumption and carbon emissions are shown below in Figure 14.

The contrast between the energy consumption and carbon emissions of the two detecting methods is shown in Figure 14. Obviously, the energy consumption and carbon emissions based on GPR detection were less than those based on traditional detection and were reduced by 792,106 MJ/km (16.94%) and 56,289 kg/km (16.91%), respectively.

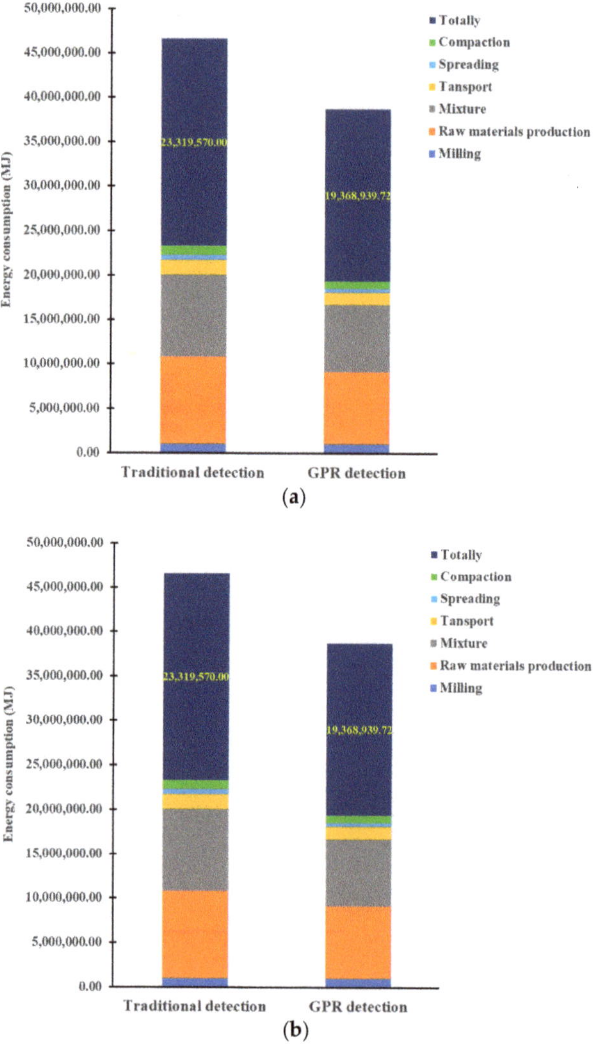

Figure 14. The contrast for energy consumption (**a**) and carbon emissions (**b**) of the two detecting methods.

5. Conclusions

This paper aims to improve the detection efficiency and increase the maintenance benefits by combining YOLO models and 3D GPR images of an asphalt road. The YOLOv5m model is selected to conduct the rapid identification of road defects according to the comparison results of six YOLO series models. Based on the analysis of economic and environmental benefits for tested-road maintenance, the advantage of GPR detection has emerged. Several conclusions can be summarized as follows:

1. The internal defects in asphalt pavement, including cracking, void zones, raveling, and settlement, were detected by 3D GPR. However, the conventional method detected only the surface conditions. Furthermore, 3D GPR detection is more nondestructive relative to the coring validation.

2. The final converged loss value of YOLOv3 was approximately 2, whereas that of YOLOv5 was lower than 0.2. Thus, the YOLOv5 models are suitable for the detection of internal defects in asphalt road, and these models provide a good training result even for a small dataset condition. The mAP values of the YOLOv5m, YOLOv5l, and YOLOv5x models were higher than 90% and the maximum was 94.45% in YOLOv5-x. It was also found with regularity that the larger a model's weights are, the higher the model's mAP will be, which suggests that an appropriate increase in model depth favors the enhancement of the training performance. Most importantly, the YOLOv5m models are the most balanced deep-learning models in terms of detection speed and actual performance of the six YOLO series models.

3. In the evaluation of the economic benefits of maintenance programs, the maintenance cost based on GPR detection was reduced by $49,398/km compared to that of traditional detection, and the economic scores based on GPR detection were higher than those of traditional detection in low-traffic and high-traffic road sections. As for environmental benefits, the energy consumption and carbon emissions of the maintenance program based on GPR detection was less than those of traditional detection by 792,106 MJ/km and 56,289 kg/km or 16.94 and 16.91 percentage points, respectively.

All these facts demonstrate that 3D GPR is effective in pavement detection and maintenance and should be recommended for the life-cycle maintenance of civil infrastructure. Future research may focus on further improving the intelligence level of GPR detection and developing the evaluation criteria of GPR detection.

Author Contributions: Conceptualization, Z.L., X.G. and S.L.; methodology, Z.L., X.G., S.L. and T.Z.; software, W.W. and S.L.; validation, Z.L., and X.G.; formal analysis, Z.L. and S.L.; investigation, Z.L., W.W., L.W. and S.L.; resources, W.W. and T.Z.; data curation, S.L. and L.W.; writing—original draft preparation, Z.L. and X.G.; writing—review and editing, Z.L. and S.L.; visualization, Z.L.; supervision, W.W.; project administration, X.G.; funding acquisition, W.W. and X.G. All authors have read and agreed to the published version of the manuscript.

Funding: This research was funded by [National Key R&D Program of China] grant number [2017YFF0205600], [Fundamental Research Funds for the Central Universities] grant number [3221008101], and [Science and Technology Project of Zhejiang Provincial Department of Transport] grant number [2020045] and [2020053].

Institutional Review Board Statement: Not applicable.

Informed Consent Statement: Not applicable.

Data Availability Statement: Restrictions apply to the availability of these data. Data was obtained from [Jinhua Highway and Transportation Management Center, China] and are available [from the authors] with the permission of [Jinhua Highway and Transportation Management Center, China].

Conflicts of Interest: The authors declare no conflict of interest.

References

1. Zou, Q.; Cao, Y.; Mao, Q.; Wang, S. Crack Tree: Automatic crack detection from pavement images. *Pattern Recogn. Lett.* **2012**, *33*, 227–238. [CrossRef]
2. Brown, S.F. Accelerated pavement testing in highway engineering. *Transport* **2004**, *157*, 173–180. [CrossRef]
3. Lai, W.L.; Dérobert, X.; Annan, P. A review of ground penetrating radar application in civil engineering: A 30-year journey from locating and testing to imaging and diagnosis. *NDT E Int.* **2017**, *96*, 58–78.
4. Zhao, S.; Al-Qadi, I.L. Development of regularization methods on simulated ground-penetrating radar signals to predict thin asphalt overlay thickness. *Signal Process.* **2017**, *132*, 261–271. [CrossRef]
5. Zhao, S.; Al-Qadi, I.L. Super-Resolution of 3-D GPR Signals to Estimate Thin Asphalt Overlay Thickness Using the XCMP Method. *IEEE Trans. Geosci. Remote Sens.* **2019**, *57*, 893–901. [CrossRef]
6. Zhang, J.; Yang, X.; Li, W.; Zhang, S.; Jia, Y. Automatic detection of moisture damages in asphalt pavements from GPR data with deep CNN and IRS method. *Autom. Constr.* **2020**, *113*, 103119. [CrossRef]
7. Zhao, S.; Al-Qadi, I.L. Algorithm development for real-time thin asphalt concrete overlay compaction monitoring using ground-penetrating radar. *NDT E Int.* **2019**, *104*, 114–123. [CrossRef]

8. Zhao, S.; Al-Qadi, I.L.; Wang, S. Prediction of thin asphalt concrete overlay thickness and density using nonlinear optimization of GPR data. *NDT E Int.* **2018**, *100*, 20–30. [CrossRef]

9. Liu, Y.; Qiao, J.; Han, T.; Li, L.; Xu, T. A 3D Image Reconstruction Model for Long Tunnel Geological Estimation. *J. Adv. Transp.* **2020**, *2020*, 8846955. [CrossRef]

10. Rasol, M.A.; Gracia, V.P.; Fernandes, F.M.; Pais, J.C.; Assuncao, S.S.; Santos, C.; Sossa, V. GPR laboratory tests and numerical models to characterize cracks in cement concrete specimens, exemplifying damage in rigid pavement. *Measurement* **2020**, *158*, 107662. [CrossRef]

11. Rodes, P.J.; Reguero, A.M.; Perez-Gracia, V. GPR Spectra for Monitoring Asphalt Pavements. *Remote Sens.* **2020**, *12*, 1749. [CrossRef]

12. Gao, J.; Yuan, D.; Tong, Z.; Yang, J.; Yu, D. Autonomous pavement distress detection using ground penetrating radar and region-based deep learning. *Measurement* **2020**, *164*, 108077. [CrossRef]

13. Kang, M.S.; Kim, N.; Im, S.B.; Lee, J.J.; An, Y.K. 3D GPR Image-based UcNet for Enhancing Underground Cavity Detectability. *Remote Sens.* **2019**, *11*, 2545. [CrossRef]

14. Gong, Z.; Zhang, H. Research on GPR image recognition based on deep learning. In *MATEC Web of Conferences*; EDP Sciences: Les Ulis, France, 2020; Volume 309, p. 03027.

15. Tong, Z.; Gao, J.; Zhang, H. Innovative method for recognizing subgrade defects based on a convolutional neural network. *Constr. Build. Mater.* **2018**, *169*, 69–82. [CrossRef]

16. Tong, Z.; Yuan, D.; Gao, J.; Wang, Z. Pavement defect detection with fully convolutional network and an uncertainty framework. *Comput. Aided Civ. Infrastruct. Eng.* **2020**, *35*, 832–849. [CrossRef]

17. Tong, Z.; Yuan, D.; Gao, J.; Wei, Y.; Dou, H. Pavement-distress detection using ground-penetrating radar and network in networks. *Constr. Build. Mater.* **2019**, *233*, 117352. [CrossRef]

18. Redmon, J.; Divvala, S.; Girshick, R.; Farhadi, A. You Only Look Once: Unified, Real-Time Object Detection. In Proceedings of the Computer Vision & Pattern Recognition, Las Vegas, NV, USA, 27–30 June 2016.

19. Redmon, J.; Farhadi, A. YOLO9000: Better, Faster, Stronger. In Proceedings of the IEEE Conference on Computer Vision & Pattern Recognition, Honolulu, HI, USA, 21–26 July 2017.

20. Redmon, J.; Farhadi, A. YOLOv3: An Incremental Improvement. *arXiv* **2018**, arXiv:1804.02767.

21. Lin, T.Y.; Gaoyal, P.; Girshick, R.; He, K.; Dollár, P. Focal Loss for Dense Object Detection. In Proceedings of the 2017 IEEE International Conference on Computer Vision (ICCV), Venice, Italy, 22–29 October 2017.

22. Liu, W.; Anguelov, D.; Erhan, D.; Szegedy, C.; Reed, S.; Fu, C.Y.; Berg, A.C. SSD: Single Shot MultiBox Detector. In *Computer Vision—ECCV 2016, ECCV 2016*; Lecture Notes in Computer Science; Springer: Cham, Switzerland; Glasgow, UK, 2016.

23. Shen, Z.; Zhuang, L.; Li, J.; Jiang, Y.; Xue, X. DSOD: Learning Deeply Supervised Object Detectors from Scratch. In Proceedings of the 2017 IEEE International Conference on Computer Vision (ICCV), Venice, Italy, 22–29 October 2017.

24. Duan, K.; Bai, S.; Xie, L.; Qi, H.; Huang, Q.; Tian, Q. CenterNet: Keypoint Triplets for Object Detection. In Proceedings of the 2019 IEEE/CVF International Conference on Computer Vision (ICCV), Seoul, South Korea, 27 October–2 November 2019.

25. Zhou, J.; Tian, Y.; Yuan, C.; Yin, K.; Wen, M. Improved UAV Opium Poppy Detection Using an Updated YOLOv3 Model. *Sensors* **2019**, *19*, 4851. [CrossRef]

26. Pang, L.; Liu, H.; Chen, Y.; Miao, J. Real-time Concealed Object Detection from Passive Millimeter Wave Images Based on the YOLOv3 Algorithm. *Sensors* **2020**, *20*, 1678. [CrossRef] [PubMed]

27. Liu, J.; Wang, X. Early recognition of tomato gray leaf spot disease based on MobileNetv2-YOLOv3 model. *Plant Methods* **2020**, *16*, 1–16. [CrossRef]

28. Sunjay, S. GPR Applications in Energy Resources Exploration. In *All India Seminar on "Sources of Planet Energy, Environmental and Disaster Science: Challenges & Strategies" (SPEEDS-2013) on Sep' 07–08, 2013*; The Institution of Engineers (India) State Centre, Lucknow&School of Management Sciences, Technical Campus: Lucknow, India, 2019.

29. Ukhwah, E.N.; Yuniarno, E.M.; Suprapto, Y.K. Asphalt Pavement Pothole Detection using Deep learning method based on YOLO Neural Network. In Proceedings of the 2019 International Seminar on Intelligent Technology and Its Applications (ISITIA), Surabaya, Indonesia, 28–29 August 2019.

30. Du, Y.C.; Pan, N.; Xu, Z.; Deng, F.; Shen, Y.; Kang, H. Pavement distress detection and classification based on YOLO network. *Int. J. Pavement Eng.* **2020**, 1–14. [CrossRef]

31. Bochkovskiy, A.; Wang, C.Y.; Liao, H.Y.M. YOLOv4: Optimal Speed and Accuracy of Object Detection. *arXiv* **2020**, arXiv:2004.10934.

32. Arya, D.; Maeda, H.; Ghosh, S.K.; Toshniwal, D.; Sekimoto, Y. Global Road Damage Detection: State-of-the-art Solutions. *arXiv* **2020**, arXiv:2011,08740. Available online: https://github.com/USC-InfoLab/rddc2020 (accessed on 26 September 2020).

33. Liu, Z.; Gu, X.; Dong, Q.; Tu, S.; Li, S. 3D Visualization of Airport Pavement Quality Based on BIM and WebGL Integration. Forthcoming. *J. Transp. Eng. Part B Pavements* **2021**. [CrossRef]

34. Yao, L.; Dong, Q.; Jiang, J.W.; Ni, F.J. Deep reinforcement learning for long-term pavement maintenance planning. *Comput.-Aided Civ. Infrastruct. Eng.* **2020**, *35*, 1230–1245. [CrossRef]

35. China Highway and Transportation Society, Chts. Technical Guideline for Ground Penetrating Radar Detection for Internal Conditions of Highway Asphalt Pavement. China. 2020. Available online: http://www.chts.cn/upload/file/20201027/6373939 111170418695360217.pdf (accessed on 20 October 2020).

36. Li, S.; Gu, X.; Xu, X.; Xu, D.; Zhang, T.; Liu, Z.; Dong, Q. Detection of concealed cracks from ground penetrating radar images based on deep learning algorithm. *Constr. Build. Mater.* **2021**, *273*, 121949. [CrossRef]
37. Tzutalin. LabelImg. Git Code. 2015. Available online: https://github.com/tzutalin/labelImg (accessed on 27 October 2017).
38. YOLOv5. 2020. Available online: https://github.com/ultralytics/yolov5 (accessed on 30 May 2020).
39. Rezatofighi, H.; Tsoi, N.; Gwak, J.Y.; Sadeghian, A.; Reid, I.; Savarese, S. Generalized Intersection Over Union: A Metric and a Loss for Bounding Box Regression. In Proceedings of the 2019 IEEE/CVF Conference on Computer Vision and Pattern Recognition, Long Beach, CA, USA, 15–20 June 2019; pp. 658–666.
40. Yosinski, J.; Clune, J.; Bengio, Y.; Lipson, H. How transferable are features in deep neural networks? In Proceedings of the 27th International Conference on Neural Information Processing Systems, Montreal, QC, Canada, 8–13 December 2014.
41. Yao, L.; Dong, Q.; Ni, F.; Jiang, J.; Lu, X.; Du, Y. Effectiveness and Cost-Effectiveness Evaluation of Pavement Treatments Using Life-Cycle Cost Analysis. *J. Transp. Eng. Part B Pavements* **2019**, *145*, 04019006. [CrossRef]
42. Yu, B.; Meng, X.; Liu, Q. Multi-objective optimisation of hot in-place recycling of asphalt pavement considering environmental impact, cost and construction quality. *Int. J. Pavement Eng.* **2018**, *21*, 1576–1584. [CrossRef]
43. Yu, B.; Wang, S.; Gu, X. Estimation and uncertainty analysis of energy consumption and CO_2 emission of asphalt pavement maintenance. *J. Clean. Prod.* **2018**, *189*, 326–333. [CrossRef]
44. Lei, M. Research on Comprehensive Evaluation System of Highway Maintenance Cost Benefit. Master's Thesis, School of Transportation, Southeast University, Nanjing, China, 2019. (In Chinese).
45. Xie, S. Research on Evaluation Method of Maintenance Benefit of Highway Asphalt Pavement. Master's Thesis, School of Transportation, Southeast University, Nanjing, China, 2016. (In Chinese).

Technical Note

A GPR-Based Pavement Density Profiler: Operating Principles and Applications

Nectaria Diamanti [1,*], A. Peter Annan [2], Steven R. Jackson [2] and Dylan Klazinga [2]

[1] Department of Geophysics, Aristotle University of Thessaloniki, 541 24 Thessaloniki, Greece
[2] Sensors & Software Inc., Mississauga, ON L4W 2X8, Canada; Peter.Annan@spx.com (A.P.A.); Steven.Jackson@spx.com (S.R.J.); Dylan.Klazinga@spx.com (D.K.)
* Correspondence: ndiamant@geo.auth.gr

Abstract: Density is one of the most important parameters in the construction of asphalt mixtures and pavement engineering. When a mixture is properly designed and compacted, it will contain enough air voids to prevent plastic deformation but will have low enough air void content to prevent water ingress and moisture damage. By mapping asphalt pavement density, areas with air void content outside of the acceptable range can be identified to predict its future life and performance. We describe a new instrument, the pavement density profiler (PDP) that has evolved from many years of making measurements of asphalt pavement properties. This instrument measures the electromagnetic (EM) wave impedance to infer the asphalt pavement density (or air void content) locally and over profiles.

Keywords: air void content; asphalt pavement; compaction; density; ground penetrating radar (GPR); non-destructive testing (NDT); quality assurance/quality control (QA/QC); pavement density profiler (PDP)

Citation: Diamanti, N.; Annan, A.P.; Jackson, S.R.; Klazinga, D. A GPR-Based Pavement Density Profiler: Operating Principles and Applications. *Remote Sens.* **2021**, *13*, 2613. https://doi.org/10.3390/rs13132613

Academic Editors: Mercedes Solla, Vega Perez-Gracia and Simona Fontul

Received: 11 June 2021
Accepted: 27 June 2021
Published: 3 July 2021

Publisher's Note: MDPI stays neutral with regard to jurisdictional claims in published maps and institutional affiliations.

1. Introduction

Hot mix asphalt (HMA) pavement industry best practices state that asphalt density (or compaction) is a key indicator in the quality and longevity of pavement in new and refinished HMA driving surfaces. Previous work has shown that improper asphalt compaction during pavement construction can lead to early pavement degradation through excessive rutting, cracking, raveling, potholes, and water infiltration [1–9].

The current methodologies for assessing pavement asphalt density/compaction have a number of shortfalls [10] which include:

- Coring at several locations and conducting air void tests in the laboratory as indicated in [11] is a time consuming, costly, and destructive process;
- Desire for real-time feedback on compaction with devices that would increase productivity of the construction, facilitate shorter construction times, and reduce construction costs;
- Existing methods for density measurements such as nuclear gauges have added complexities relative to licensing, equipment handling, and storage;
- Existing methods all provide only point measurements at spatial limited pavement locations;
- Safety concerns for any operator in trafficked areas.

These shortfalls could be mitigated by new technologies integrated onto asphalt rollers incorporating real-time asphalt density/compaction measurement data. The American Association of State Highway and Transportation Officials (AASHTO) has published preliminary standard practice for using ground penetrating radar (GPR) to measure asphalt surface dielectric—a proxy for density—profiles [12].

An example of such a technology is the pavement density profiler (PDP), a new asphalt density measuring instrument recently launched by Sensors & Software Inc. [13,14].

171

The concept behind this instrument is to automatically and non-destructively determine asphalt pavement density (or void content). PDP provides either continuous density profiles as a function of the distance traversed over the pavement, as well as local density measurements at specific points. PDP can also produce area mapping coverage providing that PDP profiles are collected with controlled positioning such as geo-referenced data obtained from GPS. Continuous profile data can be acquired when PDP is vehicle mounted (Figure 1, top-right) or when the instrument is cart mounted (Figure 1, top-left). The speed of traverse is normally slow (walking speed is typical) since the device is normally used to measure at the speed of the paving train machinery. While measurements can be made at higher traverse speeds on open areas days or weeks after pavement placement, this approach defeats the ability to obtain immediate feedback to the paving operation.

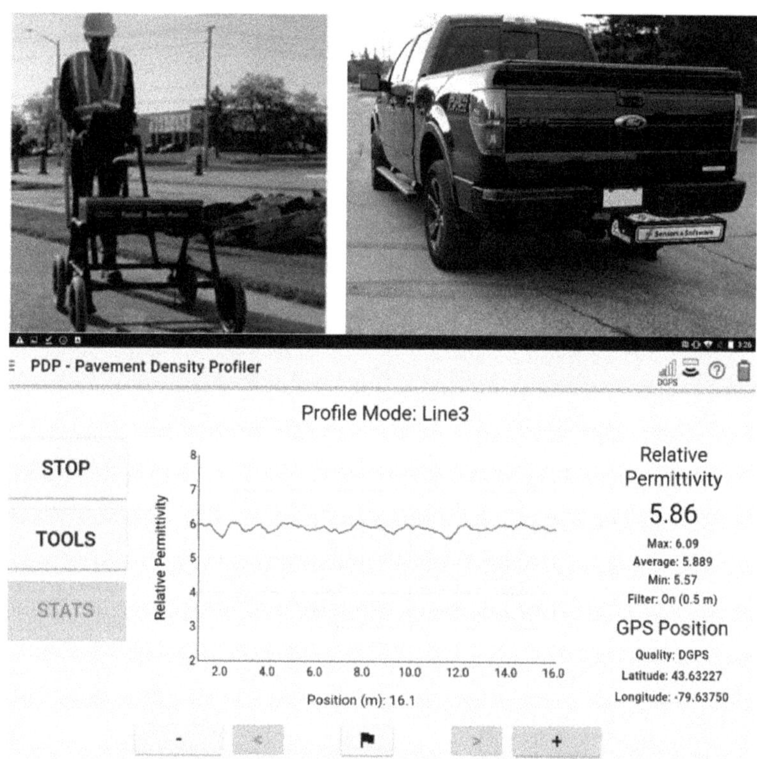

Figure 1. A cart-mounted (**top-left**) and a vehicle-based (**top-right**) PDP system. The bottom plot illustrates the PDP real-time display showing an output of the relative permittivity and its statistics.

The PDP sensor is cableless and completely self-contained including the data acquisition hardware, software, data storage, a rechargeable lithium-ion battery, built-in GPS, and Wi-Fi communications (through which the system is controlled, and data are displayed with common devices—for example, laptops, tablets, smartphones, etc.). There is also a height sensor embedded into the PDP unit which records the system height above the pavement surface. The height sensor data are logged and synchronized with PDP data and could be used in the automatic data processing procedure to compensate for height variations if significant variations occur. The device is factory calibrated and has long-term stability eliminating the need for user calibration. The system electronic components use dynamic temperature compensation over a wide range of temperatures (i.e., from -50 to $+50$ °C). Factory calibration is performed over controlled materials such as large metal sheets and thick layers of high density polyethylene (HDPE). The metal sheet is a perfect reflector, while the HDPE is effectively a half-space with relative permittivity (k_r) equal to 2.3. In many follow-up tests over HDPE, the PDP obtained an average k_r of 2.301 and a standard deviation of 0.017 using data sets with ~1000 measurements.

PDP provides immediate on-site feedback through a simple, user-friendly software interface which delivers a real-time graphical display of the selected density indicator (relative permittivity, normalized density, air-void ratio, etc.; Figure 1, bottom) while the user collects data over a profile. This methodology reduces (or eliminates) the time consuming, expensive, and most importantly invasive testing methods of coring and laboratory density measurements which are currently used for pavement quality assurance and quality control (QA/QC).

2. Asphalt Pavement Overview

Paving asphalt—bitumen or tar—is a petroleum product and it is used as a binder —called asphalt binder—to create mixtures which are then used to form driving surfaces such as roads, runways, and parking lots. The goal is to provide a smooth surface which can withstand substantial loads without deforming or fracturing. These mixtures also contain aggregate (which is typically crushed rock, clean gravel or possibly even crushed glass—the key aspect of the aggregate is that it should be impermeable and have long-term chemical stability). The asphalt binder and aggregate are combined in different proportions to create the hot mix asphalt (HMA) which is a very pliable, hot material and can be poured as a viscous fluid during construction. This hot mixture is placed into a paving machine which then extrudes the pliable material in a uniform layer on the road or area to be covered. After laying a uniform, relatively flat HMA layer, a roller or similar compacting device compresses the malleable, pliable material. As the material eventually cools, a solid impermeable surface forms—the asphalt pavement [15].

Two critical factors impacting the quality of asphalt pavements are surface smoothness and life expectancy. A smooth surface generally produces low vibration noise from passing vehicles and decreases the amount of hammering and impacting on the surface when high-speed vehicles move over undulations. Life expectancy is an important factor of pavement engineering as the fewer times paving materials must be replaced, the better. Construction design and material stability are the two controlling factors of asphalt pavement life expectancy. At the construction stage, the asphalt mix must be properly selected for the expected environment and loading conditions and then, properly installed. The material stability for the long-term variation depends on the construction practice as well as the material itself. The asphalt mixture is usually comprised by some volatile materials that burn off with time and exposure to weathering conditions. Moreover, brittleness or friability of the material heavily depends on how well the asphalt mix has been compacted during construction. Compaction depends on the mixture temperature (which controls malleability) and the compaction method. When the HMA is poorly (under) compacted, it becomes friable and crumbles and also, there is the risk of water intrusion. When it is over compacted, it becomes brittle and will crack under heavy loading. Over-compaction can also lead to early pavement degradation through excessive rutting. There is an optimal level of compaction which is expected to provide the best asphalt pavement life expectancy.

3. PDP Operating Principles and Technical Specifications

PDP belongs to the group of indirect methods for measuring density, as opposed to the direct methods that require acquiring material samples from the road surface via physical coring. The main challenges when employing the direct methods for measuring pavement density are that: (a) They are destructive, (b) the samples are of limited number and come from specific pavement locations, and (c) performing the final density measurement at the lab is a time-consuming, expensive process that does not provide real-time results. The indirect methods for measuring density, often referred to as non-destructive testing (NDT) methods, measure a material property that is related to the density. The goal is a methodology that does not damage the road structure, can generate density estimates over points and/or continuous stretches of pavement, if needed, and immediately provide readings on site. PDP, being part of the indirect methods group, does not measure density directly. It measures the electromagnetic (EM) wave impedance which is closely correlated with density (since the EM reflectivity of an asphalt pavement road surface is an indicator of the material density). The surface reflectivity method used for deriving a physical material property from the electrical properties measured with GPR is widely used for other applications such as for soil water content estimation [16–20].

PDP is an air-launched GPR, which follows the established principles of operation for GPR [21]. The PDP differs from a traditional GPR in that the system does not display GPR raw data in an image form but processes the data to obtain a real-time measure of surface reflectivity which is displayed as an apparent permittivity value. The PDP is designed to be a fit for purpose system that automatically provides a real-time graphical display of the selected density indicator while the user collects data. The various PDP output display options are discussed in the next section.

The instrument is placed at a height above the road surface and the amplitude and two-way travel time of the signal reflected off that surface are recorded, as shown in Figure 2. PDP operates at a nominal height of ~0.5 m so that the travel time to the pavement surface and back allows for the direct and reflected signals to be clearly separated in time. Its maximum height limit is ~1.0 m as dictated by emissions standards and regulations [22,23]. PDP data positions are normally acquired at equal spatial intervals, using an odometer, which are georeferenced using either the internal unit GPS or a high accuracy external GPS.

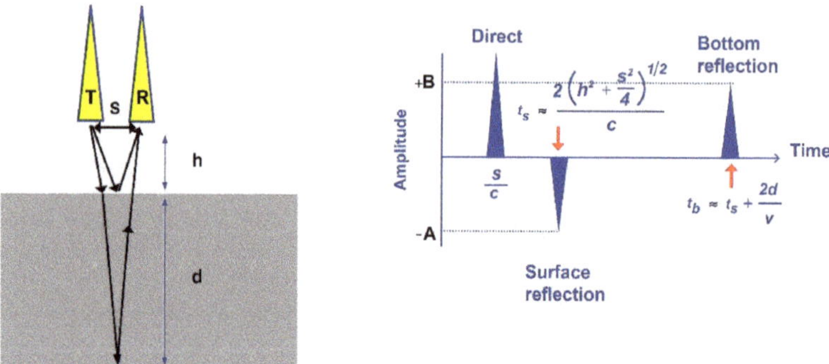

Figure 2. PDP instrument background principle of operation. It is an air-launched GPR that transmits through a transmitter (T) a radio wave pulse, which is reflected from the air/ground interface. The receiver (R) records the transient signal, and the surface reflection event (annotated on the right part of the figure) is identified and automatically processed to estimate the asphalt pavement dielectric permittivity, k_r. The GPR wave velocity in the subsurface is annotated with v while the free space propagation velocity, c, is equal to 0.3 m/ns.

The normal incident reflection coefficient R of an EM wave at the air/ground interface is described by Equation (1) and it is determined by the contrast in EM impedance Z (and hence, in k_r) between the air and the ground:

$$R = \frac{Z - Z_0}{Z + Z_0} = \frac{1 - \sqrt{k_r}}{1 + \sqrt{k_r}},\tag{1}$$

where Z_0 is the EM impedance of air, $Z = Z_0/\sqrt{k_r}$ is the impedance of the ground (pavement) material, and k_r is the dielectric permittivity of the ground material. It must be noted that R is always less than zero and that for this expression to be strictly valid the following assumptions are made: (a) Ground conductivity should be sufficiently small to be ignored; (b) surface should be planar and smooth; (c) subsurface should be homogeneous; and (d) EM waves should be vertically incident on the air/ground interface.

The amplitude, A_r, of the reflected wavelet from the ground surface is dependent on the magnitude of R. By comparing the reflected amplitude with the amplitude A_m of a wavelet measured at the same elevation over a metal plate target, A_m (which theoretically has an R equal to –1), we can calculate the dielectric permittivity (or else, relative permittivity or dielectric constant) of the material, k_r, using Equation (2):

$$k_r = \left(\frac{1-R}{1+R}\right)^2 = \left(\frac{1+\frac{A_r}{A_m}}{1-\frac{A_r}{A_m}}\right)^2, \tag{2}$$

For the k_r (or else, k_{mix} as it will be referred to later) computation, this relationship assumes normal incident signals but can be modified if non-normal incidence is a factor in system design.

A common question arising with air-launched GPR systems is the sensing depth of the surface reflection signal. We have employed numerical modelling to produce an estimate of the effective depth of investigation (or else, sampling depth) using an air-launched GPR system, which was then confirmed with field tests with the PDP [24]. For the k_r estimation, our data analysis showed that with appropriate signal time gating we could obtain k_r values that are closer to the top layer relative permittivity and sampling depth has its minimum value in this case. The parts of the signals arriving later in time, carry information from deeper layers and hence, the depth of investigation increases. For a physical model that resembles PDP, sampling depth varies from ~20 to 80 mm and depends on the signal analysis approach that has been employed. It should be noted that the analysis method can be altered depending on the application. Various methods are used for the amplitude quantification such as the peak signal or the RMS signal amplitude over a time window.

For PDP's default configuration, the volume it senses is ~300 mm in diameter on the pavement surface (this is a product of the antenna beam and the zone of influence at its default operational height which is equal to 0.5 m), and ~80 mm in exploration/sampling depth. The standard spatial sampling interval is 100 mm, meaning that for every PDP footprint on the road surface (i.e., 300 mm), three samples are acquired. Consistent spatial sampling is controlled by a calibrated odometer (sometimes referred to as distance measurement indicator (DMI)). This way, footprints overlap and adequate sampling occurs so as there are no gaps in the measured data. The PDP frequency band and height were selected such that the antenna beam and the Fresnel zone or zone of influence (~300 mm in diameter) would be similar to the minimum size of pavement heterogeneity that seems to be of concern during paving jobs. All the above parameters are optimized for the instrument's normal deployment at approximately walking speed (~1 m/s) for this specific application. However, the instrument's core elements are extremely flexible and many are software controlled, meaning the system could be customized for other applications (such as asphalt thickness, road-base evaluation, etc.). The goal in the PDP development was to make a specific fit for purpose device. Other general-purpose GPR products already exist and are available to address other applications.

As PDP is an air-launched GPR system, surface roughness can be an important factor affecting the surface reflection amplitude data. When new pavement surfaces are rolled, their surface is considered flat (surface roughness is in the order of a few millimeters) and the PDP measurements which are in the ~1.0 to 2.0 GHz range are not greatly affected. When the asphalt pavement surface is roughly milled or pitted with age, roughness in the 10-to-20-millimeter scale is observed to affect the measured response and reflection amplitudes are substantially reduced due to the energy being scattered [16,17].

Water has a major impact on the permittivity of materials and surface water can be present during paving. Some compactors spray water on the asphalt pavement surface to avoid asphalt particles sticking to the roller. The presence of a thin layer of water on the asphalt pavement surface can affect the k_r estimation and hence, the derived density value [25,26]. We have performed modelling to look at the magnitude of the water effect and the results show that the impact of surface water is very substantial (i.e., even 1 mm thick free water layer on the asphalt pavement surface alters the measured k_r value more than 150%). Therefore, PDP surveys should be performed when the pavement surface is visually dry. Fortunately, asphalt is emplaced at high temperatures and surface water from compaction evaporates quickly.

4. Density Derivation from PDP Responses

At this point, it is important to stress the need to separate (a) instrument stability in providing correct dielectric permittivity values from (b) the role of the "interpretation model" that transforms k_r values into density (or air void content). Regarding (a), if the individual instrument being used suffers from calibration errors or stability issues, inconsistencies will make their way into the dielectric permittivity values being measured and hence, the permittivity to density transformation process becomes instrument dependent. PDP is a factory calibrated instrument designed to make the permittivity estimate instrument independent. When it comes to (b), the permittivity to density transformation depends on the interpretation model being used. Not all asphalt mixes are the same and the relationship is to some degree mix dependent [27,28]. It is important to understand the need to avoid merging instrument errors and biases with the interpretation model differences within the permittivity to density translation and to be careful not to mix these two.

Regarding the interpretation model, there has been substantial previous work showing that the electrical properties of HMA (i.e., its electrical relative permittivity, k_r) are closely related to its density, ρ or similarly, to its air void content V_{air} [25–36]. This conversion is referred to as the "k_r to ρ" transformation and there are two main approaches to it. The first approach is using empirical relationships to connect ρ and k_r (i.e., these relationships are derived from the correlation of permittivity and core density data obtained from various pavement locations and types of asphalt pavements). The second approach employs EM mixing theory according to which there is a relationship between the dielectric permittivity of a mixture and the homogeneous dielectric and the volumetric proportions of its components.

The relative permittivity to density conversion is a subject that is constantly being assessed. To demonstrate the concept of the "k_r to ρ" conversion, we use a simple model which bases the computation on a geometrical mean of the components, historically known as Lichtenecker's formula [37]. The form for predicting permittivity k_r for a known density ρ is expressed as

$$k_r = \beta^\rho \tag{3}$$

Alternately, density can be expressed as

$$\rho = \frac{\ln k_r}{\ln \beta} \tag{4}$$

From the assessment of geologic materials [38], β is found to be in the range of 1.9 to 2.2. Since an asphalt mix is primarily aggregate, this model is useful for asphalt pavement. To provide some flexibility for accommodating the effects of mixtures, the value of β can be adjusted.

While such a relationship is helpful for qualitative analysis, details on a specific asphalt mix are needed to determine the maximum density, ρ_{max}, when no air is present in the fully compacted material. This requires a controlled sample of material and a true density measurement. If the fully compacted pavement density is known, then, the relative (or normalized) density, ρ_n, and air void ratio, V_a, can be computed and displayed using the following simple relationships:

$$\rho_n = \frac{\rho}{\rho_{max}} \tag{5}$$

$$V_a = 1 - \rho_n \tag{6}$$

where ρ_{max} is reduced to a value equal to ρ when there is air in the mix. Note that density prediction errors are shown to be higher when the air void content in the asphalt mix is more than ~11% and also, the density prediction accuracy greatly depends on the mix type being used (i.e., binder mix, aggregate properties, surface layer type, etc.) [27,28].

Many other alternatives for the "k_r to ρ" conversion can be employed if desired [25–36]. The complex refractive index model (CRIM) formula [39] is one of the more mainstream approaches for estimating the bulk relative permittivity of heterogeneous materials but requires more in-depth knowledge of the individual material properties. The references provided indicate the various methodologies.

More specifically, PDP can provide a display of one of the following five outputs on the fly, while the user is collecting data. The available displays and calculations described below can be found in [40], which describes the PDP toolkit, a dedicated software that allows users to import and reprocess PDP data. However, we would like to stress the ability of the instrument to provide real time feedback:

1. Relative dielectric permittivity (k_r): This is the initial value calculated by PDP. k_r is expressed as a unitless quantity relative to the permittivity of free space. All other parameters below are derived from the relative permittivity.
2. Density (ρ): This is a display for an absolute density of the asphalt, expressed in units of g/cm^3, calculated from the observed relative permittivity. If the user has a core sample with a known density value, a density offset can be applied, such that the measured PDP parameter at the core location equals the known density of the core. This offset is then applied to all the PDP data.
3. Density (ρ)—site specific: Measurements of the asphalt properties at the survey site are used to create a unique, site-specific means of translating relative permittivity to density. When the information is available (either from direct density measurements done on cores or from indirect measurements such as using nuclear density gauges), other parameters such as relative density can be displayed in addition to the absolute density of the asphalt, expressed in units of g/cm^3. This is a more complex calculation that relies on inputting the coefficients of a parametric relationship as well as the maximum density (ρ_{max}). These values can be obtained from a core sample. While more complex, this is a more accurate representation of the true density at the survey site.
4. Relative density (ρ_n): As stated already, this quantity is sometimes called normalized density or percentage compaction. This output expresses the density measured as a percentage of the site-specific maximum density. To calculate ρ_n, the user must provide ρ_{max}. This is usually obtained from a core sample via a testing lab (note that: $\rho_n = 1 - V_a$).
5. Air void content (V_a): It is expressed as a percentage of how much of the volume of the asphalt is air. This also requires inputting ρ_{max} (note that: $V_a = 1 - \rho_n$).

Regardless of the specifics defined above, our intent here is to show the step of transitioning from the electrical property measured to the inference of density. As this subject is an area of advancement, there is no "right" or unique transition. Practical field methodologies are normally used to get to a desired result and different groups use differing end results for their asphalt compaction quality metric.

5. Field Examples

5.1. Repeatability Demonstration

The first example is shown to demonstrate the repeatability and reliability of PDP data. Testing has been performed in Minnesota, USA, on a road surface that had been shaved and a single lift of about 25 mm of asphalt had been placed on the milled surface. The testing location was a recently resurfaced road. After the old asphalt was shaved, the road was covered with a single layer of new asphalt that was laid in two strips with a longitudinal joint near the center of the lane. Five PDP repeat profiles were collected on a ~90 m long test line parallel to the road centerline. The PDP device was placed on a cart and data were collected with an odometer triggering to acquire data at ~10 cm intervals. The PDP repeat data for the 90 m long line for all the five passes are shown in Figure 3 (top) in the form of the relative permittivity versus position. In addition, the bottom part of Figure 3 illustrates the average k_r value of the five passes at each position along the line, plotted together with their $\pm$ standard deviation at each location. All the profiles show strong repeatability. Minor positioning errors might occur due to the odometer or small variations in the PDP lateral position from pass to pass. However, the overall location of high and low k_r values is remarkably good.

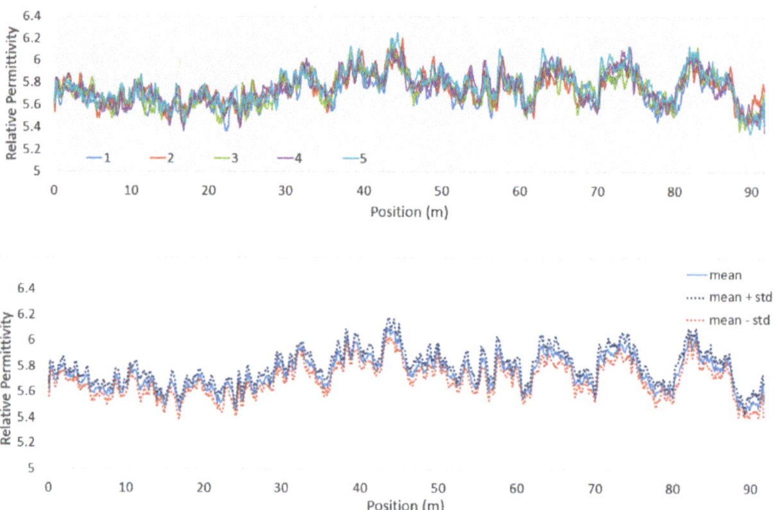

Figure 3. Plot of the k_r values along a 90 m test line with five repeat passes shown to illustrate the PDP data repeatability (top), where numbers 1 to 5 indicate the number of repeats of the test. Plot of the mean k_r value at each position along the line, plotted together with the $\pm$ standard deviation (bottom). Relative permittivity values range from about 5.40 to 6.20.

To enable the comparison of nuclear density and k_r values, five locations were identified along the ~90 m long repeatability line as high and low relative permittivity spots and were tested with a nuclear density gauge (details on this test can be found in [14]). Figure 4 presents the correlation of the density with the relative permittivity data obtained from PDP.

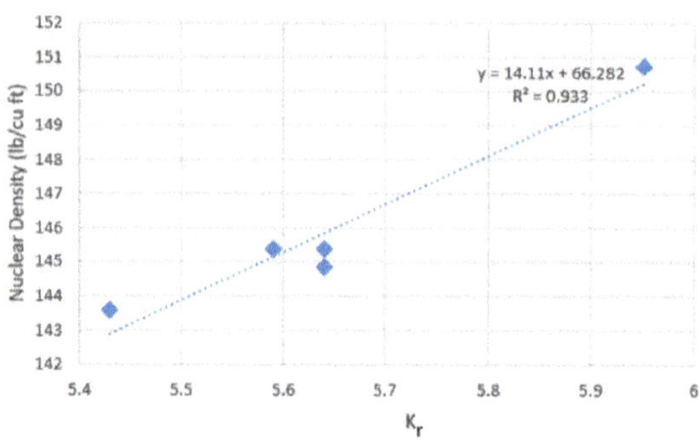

Figure 4. Correlation between nuclear density and PDP k_r data [14]. A linear model is fitted to the data.

In HMA pavements, a weak spot appears to be the joint between adjacent mats. At the test site, the centerline joint was clearly visible, as shown in Figure 5 (left). A cross-lane test was set up and traverse lines to the pavement's long dimension were collected with PDP and repeated five times. The transect crossed the joint at ~4 m, as it can be seen also from the PDP cross-lane data which are presented in the form of k_r versus position (Figure 5, right). The centerline joint appears as a low dielectric permittivity area. With regards to the data consistency, PDP data are again shown to be repeatable and the overall system's performance is reliable.

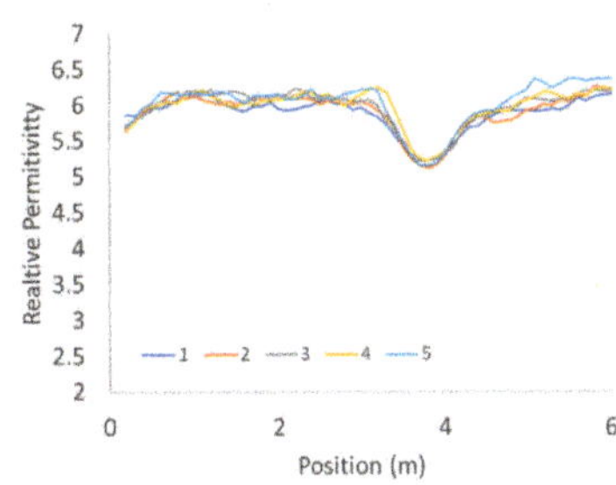

Figure 5. View of the new pavement (**left**) where the centerline joint response is clearly visible at the ~4 m position during a cross-lane test (**right**). Numbers 1 to 5 in the right plot indicate the number of repeats of the cross-lane test.

5.2. Full Lane Plan Map

PDP was employed in a pavement quality control project where data were collected during an active paving job in Brampton, ON, Canada (Figure 6, left). The roadway was a two-lane road where the asphalt pavement was laid in three lifts. The primary goal of the survey was to map the full lane width to obtain a good indication of k_r (or density) variability. A grid was established to survey both lanes of the road, consisting of lines perpendicular to the road alignment. For the example shown here, only a part of the grid is presented (i.e., an area of ~7 by 8 m^2) to show the variability more clearly. Nine parallel lines were surveyed at 1 m spacing. In order to provide a more comprehensive view, the multiple line data were merged to form a color map to show the measurement variability over the area covered. The color map of the relative permittivity is shown in Figure 6 (right). The key observation is the low *kr* values near the center of the x–axis of the grid that goes all the way along the *y*–axis. This low *kr* (hence, low ρ) area is due to the joint between the two lanes present at the middle of the pavement.

5.3. Various Output Displays

As discussed earlier, the translation from relative permittivity values to density values is a topic that still requires further research. In most cases, the translation methodology occurs by correlating core density data to k_r data. Various groups have a preference on the output they would like to see which could be any amongst density, relative or normalized density, air void ratio, etc. Figure 7 displays various output formats of PDP data that were collected along an asphalt bicycle path that was repaved. To obtain density values from the k_r data, we used Lichtenecker's formula [37]. The air void content values are higher than the values encountered in most asphalt pavements. This is reasonable due to the fact that bicycle paths do not carry as heavy loads as pavements do, so, they are not compacted as much.

Since PDP—or any other air-launched GPR system—measures relative permittivity and also, given that the k_r transformation to density is material specific in most cases, displaying the k_r variations is often the most practical display option for evaluating the asphalt pavement placement uniformity in the field, rather than absolute density values.

Each user group has specific design specifications and needs to develop their specific data presentation form that meets their workflow and validation requirements.

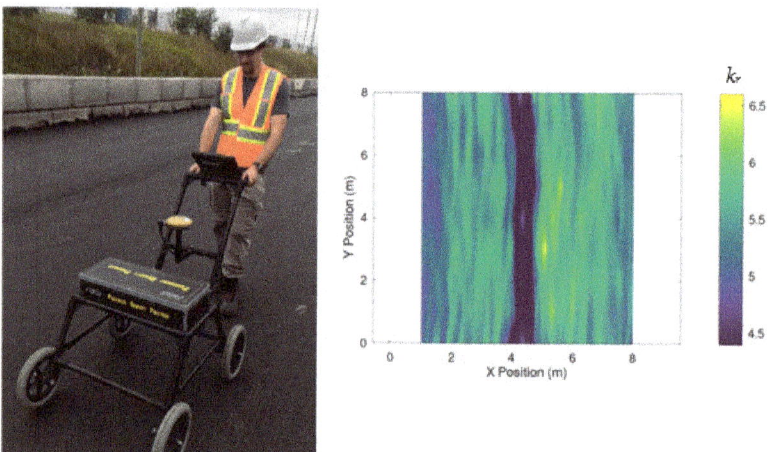

Figure 6. A cart-based PDP (**left**) and a plan map of k_r variation in the grid area (right). The darker areas are of low dielectric permittivity and are most prevalent along the joint between the two lanes (**right**).

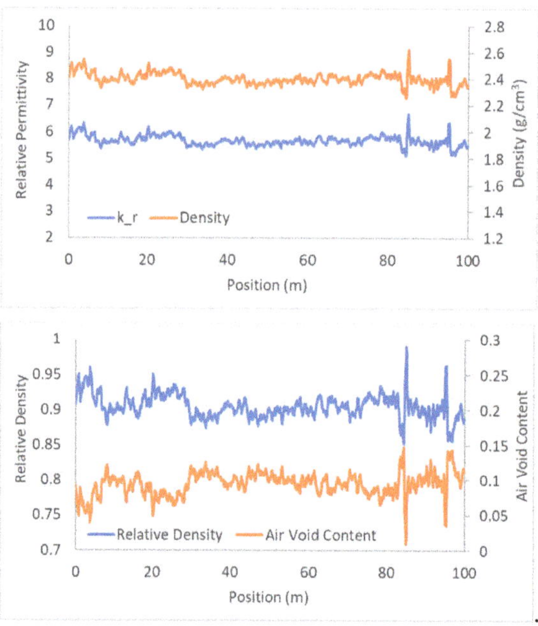

Figure 7. PDP various display output plots with distance traversed along an asphalt bicycle path.

6. Discussion and Conclusions

The asphalt pavement density is an important factor in indicating if the asphalt mix has been rolled to achieve the pavement design specifications. The goal of the PDP device is to accurately determine permittivity from the surface reflection and provide translation tools to estimate asphalt pavement density or air void content from the measured permittivity.

Extensive field testing of this application to date indicates that the permittivity to density translation is site specific (which is relative to the asphalt mix used and how it is installed).

It is important to separate instrument calibration and stability from the derived density values. PDP produces repeatable, stable results when it comes to determining the dielectric permittivity of the pavement. The accuracy of the absolute density value strongly depends on the interpretation model being used for the "k_r to ρ transformation". As a result, using relative density variability rather than density absolute values is more practical. Given that determining absolute density may be challenging, relative variations in either density or permittivity provide a powerful means for assessing the uniformity of compaction at a site. The natural conclusion is using the permittivity values directly and a physical sample (i.e., core and measure density) in anomalous areas as a most efficient way of assessing pavement placement consistency.

The interpretation model that helps derive density values from dielectric permittivity data is typically simplified, as most interpretation models to date have been focused on estimating the asphalt permittivity assuming a uniform half-space. More advanced interpretation approaches are available [41] that can handle layered and/or graded media using full wave inversion resulting in a suitably weighted average permittivity.

Author Contributions: Conceptualization, N.D., A.P.A., S.R.J. and D.K.; methodology, N.D., A.P.A., S.R.J. and D.K.; validation, N.D., A.P.A., S.R.J. and D.K.; formal analysis, N.D., A.P.A., S.R.J. and D.K.; investigation, N.D., A.P.A., S.R.J. and D.K.; writing—original draft preparation, N.D.; writing—review and editing, N.D., A.P.A., S.R.J. and D.K.; supervision, A.P.A.; project administration, A.P.A. All authors have read and agreed to the published version of the manuscript.

Funding: This research received no external funding.

Institutional Review Board Statement: Not applicable.

Informed Consent Statement: Not applicable.

Data Availability Statement: The data used in this research are available from the authors upon reasonable requests.

Conflicts of Interest: The authors declare no conflict of interest.

References

1. Zube, E. Compaction Studies of Asphalt Concrete Pavement as Related to the Water Permeability Test. In Proceedings of the 41st Annual Meeting of the Highway Research Board, Washington, DC, USA, 8–12 January 1962.
2. Santucci, L.E.; Allen, D.D.; Coats, R.L. The Effects of Moisture and Compaction on the Quality of Asphalt Pavements. *Assoc. Asph. Paving Technol.* **1985**, *54*, 168–208.
3. Ford, M.C. Pavement Densification Related to Asphalt Mix Characteristics. In Proceedings of the 1988 Annual Meeting of the Transportation Research Board, Washington, DC, USA, January 1988.
4. Brown, E.R.; Cross, S. A Study of In-Place Rutting of Asphalt Pavements. In Proceedings of the 1989 Annual Meeting of the Association of Asphalt Paving Technologist, Nashville, TN, USA, 22–26 February 1989.
5. Brown, E.R.; Collins, R.; Brownfield, J.R. Investigation of Segregation of Asphalt Mixtures in State of Georgia. In Proceedings of the 68th Annual Meeting of the Transportation Research Board, Washington, DC, USA, 22–26 January 1989.
6. Brown, E.R. Density of asphalt concrete: How much is needed? *Transp. Res. Rec. J. Transp. Res. Board* **1990**, *1282*, 27–32.
7. Huber, G.A.; Heiman, G.H. Effect of Asphalt Concrete Parameters on Rutting Performance: A Field Investigation. In Proceedings of the Association of Asphalt Paving Technologists 56, Reno, NV, USA, February 1987; pp. 33–61.
8. Roberts, F.L.; Kandhal, P.S.; Brown, E.R.; Lee, D.Y.; Kennedy, T.W. *Hot Mix Asphalt Materials, Mixture Design, and Construction*; National Asphalt Paving Association Education Foundation: Lanham, MD, USA, 1996.
9. Killingsworth, B.M. *Quality Characteristics for Use with Performance Related Specifications for Hot Mix Asphalt*; NCHRP Project 9-15 Research Results Digest 291; Transportation Research Board, National Research Council: Washington, DC, USA, 2004.
10. Commuri, S. *Intelligent Asphalt Compaction Analyzer*; Final Report FHWA-HIF-12-019; University of Oklahoma: Norman, OK, USA, 2011.
11. AASHTO T 166. *Standard Method of Test for Bulk Specific Gravity (G mb) of Compacted Asphalt Mixtures Using Saturated Surface-Dry Specimens*; WSDOT Materials Manual M 46-01.37; American Association of State Highway and Transportation Officials (AASHTO): Washington, DC, USA, 2020.

12. AASHTO. *Asphalt Surface Dielectric Profiling System Using Ground Penetrating Radar*; American Association of State Highway and Transportation Officials (AASHTO) Standard Practice; Designation: PP 98-20[1]; AASHTO: Washington, DC, USA, 2020; pp. 98.1–98.9.
13. Diamanti, N.; Redman, J.D.; Annan, A.P. A GPR-based Sensor to Measure Asphalt Pavement Density. In Proceedings of the 17th International Conference on Ground Penetrating Radar, Rapperswil, Switzerland, 18–21 June 2018.
14. Diamanti, N.; Annan, A.P.; Jackson, S.R. Measuring Asphalt Pavement Density with a GPR-based Sensor: A Case Study. In Proceedings of the 10th International Workshop on Advanced Ground Penetrating Radar, The Hague, The Netherlands, 8–12 September 2019.
15. Mallick, R.B.; El-Korchi, T. *Pavement Engineering: Principles and Practice*; CRC Press: Boca Raton, FL, USA, 2009.
16. Chanzy, A.; Tarussov, A.; Bonn, F.; Judge, A. Soil Water Content Determination Using a Digital GPR. *Soil Sci. Soc. Am. J.* **1996**, *60*, 1318–1326. [CrossRef]
17. Redman, J.D.; Davis, J.L.; Galagedara, L.W.; Parkin, G.W. Field Studies of GPR Air Launched Surface Reflectivity Measurements of Soil Water Content. In Proceedings of the 9th International Conference on Ground Penetrating Radar, Santa Barbara, CA, USA, 29 April–2 May 2002; pp. 156–161.
18. Redman, J.D.; Galagedara, L.; Parkin, G. Measuring soil water content with the ground penetrating radar surface reflectivity method: Effects of spatial variability. In Proceedings of the ASAE Annual International Meeting, Las Vegas, NV, USA, 27–30 July 2003. Paper Number: 032279.
19. Serbin, G.; Or, D. Near-surface water content measurements using horn antenna radar: Methodology and overview. *Vadose Zone J.* **2003**, *2*, 500–510. [CrossRef]
20. Huisman, J.A.; Hubbard, S.S.; Redman, J.D.; Annan, A.P. Measuring soil water content with ground penetrating radar. *Vadose Zone J.* **2003**, *2*, 476–491. [CrossRef]
21. Annan, A.P. Ground-Penetrating Radar. In *Near-Surface Geophysics*; Butler, D.K., Ed.; Society of Exploration Geophysicists (SEG): Tulsa, OK, USA, 2005; No. 13, Chapter 11; pp. 357–438.
22. FCC. *USA Regulations on UWB GPR*; Federal Communications Commission (FCC) Regulations, Part 15.503; FCC: Washington, DC, USA, 2003.
23. RSS-220 Industry Canada. *Devices Using UWB Technology*; Radio Standards Specification RSS-220; Industry Canada, 2009; Issue 1; p. 19.
24. Diamanti, N.; Redman, J.D.; Hogan, C.M.; Annan, A.P. Air-launched GPR depth of investigation. In Proceedings of the 18th International Conference on Ground Penetrating Radar, SEG, Golden, CO, USA, 14–19 June 2020.
25. Shangguan, P.; Al-Qadi, I.; Lahouar, S. Pattern recognition algorithms for density estimation of asphalt pavement during compaction: A simulation study. *J. Appl. Geophys.* **2014**, *107*, 8–15. [CrossRef]
26. Shangguan, P.; Al-Qadi, I.L. Calibration of FDTD Simulation of GPR Signal for Asphalt Pavement Compaction Monitoring. *IEEE Trans. Geosci. Remote Sens.* **2015**, *53*, 1538–1548. [CrossRef]
27. Leng, Z.; Al-Qadi, I.; Lahouar, S. Development and validation for insitu asphalt mixture density prediction models. *NDT & E Int.* **2011**, *44*, 369–375.
28. Wang, S.; Al-Qadi, I.L.; Cao, Q. Factors impacting monitoring asphalt pavement density by ground penetrating radar. *NDT & E Int.* **2020**, *115*, 102296.
29. Saarenketo, T.; Roimela, P. Ground Penetrating Radar Technique in Asphalt Pavement Density Quality Control. In Proceedings of the 7th International Conference on Ground Penetrating Radar, Lawrence, KS, USA, 27–30 May 1998; pp. 461–466.
30. Saarenketo, T.; Scullion, T. Road evaluation with ground penetrating radar. *J. Appl. Geophys.* **2000**, *43*, 119–138. [CrossRef]
31. Maser, K. *Infrasense—Development of a Pavement Thickness Density Meter*; NCHRP-61 Final Report; Infrasense Inc.: Arlington, MA, USA, 2002.
32. Sebesta, S.; Wang, F.; Scullion, T.; Liu, W. *New Infrared and Radar Systems for Detecting Segregation in Hot-Mix Asphalt Construction*; Report: FHWA/TX-05/0-4577-2; Texas Transportation Institute, The Texas A&M University System: Austin, TX, USA, 2002.
33. Saarenketo, T. Electrical Properties of Road Materials and Subgrade Soils and the Use of Ground Penetrating Radar in Traffic Infrastructure Surveys. Ph.D. Thesis, University of Oulu, Oulu, Finland, 2006.
34. Popik, M.; Lee, H.; Aho, B.; Maser, K.; Holzschuher, C. Using ground penetrating radar for evaluation of asphalt density measurements. In Proceedings of the 89th Annual Meeting of the Transportation Research Board, Washington, DC, USA; 2010.
35. Al-Qadi, I.; Leng, Z.; Larkin, A. In-place hot mix asphalt density estimation using ground penetrating radar. In *Nondestructive Testing and Evaluation (NDTE) Technologies for Airport Pavement Acceptance and Quality Assurance Activities*; ICT Report No. 11-096; University of Illinois: Champaign, IL, USA, 2011.
36. Plati, C.; Loizos, A. Estimation of in-situ density and moisture content in HMA pavements based on GPR trace reflection amplitude using different frequencies. *J. Appl. Geophys.* **2013**, *97*, 3–10. [CrossRef]
37. Lichtenecker, K. Die Dielektrizitätskonstante natürlicher und künstlicher Mischkörper. *Phys. Zeitschr.* **1926**, *XXVII*, 115–158.
38. Olhoeft, G.R. Electrical properties of rocks. In *Physical Properties of Rocks and Minerals*; Touloukian, Y.S., Judd, W.R., Roy, R.F., Eds.; McGraw-Hill: New York, NY, USA, 1981; pp. 257–330.
39. Birchak, J.R.; Gardner, C.G.; Hipp, J.E.; Victor, J.M. High dielectric constant microwave probes for sensing soil moisture. *Proc. IEEE* **1974**, *62*, 93–98. [CrossRef]

40. Sensors & Software Inc. *PDP Toolkit User's Guide. v. 2020-00058-00*; Sensors & Software Inc.: Mississauga, ON, Canada, 2020.
41. Kalogeropoulos, A.; van der Kruk, J.; Hugenschmidt, J.; Bikowski, J.; Bruhwiler, E. Full-waveform GPR inversion to assess chloride gradients in concrete. *NDT & E Int.* **2013**, *57*, 74–84.

Technical Note

Displacement Monitoring in Airport Runways by Persistent Scatterers SAR Interferometry

Luca Bianchini Ciampoli [1,*], Valerio Gagliardi [1], Chiara Ferrante [1], Alessandro Calvi [1], Fabrizio D'Amico [1] and Fabio Tosti [2]

[1] Department of Engineering, Roma Tre University, Via Vito Volterra 62, 00146 Rome, Italy; valerio.gagliardi@uniroma3.it (V.G.); chiara.ferrante@uniroma3.it (C.F.); alessandro.calvi@uniroma3.it (A.C.); fabrizio.damico@uniroma3.it (F.D.)

[2] School of Computing and Engineering, University of West London, St. Mary's Road, London W5 5RF, UK; fabio.tosti@uwl.ac.uk

* Correspondence: luca.bianchiniciampoli@uniroma3.it; Tel.: +39-06-5733-3617

Received: 31 August 2020; Accepted: 26 October 2020; Published: 30 October 2020

Abstract: Deformations monitoring in airport runways and the surrounding areas is crucial, especially in cases of low-bearing capacity subgrades, such as the clayey subgrade soils. An effective monitoring of the infrastructure asset allows to secure the highest necessary standards in terms of the operational and safety requirements. Amongst the emerging remote sensing techniques for transport infrastructures monitoring, the Persistent Scatterers Interferometry (PSI) technique has proven effective for the evaluation of the ground deformations. However, its use for certain demanding applications, such as the assessment of millimetric differential deformations in airport runways, is still considered as an open issue for future developments. In this study, a time-series analysis of COSMO–SkyMed satellite images acquired from January 2015 to April 2019 is carried out by employing the PSI technique. The aim is to retrieve the mean deformation velocity and time series of the surface deformations occurring in airport runways. The technique is applied to Runway 3 at the "Leonardo da Vinci" International Airport in Rome, Italy. The proposed PSI technique is then validated by way of comparison with the deformation outcomes obtained on the runway by traditional topographic levelling over the same time span. The results of this study clearly demonstrate the efficiency and the accuracy of the applied PSI technique for the assessment of deformations in airport runways.

Keywords: interferometric synthetic aperture radar; InSAR; permanent scatterers; PS-InSAR; transport infrastructure maintenance; airport runway; airport monitoring

---

1. Introduction

Runways are central elements in airport infrastructures, as they are mostly dedicated to the two fundamental and most critical manoeuvres of taking-off and landing of the aircrafts. In light of this, runways must comply with very strict requirements in terms of the construction [1] and the maintenance standards [2]. In this regard, it is worthy of mention that a continuously regular surface must be ensured in runways over their entire life cycle, and no superficial (e.g., cracking or rutting) or deep (e.g., subgrade subsidence) damage is acceptable to compromise safe manoeuvring of the aircrafts.

Although the provision of proper design methods for runways is crucial to minimise future maintenance and rehabilitation of the infrastructure, this can be compromised by other critical factors, such as construction site related issues (e.g., geotechnical instability of the subgrade) or the non-linear action of different heavy loads over the infrastructure life cycle. The occurrence of these events requires a dedicated and deep monitoring of differential settlements and deformations of the surface, in order to plan targeted interventions on time [3,4] and comply with budget constraints [5].

Within this context, it is important to mention that innovative inspection techniques are in demand by infrastructure agencies and administrators to be incorporated into more effective infrastructure management systems. Scope is to enhance the reliability of the decay prediction models by increasing the productivity of the system at reduced inspection costs. Nowadays, various methodologies have been reported as effective in the detection and monitoring of deformations in runways. Among these, the most acknowledged techniques are the point-based geodetic methods, such as the topographic levelling [6], often coupled to Global Navigation Satellite Systems (GNSS) for referencing purposes [7], or the LiDAR technique [8]. These methods work with different resolutions, and they all rely on the comparison between the positions of points measured across successive surveys. This approach is necessary to allow the monitoring of the deformation trend of the targets.

More recently, satellite-based techniques are gaining momentum for the monitoring of transport infrastructures. One of their major advantages is on the fact that measurements of targets can be repeated over an investigated area, with a fixed revisiting time that is related to the orbit of the satellite. This allows to collect a significant amount of information regularly distributed in time, without the requirement to close the infrastructure to traffic. Hence, tremendous benefits can be brought forward in terms of safety and costs in the asset management process [9]. Among the satellite remote sensing techniques, the Synthetic Aperture Radar Interferometry (InSAR) has proven effective in the analysis of subsidence effects in a wide range of structures [10] and infrastructures [11].

This study focuses on the application of the InSAR technology and, more specifically, of the Persistent Scatters Interferometry (PSI) [12,13] to the monitoring of differential settlements in airport runways. To demonstrate the viability of the proposed technique, a case study is presented with an application to Runway 3 at the "Leonardo da Vinci" International Airport in Rome, Italy. The investigated area is known to be affected by subsidence due to soft subgrades. Results of the InSAR acquisitions and processing are finally compared to the deformations obtained on the runway by traditional topographic levelling over the same time span.

2. Aim and Objectives

The main aim of the investigation reported in this paper is to verify the effectiveness of the satellite remote sensing technology in gaining vital information about the functionality of airport runways for inclusion in airport asset management systems. To achieve this aim, the following objectives are identified:

- to assess runway displacements at the millimetre scale and evaluate their trend on a multi-year scale using the PSI monitoring technique;
- to compare the results obtained with the PSI technique and the traditional topographic levelling method.

3. Runway Monitoring Techniques

Relevant methods employed for the monitoring of airport runways are presented in the following sections. The topographic levelling and the LiDAR techniques are here referred to as "established techniques", as opposed to the PSI method, referred to as an "innovative technique".

3.1. Established Techniques

3.1.1. Topographic Levelling

The process of measuring variations in elevation is a relatively basic operation in topographical surveys, and it is typically referred to as levelling. Various levelling techniques have been developed over the time. In regard to infrastructure surveying, geometric levelling is the most adopted [14]. The height difference is here obtained from readings on levelling staffs where the level's horizontal sightline intersects them. The level is mostly used in the middle between two levelling staffs (differential levelling), with the objective of the survey being the definition of the difference in vertical distance between the two staff positions.

As demonstrated by [6], the topographic levelling procedure allows us to monitor the long-term settlements of an infrastructure by measuring the variation in the vertical position of the target over multiple surveys. This is performed with respect to a single or multiple stable point, typically referred to as topographic benchmarks.

In recent decades, various levelling investigations on civil infrastructures have been reported in the literature [15,16]. In regard to the monitoring of the airport runways, both advantages and drawbacks can be mentioned on the use of the topographic levelling. As for the advantages, the high accuracy of the measurements [17] and the possibility to perform tests independently from indoor or outdoor environments once a reference point is defined [18] is worthy of mention. The drawbacks include a) a limited productivity [19,20], due to the need of measuring each target separately, b) the necessity to close the runway during testing, c) the impossibility to perform the survey in adverse weather conditions [15], and d) the provision of a clear line of sight between consecutive targets of the survey, implying several practical constraints. Lastly, the observed displacements are relative measurements that might be affected by potential settlements levelling benchmarks, which may result in a distortion of the results.

3.1.2. LiDAR Surveys

LiDAR is a surveying method that measures the distance to a target by illuminating the target with a laser light and measuring the reflected light with a sensor [21]. The differences in the laser return, times, and wavelengths can then be used to create digital 3D representations of the target. Since this technology has started to spread in several scientific and professional fields, successful applications in transport infrastructures monitoring have been reported [22] for both the static [23] and the mobile [24] configurations of the equipment.

With regard to airport runway monitoring, static terrestrial laser scanners are being mostly used for reconstructing the geometry of the pavement surface in order to detect defects and decayed areas [8,25]. This is carried out by illuminating the whole runway surface by means of different scans performed at different and distributed survey stations. A certain rate of superposition of the point clouds must also be provided for matching purposes [26]. The georeferencing operation between different scans is ensured by the use of common ground targets working as reference objects. However, by coupling a GNSS receiver to this ground targets, it is possible to skip from relative to global coordinates and compare successive surveys, in order to monitor the evolution of differential settlements.

Among the advantages of this technique, it is worth mentioning the rapidity and the high resolution of the surveys, as millions of point clouds can be collected in limited time. Furthermore, the use of global georeferenced ground targets allows us to prevent the measurements from relative errors due to the instability of the reference. In turn, LiDAR surveys require the runaway to be closed to traffic during the tests. In addition, the application of this technique, as well as the levelling, requires the presence of operators on site. This represents a considerable safety concern as airport runways and aprons are considered as high-risk environments.

3.2. Persistent Scatterers Interferometry (PSI)

The InSAR technique, or SAR interferometry, relies on the measurement of the signal phase variation between images acquired by a satellite orbiting over the same area [9]. Indeed, once the phase contribution related to atmospheric conditions and both temporal and spatial decorrelations are adequately accounted for, it is possible to detect a time sequence of the sensor–target distance along the sight direction of the satellite. This can be related to the surface deformations, e.g., the subsidence. [11].

For this purpose, various processing techniques have been proposed over time. Among these, PSI is one of the most acknowledged [12,13]. PSI is based on the statistical analysis of the signals back-scattered from a network of phase-coherent targets, namely the Persistent Scatterers (PS), which are defined as the points on the ground returning stable signals to the satellite sensor. Indeed, the constant scattering properties of the PS over time and the reflection dominance within a pixel cell are effective in reducing the temporal and geometric decorrelations. In addition, the atmospheric contribute can be estimated and removed using the series of images acquired at different times.

SAR sensors operate at different bands of the microwave domain, namely X, C, and L, corresponding to wavelengths (λ) ranging from 2.4 to 30 cm. Each wavelength is associated to a differently sized resolution cell on the ground, whose displacement trend is described by a single PS.

By means of the PSI technique, satellite remote sensing can provide a continuous monitoring of the overall stability of structures and infrastructures, as well as the surrounding environment. This is obtained by the analysis of multiple SAR images collected at different time stages. Subject-related literature reports the SAR techniques as viable tools for the monitoring of ground deformations, landslides, subsidence, and tectonic motions [27–29]. Similarly, in the last few years, various successful applications of the PSI technique have been presented, proving the feasibility of this technique for the assessment of transport infrastructures and surveillance areas, such as highways [30–32], bridges [33–37], subways and tunnels [31,38,39], railways [40–43], and airport runways [44–46]. This evidence confirms the wide applicability of these techniques within these specific areas of endeavour.

The use of the InSAR technique in transport infrastructure monitoring holds several advantages [9]. InSAR data can be collected regardless of the atmospheric and lighting conditions. In addition, a single InSAR survey permits the analysis of extended areas, due to the wide footprint of the sensor. The continuous motion of the satellites ensures the availability of regularly spaced images, which allows us to perform very dense analyses as opposed to on-site and low-frequency inspections. The acquisition and processing of SAR images do not require on-site operations, thereby preventing both the closure of the runway to air traffic and the presence of operators on the site, with related economic and safety benefits.

On the other hand, the applicability of the PSI analysis is by definition limited to areas where an adequate number of PSs can be observed. This implies that, in case of varying surface conditions (e.g., frequent repaving or accelerated degradation), the stability of the scattering properties of the target may be compromised, with a reduced number of PSs being detected. Furthermore, reliable InSAR assessments require the processing of various SAR images in order to detect statistically stable PSs. This occurrence involves potential computational-related issues, due to the size of the database required at each survey. Lastly, according to the frequency of the adopted sensor, uncertainties may arise in the scattering source, as it is impossible to recognise the scattering object within the resolution cell.

4. Case Study

4.1. Site Description

In the present study, a case study is presented where the PSI technique is applied Runway 3 at the "Leonardo da Vinci" International Airport in Rome, Italy.

The airport is located in the area of Fiumicino, about 30 km on the west of Rome, and carries most of the intercontinental air traffic from and towards Italy, that ranks it as one of the major airports in Europe. With an excess of 43 million passengers in 2018 and over 199,000 tons of traffic, the airport is the largest in Italy by number of passengers/year and ranks second by number of cargo flights/year.

The airport has three terminals reserved for domestic, international, and intercontinental flights and three runways. Initially, the layout of the airport included two runways (Runway 1 and Runway 2) only. However, due to an increasing traffic demand, it was expanded with a new runway located in the north-east area of the airport, which develops in the north–south direction (Figure 1)

Figure 1. Satellite view of "Leonardo Da Vinci" International Airport and location area of Runway 3 (Google Earth Image, 2015).

The area where Runway 3 was realized is known to be affected by the presence of clayey and peaty soils. As a result, the runway was affected by long-term differential subsidence effects, which required accurate monitoring and had to be brought to a full-depth pavement rehabilitation in 2015. This was aimed at increasing the bearing capacity of the subgrade in the norther section of the runway, where severe differential settlements had been recorded in the previous years. Nevertheless, Runway 3 remains a critical asset that is levelled every year in order to be able to monitor the deformation trends and plan maintenance activities.

4.2. Levelling Data

In regard to the on-site topographic surveys conducted on the survey area, a classical geometric levelling survey of the runway was conducted by means of the DNA03 Digital Level system, manufactured by Leica. The elevation measurements were collected by means of a 2 m high levelling rod made of invar, i.e., a metal with a limited thermal expansion coefficient. The main features of the employed levelling system are summarised in Table 1.

Table 1. Main features of the employed levelling equipment.

	Leica DNA03
Measuring Range	Up to 110 m
Measuring Time	Operator Dependent
Levelling Accuracy (Std Dev.)	±0.3 mm/km
Compensator	Pendulum with magnetic damping
Display	LCD

In particular, in situ levelling data were collected by multiple closed-loop levelling nets covering the entire Runway 3. The collected nets were automatically compensated, with an average squared root mean error of 0.94 mm. The starting and ending point of the measure was a levelling benchmark located in the north-west corner of the runway area, which was verified to be stable in elevation. This point was connected to the high precision levelling net developed by the Istituto Geografico Militare (IGM), through a levelling line having an average accuracy of 1.0 mm/km.

Tests were performed every year from 2015 to 2019 and covered five sections along the runway with a transversal spacing of 15 metres and a length equal to the entire runway longitudinal development (Figure 2).

Figure 2. Scheme of in situ levelling data: (**a**) section in the middle region of Runway 3; (**b**) increase in measuring points density on the ground in the area interested by subsidence effects.

Furthermore, the output of these data allows us to detect and exactly quantify the displacements and the average velocity in the investigated time period. Therefore, this information is significant for a validation of the displacements detected by the InSAR technique.

4.3. SAR Imagery

In regard to the application of the InSAR technique, a multi-temporal interferogram analysis of SAR images, namely the Persistent Scatterers Interferometry technique (PSI), was applied. To this effect, two different data stacks were acquired in ascending and descending geometries using high-resolution SAR imagery acquired in X-Band, which allows the detection of displacements with a millimetre accuracy.

In more detail, a dataset of 72 Stripmap images collected in ascending and descending geometries from the COSMO–SkyMed mission (COSMO–SkyMed Product—©ASI: Italian Space Agency, 2015–2019, All Rights Reserved) were processed. The system operates in the X-band corresponding to a wavelength of 3.1 cm, with a 3 m ground-resolution cell. The radar antenna is a phased array that is 1.4 m wide x 5.7 m long. The system is capable of both single- and dual-polarisation data collection. The central frequency is 9.6 GHz with a maximum radar bandwidth of 400 MHz. The main features of the SAR dataset are reported in Table 2.

Table 2. Main features of the Synthetic Aperture Radar Interferometry (SAR) imagery dataset from the COSMO–SkyMed mission—Italian Space Agency (ASI).

	Ascending Geometry	Descending Geometry
Number of Images	35	37
Reference Period	01/2015–04/2019	03/2016–04/2019
Frequency/Wavelength	9.6 GHz/3.1 cm	
Ground-Range Resolution	3 m	
Azimuth Resolution	3 m	

4.4. Data Processing

These products have been acquired and processed using the PS technique of SARscape Interferometric Stacking Module [47], integrated in the Envi software, within the framework of the project "MOBI: MOnitoring Bridges and Infrastructures networks" (proposal ID 46829), approved by the European Space Agency (ESA).

The processing algorithm includes the following steps [12,13,48,49]:

- Generation of differential interferograms out of the stack of SAR images;
- Implementation of High definition Digital Elevation Models (DEM) for topographic phase-term removal;
- Selection of candidate PS points, through the calculation of the Amplitude Dispersion Index;

- Coherence-based filtering of the dataset;
- Phase unwrapping;
- Identification and removal of the phase values not related to the displacements: evaluation of spatial, orbital, and atmospheric decorrelations;
- Identification of the displacements and calculation of deformation time series.

In regard to the implementation of the high-definition DEM models, an SRTM v3 "Shuttle Radar Topography Mission" DEM was collected and implemented in the interferometric process [50,51] in order to identify and subtract phase-related parameters linked to the topography of the investigated area. The DEM, with a pixel resolution of 3 arc-second (90 m × 90 m), is made available by NASA in partnership with the United States Geological Survey (USGS) [50].

The outputs of the PSI processing algorithm were exported into a GIS environment and the PSs were displayed as a function of the average annual-motion velocities.

A specific procedure was adopted in order to compare each ground-truth levelled data with a single satellite-derived displacement measure, as follows:

- All the PSs in the vicinity of the observed levelled point are selected within a distance radius of 10 m (Figure 3a,b);
- Out of all the selected PSs, a single displacement time series is derived by calculating the moving average of the deformations at each acquisition date (Figure 3c);
- The displacement velocity is defined by a linear regression of the displacement against time (Figure 3c).

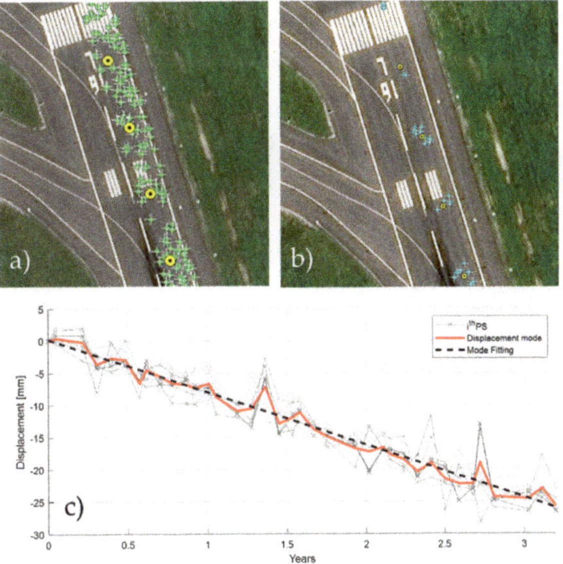

Figure 3. Scheme of SAR data processing: (**a**) visualisation of the whole PS (green) and levelling (yellow) datasets; (**b**) selection of PSs (blue) within a 10 m radius from each levelling point (yellow); (**c**) reconstruction of a single time series and deformation velocity starting from the PS sample.

Finally, in order to obtain continuous average-velocity maps representative of the runway condition, a geo-statistical gridding method was implemented both to the PS datasets and the levelling measurements.

The geo-statistical ordinary Kriging method was used for this purpose [52,53]. Kriging is applied due to the flexibility and the high accuracy in the gridding methods and the provision of representative maps applied to different types of datasets. Moreover, it can compensate for clustered data by weighting less than the cluster in the overall prediction. Each grid node value is based on the known data points neighbouring the node. Each data point is weighted by its distance away from the node, and consequently, points that are further from the node will be weighted less in the estimation of the node. To compute the $\hat{Z}(x_0)$ value at a randomly given grid node at position x_0, the following equation is used [52,53]:

$$\hat{Z}(x_0) = [w_1\ w_2\ \cdots\ w_n] \cdot \begin{bmatrix} z_1 \\ \cdots \\ z_n \end{bmatrix} = \sum_{i=1}^{n} w_i(x_0) \times Z(x_i)$$

where $\hat{Z}(x_0)$ is the estimated value of the grid node, n is the number of neighbouring data values used in the estimation, $Z(x_i)$ is the observed value at the ith location weighting $w_i(x_0)$ with i ranging between 1 and n. The values of the weights add up to 1 in order to ensure that no bias occurs towards clustered data points. The weights are intended to summarise two important procedures in a spatial inference process, i.e., (i) to reflect the structural proximity of the samples to the estimation location; (ii) in parallel, they should not have a separation effect in order to avoid bias effects caused by potential sample clusters.

5. Results

As a qualitative assessment of the reliability of the PSI technique, Figure 4 shows the velocity maps obtained from the levelling and the PSI techniques by interpolation of the displacement velocity data. The similarity between the two maps in low (Figure 4a), intermediate (Figure 4b), and severe (Figure 4c) displacing conditions demonstrates the accuracy of the prediction made using the satellite measurements.

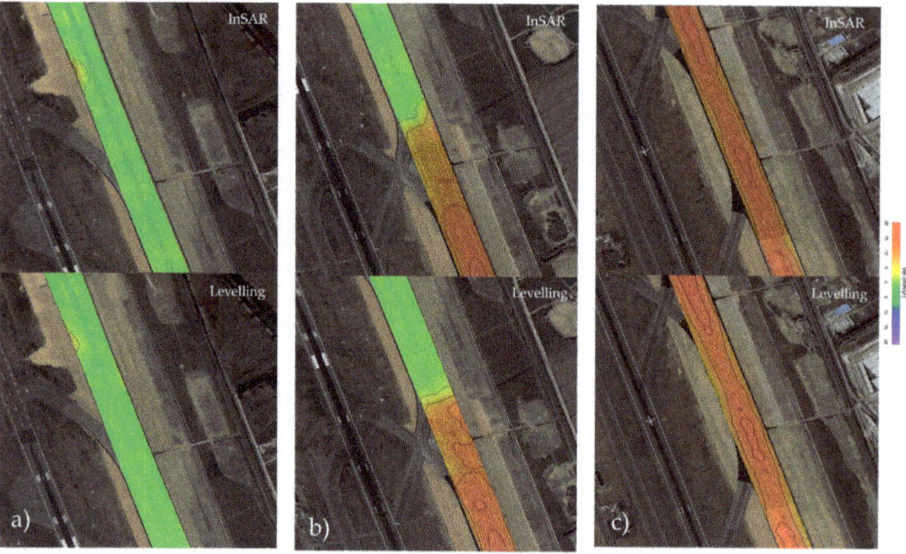

Figure 4. Comparison between the velocity maps obtained by (top) Persistent Scatterers Interferometry (PSI) datasets and (bottom) levelling, relative to (**a**) low (from 2600 m to 3200 m in N direction), (**b**) intermediate (from 1900 m to 2600 m in N direction), and (**c**) severe (from 850 m to 1620 m in N direction) displacing conditions.

A quantitative analysis of the results (Figure 5) shows a displacement velocity scatter plot of the whole ensemble of survey points, where points belonging to different longitudinal survey profiles are marked with different colours. A relatively good matching is observed between levelling and PSI across all the five survey sections acquired by levelling, which returned paired values distributed very close to the bisector, regardless of the value of velocity.

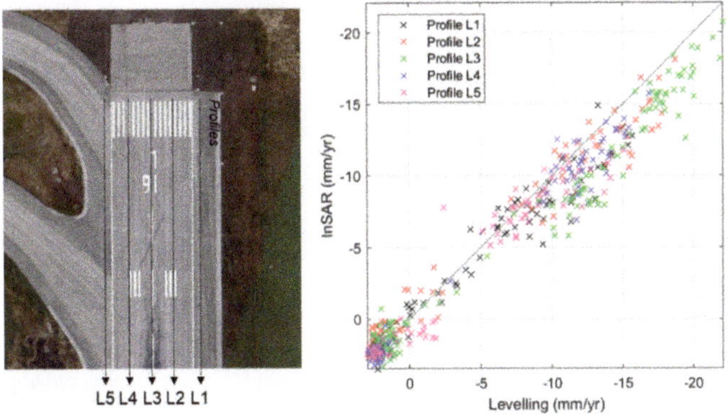

Figure 5. Displacement velocity scatter plot comparing observed (levelling) and predicted (PSI) data.

The correlation and the loss performance of the prediction are reported in Table 3. This includes Pearson's coefficients (r) and root squared mean error (RSME) values for the five survey profiles.

Table 3. Summary of the PSI potential in predicting levelling results.

Survey Profile	r (-)	RSME (mm/yr)
L1	0.9731	1.6430
L2	0.9837	1.6931
L3	0.9857	2.3477
L4	0.9907	1.3247
L5	0.9681	1.5361

It is worthy to note that the survey profile L3 returns slightly less accurate results, as shown in both Figure 5 and Table 3. Especially for the highest values of velocity (which are mainly included within L3), the PSI method seems in fact to underestimate the ground-truth levelled displacement trend. The nature of this coherent error is most likely related to the atmospheric noise contribution, which seems not to be completely filtered out through the applied processing [53,54]. However, the extent of the error is found to be quite limited (around 0.5 cm/year, approximatively). Accordingly, it is observed that the potential of the method in reconstructing the deformation behaviour of the runway is not significantly affected. This allows the provision of very useful information to airport managers for scheduling flight-lists and planning strategic on-site monitoring operations.

As a further confirmation of the above observations, Figure 6 shows the displacement velocity of the survey profiles against the space (WGS84 N Coordinate). Still, it is possible to note that as the displacement trend reaches values higher than 15 mm/year (Figure 6b,c,e), the PSI method turns out to slightly underestimate the deformation rate. However, besides this specific observation, Figure 6 succeeds in demonstrating the effectiveness of PSI in reconstructing the actual deformation pattern over the inspected infrastructure.

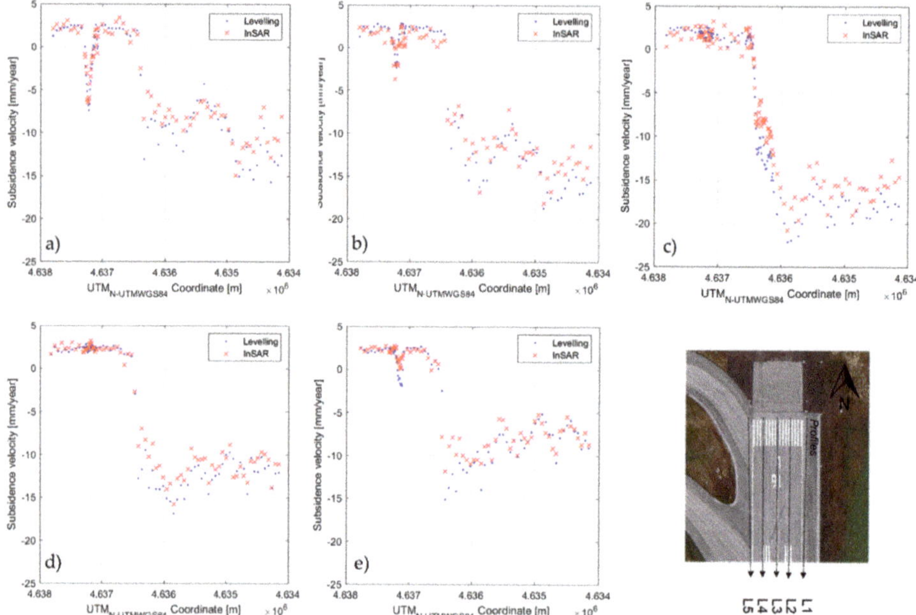

Figure 6. Displacement velocity trends for longitudinal survey profiles (**a–e**) L1 to L5.

In order to evaluate the capability of the method to reconstruct the exact deformation time series of each levelled point, a few examples are here reported, with reference to the three displacement conditions mentioned in Figure 4. More specifically, PS1, PS2, and PS3 in Figure 7 refer to low, intermediate, and severe displacement conditions, respectively.

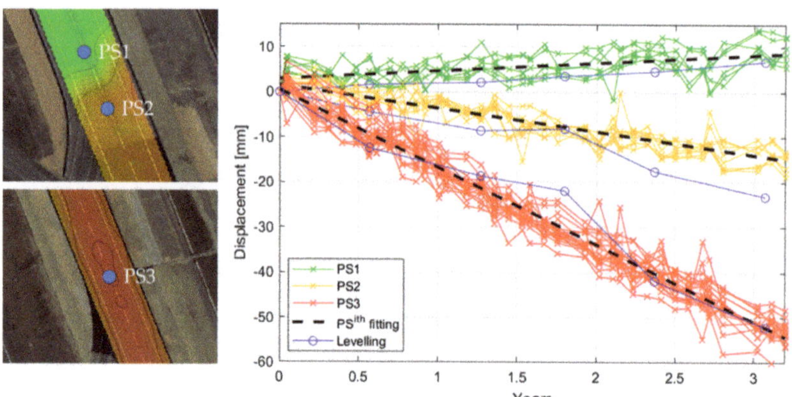

Figure 7. Deformation time series for three points related to low, intermediate, and severe displacement conditions.

A quantitative analysis of the comparison between the two methods is instead reported in Table 4. Specifically, for every levelled point, the displacement rate is reported for both the survey methodology, with the reliability of the InSAR measurement being expressed by the root squared mean error.

Table 4. Quantitative analysis of the comparison between InSAR and levelling for the three PS in the example.

Surveyed Point	Displacement Velocity by Levelling (mm/yr)	Displacement Velocity by InSAR (mm/yr)	RSME (mm)
PS1	1.68	1.76	2.4659
PS2	−6.21	−5.16	4.9875
PS3	−16.93	−17.16	3.9101

6. Conclusions

This work demonstrates the applicability of the high-resolution X-band COSMO–SkyMed mission data and the Persistent Scatters Interferometry (PSI) technique in monitoring the deformations occurring on airport runways.

To this purpose, a time-series analysis of a set of satellite images acquired from January 2015 to April 2019 is carried out by employing the PSI technique. To assess the actual reliability of the method in reconstructing the exact deformation pattern of the monitored infrastructure, the outcomes from the PSI technique were compared to those obtained by the traditional topographic levelling technique, which has been applied on the runway at the same time range as the satellite analysis.

The outcome of the analysis clearly indicates that the use of the PSI technique is reliable and accurate for deformation assessment purposes.

More specifically, the application of the PSI technique has proven to be effective in reconstructing the average trend of the annual deformation velocity (see the coloured maps in Figure 4 and the profiles reported in Figure 6) and the monthly deformation time series of each levelled point (see the examples included in Figure 7).

As opposed to the advantages related to the application of this method, it is fair to comment that in the case of the highest rates of deformation observed in the investigated runway, the PSI technique was found to slightly underestimate the subsidence velocity. Such a coherent bias was related to a limited processing of the atmospheric noise contribution, and as a consequence, it is expected to be accounted for by means of an advanced processing phase.

In general, the results presented in this research demonstrate that the PSI technique is worthy of implementation in Airport Pavement Management Systems (APMS), which may profit by a significant increase in the efficiency in the scheduling of maintenance operations.

Author Contributions: Conceptualization, L.B.C. and A.C.; methodology, A.C. and F.D.; investigation, V.G. and C.F.; data curation, L.B.C. and F.D.; writing—original draft preparation, L.B.C. and F.T.; writing—review and editing, F.D., A.C., and F.T. All authors have read and agreed to the published version of the manuscript.

Funding: This research was funded by the Italian Ministry of Education, University and Research (MIUR) under the National Project "Extended resilience analysis of transport networks (EXTRA TN): Towards a simultaneously space, aerial and ground sensed infrastructure for risks prevention", PRIN 2017, Prot. 20179BP4SM.

Acknowledgments: The license for using the software ENVI SARscape® is granted by the European Space Agency (ESA) approved project "MOBI: MOnitoring Bridges and Infrastructure networks" (EOhops proposal ID 52479). The COSMO–SkyMed Product®—©ASI: Italian Space Agency, 2015–2019, All Rights Reserved- are © of the ASI (Italian Space Agency) delivered under the license to use. In addition, the authors express their deep gratitude to G.R.S. S.r.l. for providing the topographic levelling database, and for the valuable contribution in interpreting the results.

Conflicts of Interest: The authors declare no conflict of interest.

References

1. Federal Aviation Administration (FAA). Advisory Circular AC 150/5370-10H—Standard Specifications for Construction of Airports. 2018. Available online: https://www.faa.gov/documentLibrary/media/Advisory_Circular/150-5370-10H.pdf (accessed on 10 March 2020).

2. Federal Aviation Administration (FAA). Advisory Circular AC 150/5320-6F—Airport Pavement Design and Evaluation. 2016. Available online: https://www.faa.gov/documentLibrary/media/Advisory_Circular/150-5320-6F.pdf (accessed on 10 March 2020).

3. Li, D.; Selig, E.T. Cumulative Plastic Deformation for Fine-Grained Subgrade Soils. *J. Geotech. Eng.* **1996**, *122*, 1006–1013. [CrossRef]

4. Zhou, J.; Gong, X.; Li, J. Experimental study on saturated soft clay under cyclic loading. *Ind. Constr.* **2000**, *30*, 43–47.

5. Bevilacqua, M.; Braglia, M. The analytic hierarchy process applied to maintenance strategy selection. *Reliab. Eng. Syst. Saf.* **2000**, *70*, 71–83. [CrossRef]

6. Cosser, E.; Roberts, G.W.; Meng, X.; Dodson, A.H. Measuring the Dynamic Deformation of Bridges Using a Total Station. In Proceedings of the 11th FIG Symposium on Deformation Measurements, Santorini, Greece, 25–28 May 2003.

7. Chen, W.; Yuan, J.; Li, M. Application of GIS/GPS in Shanghai Airport Pavement Management System. *Procedia Eng.* **2012**, *29*, 2322–2326. [CrossRef]

8. Barbarella, M.; De Blasiis, M.R.; Fiani, M. Terrestrial laser scanner for the analysis of airport pavement geometry. *Int. J. Pavement Eng.* **2017**, *20*, 466–480. [CrossRef]

9. Ciampoli, L.B.; Gagliardi, V.; Clementini, C.; Latini, D.; Del Frate, F.; Benedetto, A. Transport Infrastructure Monitoring by InSAR and GPR Data Fusion. *Surv. Geophys.* **2019**, *41*, 371–394. [CrossRef]

10. Bru, G.; Herrera, G.; Tomás, R.; Duro, J.; De La Vega, R.; Mulas, J. Control of deformation of buildings affected by subsidence using persistent scatterer interferometry. *Struct. Infrastruct. Eng.* **2010**, *9*, 1–13. [CrossRef]

11. Koudogbo, F.; Urdiroz, A.; Robles, J.G.; Chapron, G.; Lebon, G.; Fluteaux, V.; Priol, G. Radar interferometry as an innovative solution for monitoring the construction of the Grand Paris Express metro network—First results. In Proceedings of the World Tunnel Conference, Dubai, UAE, 21–25 April 2018.

12. Ferretti, A.; Prati, C.; Rocca, F. Nonlinear subsidence rate estimation using permanent scatterers in indifferential SAR interferometry. *IEEE Trans. Geosci. Remote Sens.* **2000**, *38*, 2202–2212. [CrossRef]

13. Ferretti, A.; Prati, C.; Rocca, F. Analysis of permanent scatterers in {SAR} interferometry. *IEEE Trans. Geosci. Remote Sens.* **2000**, *39*, 761–763.

14. Elhassan, I.M.; Ali, A. Comparative study of accuracy in distance measurement using: Optical and digital levels. *J. King Saud Univ. Eng. Sci.* **2011**, *23*, 15–19. [CrossRef]

15. Kuhlmann, H.; Glaser, A. Investigation of New Measurement Techniques for Bridge Monitoring. In Proceedings of the 2nd Symposium on Geodesy for Geotechnical and Structural Engineering, Berlin, Germany, 21–24 May 2002.

16. Uddin, W.; Gutelius, B.; Parrish, C. Airborne Laser Survey Specifications and Quality Management Protocols for Airport Obstruction Surveys. *Transp. Res. Rec. J. Transp. Res. Board* **2011**, *2214*, 117–125. [CrossRef]

17. Hill, C.D.; Sippel, K.D. Modern Deformation Monitoring: A Multi Sensor Approach. In Proceedings of the FIG XXII International Congress, Washington, DC, USA, 19–26 April 2002.

18. Radovanovic, R.S.; Teskey, W.F. Dynamic Monitoring of Deforming Structures: GPS verses Robotic Tacheometry Systems. In Proceedings of the 10th FIG International Symposium on Deformation Measurements, Orange, CA, USA, 19–22 March 2001.

19. Meng, X. Real-time Deformation Monitoring of Bridges Using GPS/Accelerometers. Ph.D. Thesis, University of Nottingham, Nottingham, UK, 2002.

20. Uddin, W.; Al-Turk, E. Airport obstruction space management using airborne LIDAR three dimensional digital terrain mapping. In Proceedings of the Federal Aviation Administration Technology Transfer Conference, Washington, DC, USA, 20–22 April 2002.

21. Kim, J.S.; Lee, J.C.; Kang, I.J.; Cha, S.Y.; Choi, H.; Lee, T.G. Extraction of geometric information on highway using terrestrial laser scanning technology. *Int. Arch. Photogramm. Remote Sens. Spat. Inf. Sci.* **2008**, *37*, 539–544.

22. Williams, K.; Olsen, M.J.; Roe, G.V.; Glennie, C. Synthesis of Transportation Applications of Mobile LIDAR. *Remote Sens.* **2013**, *5*, 4652–4692. [CrossRef]

23. Laurent, J.; Hébert, J.F.; Lefebvre, D.; Savard, Y. Using 3D Laser Profiling Sensors for the Automated Measurement of Road Surface Conditions. In Proceedings of the 7th RILEM International Conference on Cracking in Pavements, Delft, The Netherlands, 20–22 June 2012; Volume 4, pp. 157–167.

24. Pu, S.; Rutzinger, M.; Vosselman, G.; Elberink, S.O. Recognizing basic structures from mobile laser scanning data for road inventory studies. *ISPRS J. Photogramm. Remote Sens.* **2011**, *66*, S28–S39. [CrossRef]

25. De Blasiis, M.R.; Di Benedetto, A.; Fiani, M. Mobile Laser Scanning Data for the Evaluation of Pavement Surface Distress. *Remote Sens.* **2020**, *12*, 942. [CrossRef]

26. Barbarella, M.; D'Amico, F.; De Blasiis, M.R.; Di Benedetto, A.; Fiani, M. Use of Terrestrial Laser Scanner for Rigid Airport Pavement Management. *Sensors* **2017**, *18*, 44. [CrossRef] [PubMed]

27. Colesanti, C.; Ferretti, A.; Prati, C.; Rocca, F. Monitoring landslides and tectonic motions with the Permanent Scatterers Technique. *Eng. Geol.* **2003**, *68*, 3–14. [CrossRef]

28. Colombo, D.; Farina, P.; Moretti, S.; Nico, G.; Prati, C. Land subsidence in the Firenze-Prato-Pistoia basin measured by means of spaceborne SAR interferometry. In Proceedings of the 2003 IEEE International Geoscience and Remote Sensing Symposium (IGARSS 2003), Toulouse, France, 21–25 July 2003; Volume 4, pp. 2927–2929.

29. Frattini, P.; Crosta, G.B.; Allievi, J. Damage to Buildings in Large Slope Rock Instabilities Monitored with the PSInSAR™ Technique. *Remote Sens.* **2013**, *5*, 4753–4773. [CrossRef]

30. Yu, B.; Liu, G.; Zhang, R.; Jia, H.; Li, T.; Wang, X.; Dai, K.; Ma, D. Monitoring subsidence rates along road network by persistent scatterer SAR interferometry with high-resolution TerraSAR-X imagery. *J. Mod. Transp.* **2013**, *21*, 236–246. [CrossRef]

31. Perissin, D.; Wang, Z.; Lin, H. Shanghai subway tunnels and highways monitoring through Cosmo-SkyMed Persistent Scatterers. *ISPRS J. Photogramm. Remote Sens.* **2012**, *73*, 58–67. [CrossRef]

32. Fárová, K.; Jelének, J.; Kopačková-Strnadová, V.; Kycl, P. Comparing DInSAR and PSI Techniques Employed to Sentinel-1 Data to Monitor Highway Stability: A Case Study of a Massive Dobkovičky Landslide, Czech Republic. *Remote Sens.* **2019**, *11*, 2670. [CrossRef]

33. Milillo, P.; Giardina, G.; Perissin, D.; Milillo, G.; Coletta, A.; Terranova, C. Pre-Collapse Space Geodetic Observations of Critical Infrastructure: The Morandi Bridge, Genoa, Italy. *Remote Sens.* **2019**, *11*, 1403. [CrossRef]

34. D'Amico, F.; Gagliardi, V.; Ciampoli, L.B.; Tosti, F. Integration of InSAR and GPR techniques for monitoring transition areas in railway bridges. *NDT E Int.* **2020**, *115*, 102291. [CrossRef]

35. Alani, A.M.; Tosti, F.; Bianchini Ciampoli, L.; Gagliardi, V.; Benedetto, A. Integration of GPR and InSAR methods for the health monitoring of masonry arch bridges. *NDT E Int.* **2020**, in press. [CrossRef]

36. Jung, J.; Kim, D.-J.; Vadivel, S.K.P.; Yun, S.-H. Long-Term Deflection Monitoring for Bridges Using X and C-Band Time-Series SAR Interferometry. *Remote Sens.* **2019**, *11*, 1258. [CrossRef]

37. Ciampoli, L.B.; Gagliardi, V.; Calvi, A.; D'Amico, F.; Tosti, F. Automatic network level bridge monitoring by integration of InSAR and GIS catalogues. *Multimodal Sens. Technol. Appl.* **2019**, *11059*, 110590I. [CrossRef]

38. Roccheggiani, M.; Piacentini, D.; Tirincanti, E.; Perissin, D.; Menichetti, M. Detection and Monitoring of Tunneling Induced Ground Movements Using Sentinel-1 SAR Interferometry. *Remote Sens.* **2019**, *11*, 639. [CrossRef]

39. Barla, G.; Tamburini, A.; Del Conte, S.; Giannico, C. InSAR monitoring of tunnel induced ground movements. *Géoméch. Tunnelbau* **2016**, *9*, 15–22. [CrossRef]

40. Yang, Z.; Schmid, F.; Roberts, C. Assessment of Railway Performance by Monitoring Land Subsidence. In Proceedings of the 6th IET Conference on Railway Condition Monitoring (RCM 2014), Birmingham, UK, 17–18 September 2014; pp. 1–6.

41. Chang, L.; Dollevoet, R.P.B.J.; Hanssen, R.F. Nationwide Railway Monitoring Using Satellite SAR Interferometry. *IEEE J. Sel. Top. Appl. Earth Obs. Remote Sens.* **2016**, *10*, 596–604. [CrossRef]

42. Qin, X.; Liao, M.; Zhang, L.; Yang, M. Structural Health and Stability Assessment of High-Speed Railways via Thermal Dilation Mapping With Time-Series InSAR Analysis. *IEEE J. Sel. Top. Appl. Earth Obs. Remote Sens.* **2017**, *10*, 2999–3010. [CrossRef]

43. Tosti, F.; Gagliardi, V.; D'Amico, F.; Alani, A.M. Transport infrastructure monitoring by data fusion of GPR and SAR imagery information. *Transp. Res. Procedia* **2020**, *45*, 771–778. [CrossRef]

44. Jiang, L.; Lin, H. Integrated analysis of SAR interferometric and geological data for investigating long-term reclamation settlement of Chek Lap Kok Airport, Hong Kong. *Eng. Geol.* **2010**, *110*, 77–92. [CrossRef]

45. Gao, M.; Gong, H.; Chen, B.; Zhou, C.; Chen, W.; Liang, Y.; Shi, M.; Si, Y. InSAR time-series investigation of long-term ground displacement at Beijing Capital International Airport, China. *Tectonophysics* **2016**, *691*, 271–281. [CrossRef]

46. Jiang, Y.; Liao, M.; Wang, H.; Zhang, L.; Balz, T. Deformation Monitoring and Analysis of the Geological Environment of Pudong International Airport with Persistent Scatterer SAR Interferometry. *Remote Sens.* **2016**, *8*, 1021. [CrossRef]

47. Sarmap. SARscape Technical Description. Available online: http://www.sarmap.ch/pdf/SARscapeTechnical.pdf/ (accessed on 22 July 2020).

48. Sarmap. SAR-Guidebook. Available online: http://www.sarmap.ch/pdf/SAR-Guidebook.pdf (accessed on 22 July 2020).

49. NASA. The Shuttle Radar Topography Mission (SRTM) Collection User Guide. 2015. Available online: https://lpdaac.usgs.gov/documents/179/SRTM_User_Guide_V3.pdf (accessed on 11 February 2020).

50. Rodriguez, E.; Morris, C.S.; Belz, J.E.; Chapin, E.C.; Martin, J.M.; Daffer, W.; Hensley, S. *An assessment of the SRTM Topographic Products*; Technical Report JPL D-31639; Jet Propulsion Laboratory: Pasadena, CA, USA, 2005; p. 143.

51. Giraldo, R.; Delicado, P.; Mateu, J. Ordinary kriging for function-valued spatial data. *Environ. Ecol. Stat.* **2010**, *18*, 411–426. [CrossRef]

52. Papritz, A.; Stein, A. Spatial prediction by linear kriging. In *Spatial Statistics for Remote Sensing*; Springer Science and Business Media LLC: Berlin/Heidelberg, Germany, 1999; Volume 1, pp. 83–113.

53. Doin, M.-P.; Lasserre, C.; Peltzer, G.; Cavalié, O.; Doubre, C. Corrections of stratified tropospheric delays in SAR interferometry: Validation with global atmospheric models. *J. Appl. Geophys.* **2009**, *69*, 35–50. [CrossRef]

54. Fattahi, H.; Amelung, F. InSAR bias and uncertainty due to the systematic and stochastic tropospheric delay. *J. Geophys. Res. Solid Earth* **2015**, *120*, 8758–8773. [CrossRef]

Publisher's Note: MDPI stays neutral with regard to jurisdictional claims in published maps and institutional affiliations.

Article

Quantification of the Mechanized Ballast Cleaning Process Efficiency Using GPR Technology

Anna Borkovcová [1], Vladislav Borecký [1], Salih Serkan Artagan [1,2,*] and Filip Ševčík [1]

[1] Department of Transport Structures, Faculty of Transport Engineering, University of Pardubice, Studentska 95, 532 10 Pardubice, Czech Republic; anna.borkovcova@student.upce.cz (A.B.); vladislav.borecky@upce.cz (V.B.); filip.sevcik@upce.cz (F.Š.)

[2] Vocational School of Transportation, Eskişehir Technical University, Basın Şehitleri Cad. No:152, Eskişehir 26470, Turkey

[*] Correspondence: ssartagan@eskisehir.edu.tr

Abstract: Ground Penetrating Radar (GPR) has been used recently for diagnostics of the railway infrastructure, particularly the ballast layer. To overcome ballast fouling, mechanized ballast cleaning process, which increases track occupancy time and cost, is usually used. Hence it is of crucial significance to identify at which stage of track ballast life cycle, and level of fouling, ballast cleaning should be initiated. In the present study, a series of in situ GPR surveys on selected railway track sections in Czechia was performed to obtain railway granite ballast relative dielectric permittivity (RDP) values in several phases of railway track lifecycle. GPR data were collected in the form of B-scan, and time-domain analysis was used for post-processing. The results indicate (i) change of railway ballast RDP in time (long term); (ii) a dependency of ballast fouling level on RDP; and (iii) the RDP change during the ballast cleaning process, thus its efficiency. This research aimed to provide new perspectives into the decision-making process in initiating the mechanized ballast cleaning intervention based on the GPR-measured data.

Keywords: ground penetrating radar (GPR); relative dielectric permittivity (RDP); railway ballast fouling; granite ballast; mechanized ballast cleaning; railway infrastructure

Citation: Borkovcová, A.; Borecký, V.; Artagan, S.S.; Ševčík, F. Quantification of the Mechanized Ballast Cleaning Process Efficiency Using GPR Technology. *Remote Sens.* **2021**, *13*, 1510. https://doi.org/10.3390/rs13081510

Academic Editor: Susana Lagüela López

Received: 6 March 2021
Accepted: 8 April 2021
Published: 14 April 2021

Publisher's Note: MDPI stays neutral with regard to jurisdictional claims in published maps and institutional affiliations.

1. Introduction

Increasing service frequencies for railway freight and passenger along with the high safety, sustainability, and reliability requirements, railways necessitate optimized strategic planning and decision-making for the maintenance activities, which should be figured out based on extensive and reliable fault detection methodologies. Conventional diagnostic methods are destructive, economically inefficient, and cause traffic constraints increasing the track occupation time. In contrast, railway infrastructure administrators target the task of optimizing the diagnostic inspection intervals and following maintenance works, to diminish the financial costs and track occupation time. Therefore, non-destructive testing (NDT) techniques have been increasingly used lately.

Qian et al. [1] developed a non-intrusive technique, as a fundamental element of Railroad Infrastructure 4.0, to determine the distribution of ballast pressure under the sleeper using the bending moment profile across the concrete sleeper, and the estimated rail seat loads as inputs, to compute the state of ballast support employing an optimization algorithm. Tamrakar et al. [2] used spectral analysis of surface wave technique (SASW) to estimate the mechanical properties of ballast and evaluated the complications in the implementation of the SASW method for characterization of ballast and foundation soil or subgrade. Mvelase et al. [3] employed laser technology to examine the influence of railway ballast roundness on shear strength resulting in improved in situ quality evaluations in regards to ballast layer maintenance or replacement. Schmidt et al. [4] stated that the outcomes of the constant head permeability experiments in combination with a novel

199

imaging technology for ballast will give insights to decision-makers in the determination of the time as to when ballast should be cleaned. Calibration [5] and application [6] of Time Domain Reflectometry (TDR) test probes were performed for determination of moisture variation in the railway infrastructure layers.

Ground Penetrating Radar (GPR), as another NDT technique, has steadily been in use in diagnostics of the railway infrastructure recently [7,8]. This method is still not the most routine tool implemented to evaluate the status of the railway infrastructure, however, it is being utilized to a greater extent due to its numerous benefits such as time efficiency, reduced costs, rapid fulfilling of surveys, continuous data acquisition of long sections, and the non-destructive principle [9]. Moreover, competent use of GPR brings forward the advantages of prioritizing the sections requiring urgent interventions, insights on the rates of ballast layer deformation, and enhanced strategical planning for mechanized operations (e.g., tamping and undercutter/ballast cleaning machines) [10]. GPR applications on railway infrastructure comprise detecting the thickness of the layers [11–13], moisture trapped areas [14], congested pipes, utility networks, culverts, other buried objects, and the required diagnosis of the railway infrastructure status.

The basic tasks of the ballast layer are to bear the railway axle loads from the superstructure, transmit those to the infrastructure, and facilitate water draining [15]. Railway ballast progressively degrades because of cyclic loading and weathering impacts. So-called ballast fouling (contamination of ballast) develops as the air voids between the ballast stones are replaced by finer-sized fouling materials. Ballast fouling takes place when ballast stones break down and/or ingress of other materials occurs either from the surface of ballast or from the ballast layer bottom [16]. To discharge the substandard finer materials from the ballast layer, mechanized ballast cleaning is commonly used with the machinery called ballast-cleaners and/or undercutters, which undertake screening of existing ballast and removing of the spent material with the capability of the addition of new standard clean ballast to maintain the required ballast quantity and profile. However, this vital intervention, which is generally performed before tamping, is time-consuming and expensive. Thus, it is of crucial significance to identify at which step of track ballast life cycle, level of fouling, and mechanized ballast cleaning should be initiated and is most favorable costwise. In order to attain this critical knowledge, long-term and recurring investigations of the status of railway ballast are required.

1.1. Mechanized Ballast Cleaning Activity on Railways

The ballast layer is anticipated to address several requirements for safe, secure, and sustainable railway operations, such as homogeneous load-bearing capability, high resilience to loads from each direction (vertical, transversal, and longitudinal), to keep track in place, enabling elementary fixing and flexibility of track geometry, and immediate draining of water. The pollution of the ballast layer (i.e., ballast fouling) occurs when fine particles (up to 22.5 mm) within the ballast layer fill the air voids. Among reasons for ballast fouling, there exist, mechanical breakdown of the ballast aggregates caused by dynamic forces, dirt, and dust from the surface and sublayers, railway traffic contamination, and interventions such as tamping and packing of sleepers. All these mechanisms lead to water drainage disability, reduction in the internal friction, differential settlements, stress increment in the ballast layer, and the formation of clogged portions in the ballast layer. The aim of mechanized ballast layer cleaning is to screen and eliminate finer material from the ballast layer and to reestablish the required characteristics of the ballast layer. When ballast fouling level reaches over 30% of the ballast aggregates' total weight, the ballast layer cleaning process should be initiated [17].

Ballast cleaning intervention can be regarded as one of the physically most challenging activities in maintaining and reconstructing the railway infrastructure [17]. The efficiency of the mechanized cleaning process has an explicit impact on the overall standard of maintenance and reconstruction works. Therefore, considering the ballast cleaning activity as a complex intervention rather than the cleaning activity itself might enable analyzing it better. This complicated process involves ballast excavation, ballast screening and removing the fine grains and spent ballast, putting the reusable, good performing ballast stones back in track, and clearing of obstructed areas with fine grains along with adjustment of track alignment [18]. Recently, noticeable consideration is focused on the efficiency and capacity of the ballast cleaning and undercutter machinery. The ever-increasing output requirements resulted in an enhancement in the machinery design (particularly the ballast screening capability has been enhanced and dimensions and speed of the ballast excavation chains have been increased) with respect to ergonomic and environmental parameters [17]. Schilling [18] presented the track possession optimization program for track reconstruction with a comparative analysis of three different ballast cleaning technologies, whereas Korolev et al. [19] discussed the technologies of ballast cleaning and analyzed the impact of miscellaneous parameters on the implementation of the ballast cleaning technology for reducing the operating costs for Russian railways.

According to Tzanakakis [20], ballast cleaning process is implemented in a range varying from 12 to 15 years on a common railway mainline under intensive operation and is performed along with other large-scale railway maintenance and reconstruction interventions. However, the necessity of ballast cleaning process is usually decided subjectively by the chief district maintenance officer in real cases.

1.2. Ground Penetrating Radar

The term RADAR, dating back to the 1930s, is generated as an acronym for Radio Detection and Ranging. GPR, which is a particular sort of radar system, has been employed for over 50 years in probing, detecting, and visualizing underground and construction materials [21]. Through the combination of Maxwell's formulation of electromagnetic (EM) theory and constitutive equations, one can attain the quantitative characterization of GPR signals [22]. From the physical point of view, the prospects of employing the GPR method are affected by the EM characteristics of the material under survey, which are basically Relative Dielectric Permittivity (RDP) (alias dielectric constant, ε_r), magnetic permeability μ, and electrical conductivity σ.

RDP portrays the capacity of the material to store and emit the electric charge generated by the EM field. RDP of a material can be characterized as the measure of electrostatic energy reserved per volume. It is a figure demonstrating the speed of radar energy as it goes through the medium. RDP values of particular materials might be found in the literature, however, the composition of the in situ material in the lossy environment has to be considered [23]. Electromagnetic Wave Velocity (EMWV), i.e., v_r (m/s), is computed from the two-way travel time (twt) of the EM signal (to and from the target). As per the construction materials, it might be assumed that the GPR signal interacts negligibly with the magnetic field in general. Based on this assumption, EMWV can be calculated from Equation (1),

$$v_r = c \: / \: \sqrt{\varepsilon_r} \; [\text{m/s}], \tag{1}$$

where v_r is relative EMWV, c is the speed of light, and ε_r is RDP [24]. Once the EMWV of the material is known, the thickness/depth of the material of interest can be obtained via Equation (2),

$$d = v_r * \text{twt}/2 \; [\text{m}], \tag{2}$$

where d is the thickness/depth of the object/interface and twt is the two-way travel time of EM signal to and from the object/interface of interest [24].

Figure 1 illustrates a radargram (both A-scan and B-scan) where twt differences between the maximum reflection amplitudes of air/ballast interface and the ballast bottom can be observed. (In Figure 1, a metal plate interface was used to collect a distinctive reflection from the ballast base). Those time intervals are used in Equation (2) in order to obtain EMWV in the case of known depth. Then using Equation (1), one can back-calculate the RDP value.

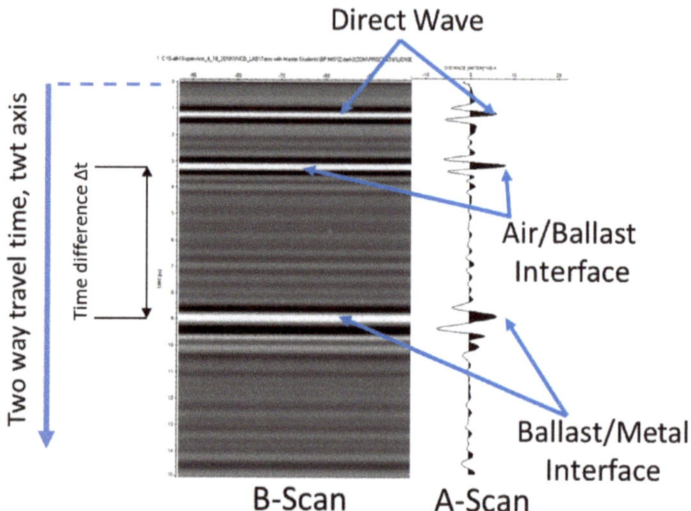

Figure 1. Schematic presentation of air/ballast interface and ballast/bottom interface (from 2 GHz air-coupled antenna) (adapted from [25]).

1.3. Railway Ballast Assessment by GPR

According to Selig and Waters [26] ballast layer might be grouped in four zones as follows:

- Crib—the portion of ballast between the sleepers;
- Shoulder—the ballast aggregates between the edge of the sleeper and down to the top of the sub-ballast layer;
- Top ballast—the upper portion of the ballast aggregates where tamping intervention is often performed.
- Bottom ballast—the lower portion of the ballast aggregates supporting the whole structure.

Ballast fouling can display variations along the track, introducing changes in both RDP values and layer thicknesses. Clean ballast possesses a lower value of the average RDP since it has a higher volume of air voids (RDP value of air equals 1). Ballast fouling can be caused by five mechanisms as follows (the percentage of occurrences are given for each mechanism in parenthesis): ballast breakdown (76%), ingress of granular layers beneath the ballast layer (13%), ingress of fines from ballast surface (7%), penetration of materials from subgrade (3%), and degradation in sleepers (1%) [26].

Although there exist many fouling indicators, the fouling index proposed by Selig and Waters [26] is the most commonly applied one. Percentage passing values of certain sieves are summed up to find out the fouling index using Equation (3).

$$F_I = P_{0.075} + P_{4.75},$$

(3)

where *FI* is the fouling index, $P_{0.075}$ and $P_{4.75}$ symbolizes the percentage of materials, respectively, passing the sieve sizes of 0.075 and 4.75 mm.

Foun [27] used another fouling index called percentage void contamination (PVC) obtained from laboratory GPR testing to develop a method for detailed ballast cleaning plans. Moreover, other fouling indexes, such as the effective degree of fouling, D-bar method, relative ballast fouling, the percentage of fouling, and PVC were comparatively referred to by Anbazhagan [28].

The information on RDP values addresses a significant part of determining ballast thickness and status. RDP value is hugely reliant on the water amount as well as the fouling level of the ballast layer. However, a precise examination of these phenomena still represents an arduous task for GPR railways investigations [29]. Lalagüe [30] gave an overview of the published RDP values of granite ballast under several varying cases (clean ballast in dry and moist conditions as well as fouled ballast in dry and moist conditions) in the studies of various researchers. In Table 1, an extended version of published RDP values of granite ballast under several different conditions can be seen.

Table 1. Relative dielectric permittivity (RDP) values of granite railway ballast under several different conditions in published literature.

	Ballast Condition	References			
		Clark et al. [31]	Sussmann [32]	Leng and Al-Qadi [33]	Artagan and Borecky [34]
Clean	Dry clean	3.00	3.60	3.25	3.09
	Moist clean	3.50	4.00		4.59
	Saturated clean	26.90	26.40		25.50
Fouled	Dry fouled	4.30	3.70	3.77	3.85
	Moist fouled	7.80	5.10		5.21
	Saturated fouled	38.50			

As indicated by Maturana et al. [35], the presence of reflectors within the ballast layer, the attenuation, indistinguishable ballast bottom interface, and decline in the EMWV owing to various reasons influence the GPR signal in the ballast layer. GPR surveys repeated in regular time intervals facilitate the estimation of the status and degradation pace of the ballast layer, which might assist to optimize the overall maintenance action plan through effective scheduling of required short, medium, and long-term maintenance activities with remarkable savings associated with cost and time [35]. As also mentioned in a recent article by Artagan and Borecky [34], numerous scientists, partially or wholly, handled the evaluation of railway ballast fouling and thickness using several attributes of GPR signal in their researches [7,28,33–49]. GPR tests were performed on a model rail line track of various ballast states utilizing antennas with different frequencies in the work by Su et al. [39]. In view of the investigation of the handled data, the impact of antenna frequency, water amount in the ballast layer, the influence of geotextile discernible by radar, and the distinction in RDP values were examined and assessed.

In their case study, Keogh et al. [50] created and executed a complex framework for the aim of non-destructive investigation of railway ballast and sleeper conditions, where they modeled the effective in situ RDP of railway ballast according to the volumetric weighted sum of the RDP values of the railway ballast components. This model anticipated a decline in EMWV values (10–30%) in the ballast layer during its fouling interaction.

Jack and Jackson [11] monitored variations in the status of the ballast layer to organize the collected GPR data into the stretches associated with the corresponding variations in the ballast layer status. In another study by Gallagher et al. [12], the GPR method was utilized to determine the interface between ballast and subgrade along with the level of ballast fouling. Hugenschmidt [13] expressed that the use of GPR had an edge over the conventional methods for ballast condition evaluation, leading to a minimized quantity of test-pits required, and also identified the railway segments where material from underlying layers infiltrated into the ballast layer. In a recent experimental research by Tosti et al. [8], RDP values of clean ballast were compared between different types and frequencies of the antennas in order to identify critical elements most applicable for the railway ballast GPR surveys. To assess the degree of ballast fouling utilized in Portuguese railway lines, laboratory experiments were undertaken on various materials utilizing different antennas [29,44]. The consolidated effect of varying fouling levels and diverse water contents was examined. Sussmann et al. [51] carried out both laboratory and field tests in order to estimate RDP values of railway track ballast through a mixing model.

Several studies employed the EMWV values in converting the time axis to the depth axis in GPR data for the ballast layer. Hugenschmidt [13] suggested an average EMWV value of 14 cm/ns for the ballast layer after comparison of ground truth data with the GPR data, whereas 13 cm/ns for the ballast layer was calculated in the research by Jack and Jackson [11]. Intervals of EMVW values changing from 12 to 21 cm/ns and from 0.8 to 1.2 cm/ns were used, respectively, for clean and fouled ballast in the research by Göbel et al. [52].

Artagan et al. [53] determined the effect of water content within the granite ballast layer for the clean and fouled conditions by collecting GPR data in the laboratory. Clean granite ballast in saturated conditions was reported to have an RDP value of 25.5, while the mean RDP value of clean drained granite ballast was discovered to reflect almost identical values (3.09) with the clean dry ballast. As the fouling levels increased, so did the RDP values.

Besides, other aspects of GPR analysis of railway ballast have come to the interest of the researchers such as the use of frequency-domain analyses [7,38,40], analytical modeling [54] and numerical simulations [46,55,56]. Moreover, considerable efforts are directed into coupling GPR with other NDT techniques in railway ballast evaluation. Fontul et al. [57] expressed seismic waves, GPR, electric resistivity, and gravimetry as the fundamental NDT geophysical methods for railway infrastructure assessment. Comprehensive overviews on the GPR-based applications for condition monitoring of railway ballast can be found in a couple of works [10,58,59].

It might very well be perceived from the mentioned literature above that GPR is a favorable instrument to evaluate the status of railway ballast and the associated degree of fouling utilizing determined RDP/EMWV values. It is worth mentioning that test pits are required for calibrating the RDP/EMWV values as recommended in the literature above.

Although extensive research has been carried out on GPR-based determination of the state of the railway ballast recently, these studies are mostly limited to the then-current condition of the ballast layer. To the best of authors' knowledge, no study exists monitoring the condition change of ballast layer by GPR just before and after mechanized cleaning process. Moreover, there are a few studies [60,61] monitoring long-term change of ballast layer.

Therefore, the presented article focused on identifying the status of the railway granite ballast layer and finding out the variation in fouling levels before and after the mechanized ballast cleaning intervention by means of the GPR-measured RDP values of the ballast layer. In situ RDP values of the granite ballast layer obtained in this study provide additional literature data for that particular type of ballast material. The partial goals were to determine the long-term variation of RDP and EMWV particularly before and after the mechanized ballast cleaning activities and to describe the dependency between the collected GPR data and the fouling levels. To accomplish the set goals, repeated surveys (long-term monitoring over a total period of 4 years) at various stages of the track lifecycle in several sections of a real railway track were performed with miscellaneous GPR antenna configurations. Ground truth data were collected and analyzed in the laboratory. Time-domain analysis was used to process the data and compute the RDP values at different stages of railway ballast economic life. Then, quantification of the influence of the mechanized ballast cleaning on the condition of the railway ballast layer was obtained using GPR.

2. Survey Site and Equipment Used

Due to the particular interest in the utilization of two types of ballast cleaners, the site of Čáslav to Kutná Hora railway section in Czechia, where the maintenance and reconstruction activities were planned, was selected in this study. The picked railway section is under operation for both passenger and freight trains with a double-track railway line with ballast aggregates made of granite. The mentioned railway section constitutes a portion of a cross-country track, which additionally serves for the European TEN-T (Trans-European Transport Networks) railway line within the global passenger and freight network with a design maximum track speed of 120 km/h. At the beginning of the reconstruction activities, R 65 and S 45 rails, which were encountered partially in the sections, were all altered with 60 E2 type rails. Existing SB8 and SB8P sleepers, which were laid 60 cm apart from one another, were controlled and retained. The fastening system of type K was observed, checked, and maintained in track. Although the design thickness of the ballast layer was 350 mm under the sleepers, 300 mm ballast layer thickness was observed in some of the portions of the surveyed railway track after opening test pits. However, as a result of the reconstruction activities, the ballast layer was adjusted to be constantly 350 mm thick beneath the sleepers. Eight railway track sections, each 50 m long, were surveyed close to the four level crossings set out in the railway line direction from Čáslav to Kutná Hora, numbered as P2 (km 279,223), P3 (km 281,182), P6 (km 283,747), and P8 (km 286,468). All of these eight sections under investigation were situated in proximity and on both sides of the mentioned level crossings, as depicted in Figure 2. Table 2 provides the mileage of these sections.

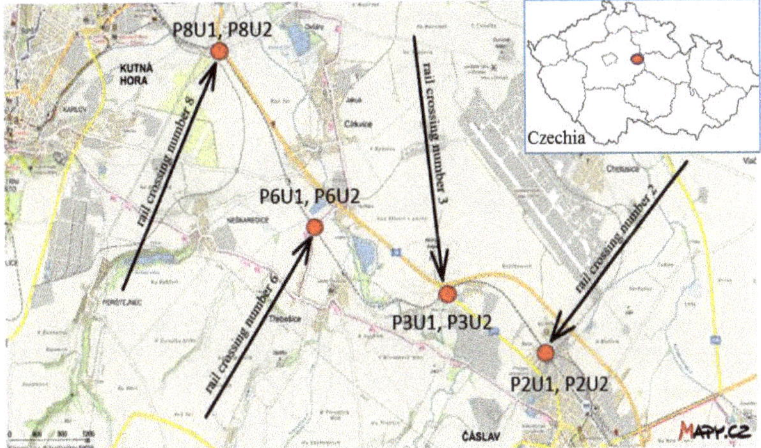

Figure 2. Surveyed railway level-crossings displayed on the map (mapy.cz overlay).

Table 2. Survey sections stationing and length.

Level Crossing	Section	Start Mileage (km)	End Mileage (km)	Length (m)
P2	U1	279.150	279.200	50
	U2	279.250	279.300	50
P3	U1	281.100	281.150	50
	U2	281.200	281.250	50
P6	U1	283.675	283,725	50
	U2	283.775	283.825	50
P8	U1	286.400	286.450	50
	U2	286.500	286.550	50

The RM 79 ballast cleaner/undercutter unit covered reconstruction and maintenance interventions at P2 and P3 sections, whereas the RM 900 unit was utilized in the reconstruction activities at P6 and P8 sections.

In 2014 and 2017, GPR data were obtained with GSSI systems. For all 2018 GPR surveys described in this study, the RIS Hi-Pave GPR system developed by IDS GeoRadar company was used. A manual GPR survey trolley, with polyamide wheels and the skeleton made of glass-fiber reinforced plastic square profiles, was employed. Two types of GPR antennas were used in the surveys; (i) 2 GHz Horn air-coupled antenna (HN-2000) and (ii) 400/900 MHz dual-frequency ground-coupled antenna (TR DUAL-F 400/900). As per data acquisition, K2 Fast Wave software from IDS was operated, whereas ReflexW software was exploited for GPR signal processing and data interpretation. Figure 3 displays the GPR assembly along with the antennas and the auxiliary equipment used in the surveys.

Figure 3. Ground Penetrating Radar (GPR) trolley designed for the surveys (air-coupled antenna in diagonal orientation).

3. Track Surveys, Laboratory Tests, and GPR Data Processing

The present study covered both field and laboratory tests. The GPR field surveys were adjusted according to the timetable of the rehabilitation activities in the construction site. The start and end points of the abovementioned sections (Figure 2 and Table 2) were determined by spraying a dye of recognizable color either on the rails or the sleepers. A digital measuring wheel was used to measure distances. Moreover, in order to distinguish the start and end points of the sections in the GPR data, aluminum foil pieces were placed in these positions. All laboratory tests were performed in the Educational and Research Centre in Transport (University of Pardubice).

3.1. In Situ GPR Surveys

GPR data for the first two stages were obtained in 2014 and 2017 by the state railway administrator Správa železniční dopravní cesty (SŽDC) and they were processed and interpreted to reveal the conditions of the railway track sections. Following GPR surveys undertaken in 2018 represent three more stages to compare the ballast layer status before and after the maintenance intervention works. At first, GPR data were collected from all the sections before any reconstruction activities were initiated. Then, the surveys were performed just after the ballast cleaning process, prior to the tamping and fixing of other geometric parameters. Finally, the ultimate conditions of all sections were surveyed when all the reconstruction activities were completed in all of the sections. During the GPR surveys (in 2018) and over 14 days prior to them, no precipitation was recorded in the site, and the temperature was changing from 20 to 25 °C. All these stages were numbered from 1 to 5 below:

1. Surveys undertaken by SŽDC in 2014
2. Surveys undertaken by SŽDC in 2017

3. Surveys carried out before any reconstruction activities began (12 October 2018)
4. Surveys carried out just after the ballast cleaning process (17 October 2018)
5. Surveys carried out in the final state after all the reconstruction activities were completed (19 October 2018)

Particular phases of the surveyed track sections are depicted in Figure 4.

Figure 4. Track condition (**a**) before (phase 3), (**b**) during, and (**c**) after construction activities (phase 5).

Data collection and configuration parameters were set in K2 Fast Wave software. Calibration of the odometer was also performed to precisely record the length of the surveyed track and to control the trace-interval distance, which was undertaken by going through a reference line of a known distance.

As per the GPR data from the SŽDC surveys (Stage 1 in 2014 and Stage 2 in 2017), profiles (B-scans) were extracted for all the eight sections from the continuous radargrams collected then. In all three stages of measurements performed in 2018 (i.e., stages 3, 4, and 5), for each one of the eight survey sections, one profile (B-scan) through the center of the track axis was collected. GPR antenna configurations and survey advancing directions were decided with respect to the selected coordinate system. Then, for each frequency of antennas, time windows, i.e., time axis ranges were picked. The estimated EMWV (cm/ns) value in the surveyed material was then entered as an initial input. Dependent on this value, the maximum thickness of the ballast layer, from where the reflected pulse could be recorded, was computed preliminarily using Equation (2). The number of samples per trace was set to 512 and 1024. Time window values were selected as 20 and 60 ns, respectively, for 2 GHz air-coupled antenna and for ground-coupled dual frequency antenna (400/900 MHz). Horizontal spacing of 0.02 m was selected for the surveys. In the SŽDC surveys, the step between traces was set to 0.25 and 1 m, respectively, in 2014 and 2017.

As an essential calibration base for GPR data processing, the existing thickness of the ballast layer under the sleepers was measured to compute the RDP values and hence the EMWV values. In this respect, two test pit locations were selected where the ballast condition was expected to be most representative for the track sections under investigation. The first in the P6U1 section was opened by an excavator, whereas the second trench was dug conventionally in the P2U2 section. Both of the test-pits are shown in Figure 5. The extracted material from the trenches has been properly marked and taken to the laboratories for relevant tests.

Figure 5. Ground truth data (in sections P2U2 & P6U1).

For the sake of clarity, the survey profiles in all of the sections were always executed in the advancing stationing direction. For each section, GPR data were collected using two different configurations of the air-coupled antenna. First, the longitudinal configuration (long axis of the antenna parallel to track axis) was used to practically determine the sleepers, which partially mask the reflections from the bottom of the ballast layer. Then, the air-coupled antenna was positioned diagonally to diminish the impact of the reinforced concrete sleepers on the GPR signal and to enhance the perceivability of the reflected signals from the ballast bottom interface. For the ground-coupled antenna, orientation was the same for all surveys.

3.2. Laboratory Tests

Two sets of tests were conducted on the collected field ballast aggregates: (i) percentage of fouling and (ii) gradation test of fouling material.

Sample #1

The first sample was extracted from section P6U1, where it was supposed to be most representative for the ballast condition in the second half of the reconstructed railway track sections. For this, 92.140 kg of material was taken from this test-pit, where Table 3 presents the weight of coarse material and finer fouling material, which were distinguished using a caliper.

Table 3. Total weight of ballast material categorized as fine and coarse material for sample #1.

Material Fraction	Weight	Percentage
Coarse fraction 31.5–63 mm	79.240 kg	86%
Fine fraction < 31.5 mm	12.900 kg	14%
Total material	92.140 kg	100%

By proportioning the weight of fouling material (grain size less than 31.5 mm) with the total material weight, the percentage of fouling according to Equation (4) was found as 14%.

$$\text{Percentage of fouling} = w_{fine} / w_{total,} \tag{4}$$

where w_{fine} and $w_{total,}$ stand, respectively, for the weight of fine fouling material and the total weight of material.

Gradation test was carried out for fine material (finer than 31.5 mm) and quotients of 19.67 % and 1.21% were obtained, respectively, for the sieve sizes of 4.75 and 0.075 mm. Hence the fouling index was computed as 20.88% according to Equation (3), matching the fouled ballast category according to Selig and Waters [26].

Sample #2

The second sample was extracted from the opened trench in the P2U2 section, where water drain ditches were not observed, which brought forward the question of whether this area is anomalously contaminated owing to the inadequate draining of water from the sections. From this test-pit, 47,331 kg of material was collected, where the weights of coarse and finer fouling material are displayed in Table 4.

Table 4. Total weight of ballast material categorized as fine and coarse material for sample #2.

Material Fraction	Weight	Percentage
Coarse fraction 31.5–63 mm	31.551 kg	66.66%
Fine fraction <31.5 mm	15.780 kg	33,34%
Total material	47.331 kg	100%

As a result of the calculation of the proportion of the weight of fouling material (grain size less than 31.5 mm) and the total weight of the material, the percentage of fouling (from Equation (4)) was obtained as 33.34%. This figure verifies the assumption with regard to the visual inspection that ballast cleaning is significantly required for this section because of the lacking water drain ditches.

Gradation tests were undertaken for fine material (size finer than 31.5 mm). Proportions of 39.92% and 2.57% were attained respectively, for the sieve sizes of 4.75 mm and 0.075 mm, yielding a fouling index of 42.49% (Equation (3)) which falls into the highly fouled ballast category according to Selig and Waters [26].

The gradation tests for fine materials were carried out thrice for sample #1 and sample #2. Average gradation curves can be seen in Figure 6 for both samples. Sieve opening sizes are marked in Figure 6 together with the input values (0.075 and 4.75 mm) for calculating the fouling index value by Selig and Waters [26].

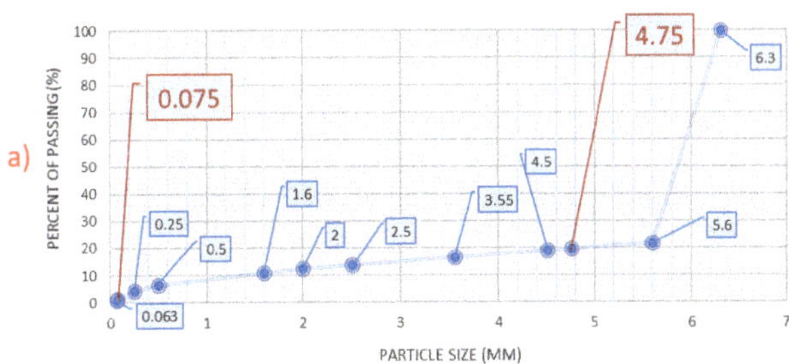

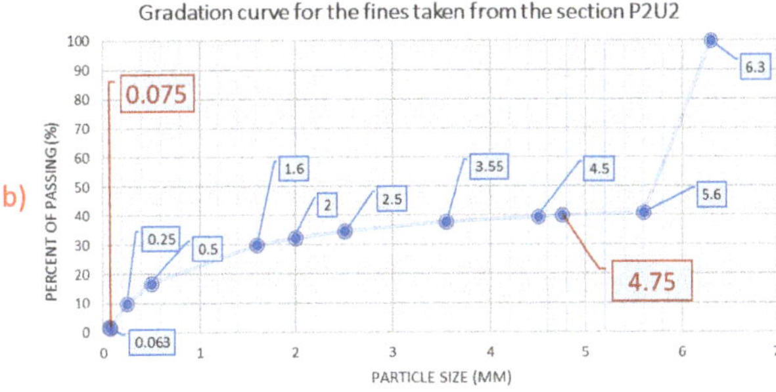

Figure 6. Gradation curves for the finer material extracted from the sections (**a**) P6U1 and (**b**) P2U2.

3.3. GPR Data Processing

Raw data are monitored and interpreted according to the in situ visual observations and the ground truth data. An example of a preliminary interpretation is given in Figure 7.

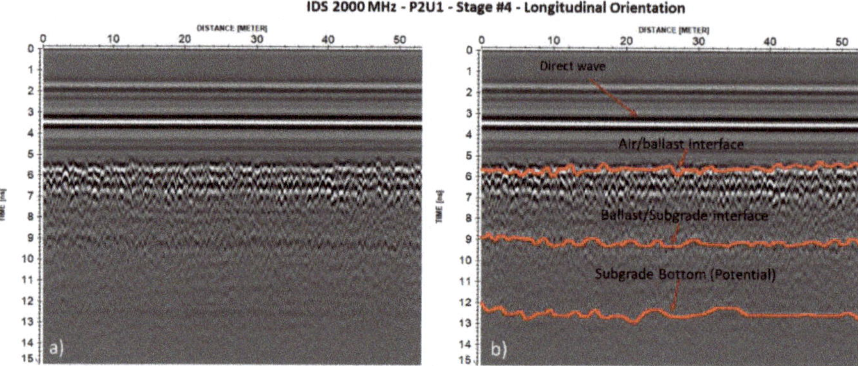

Figure 7. Preliminary interpretation of raw data from 2 GHz air-coupled antenna for section P2U1 at stage 4 with longitudinal orientation (**a**) raw data (only dewow filter is applied) (**b**) initial interface picking based on in situ visual observations and the ground truth data.

In comparison with the air-coupled antennas, ground-coupled antennas, particularly, with low frequencies provide lower resolutions to determine surface and interface positions very precisely in twt axis, especially in the case of the fouled ballast layer which corresponds to Stage 3 (measurements on 12 October 2018).

RDP and thus EMWV computations were based on the known thickness values of the ballast layer where Equations (1) and (2) were used. In the present study, used data processing functions included low frequency removal (dewow), time-zero and antenna bumping corrections, gain function, band-pass filtering, background removal, and running average filters. Dewow function was used to remove masking impacts of intrinsic lower frequency content in each antenna [62]. Move start time and correct max phase functions in ReflexW were employed for air-coupled antenna, respectively, to provide a static correction in time direction and compensate antenna bumping to bring the surface reflection in a horizontal position. Additionally, the gain function (with varying parameters) frequency was employed depending on the attenuation features of the material under survey to magnify the reflections in order to compensate for possible damping or geometric spreading losses. A band-pass filtering (band pass Butterworth in ReflexW software) was utilized within 1.5 times of the central frequency [63] for each antenna in order to enhance the signal to noise ratio by cutting off the side bands to avoid noise resulting from encompassing environment and hereditary loss of the GPR signal [7]. Background removal tool was utilized to discharge background noise with the aim of raising the signal-to-noise ratio. In this regard, the aim was to eliminate temporarily consistent noise from the whole B-scan and enable the signals visible, which were previously covered by this noise. Running average tool was then used to visualize horizontal interfaces in a more emphasized manner removing the trace-based clutter. An example of the abovementioned data processing scheme is given in Figure 8 for section P8U1 at stage 3 (before cleaning) in diagonal orientation of 2 GHz antenna.

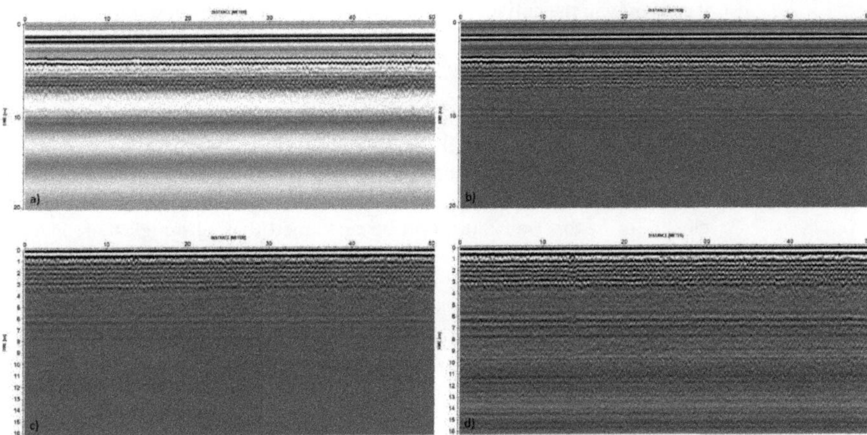

Figure 8. Data processing scheme for section P8U1 at stage 3 (before cleaning) in diagonal orientation of 2 GHz antenna: (**a**) raw GPR data, (**b**) dewow filter applied, (**c**) after move start time filter and after correct max phase applied, and (**d**) after band pass filter (Butterworth), background removal, gain, and running average processing steps.

Table 5 present the post-processing parameters used for each nominal frequency of the antenna used in the surveys.

Table 5. Post-processing parameters used for different frequencies.

Post-processing Step Parameters According to the Nominal Frequency	400 MHz	900 MHz	2000 MHz
Time-zero correction	According to the surface reflection twt		
De-wow (ns)	2.5	1.1	0.5
Correct max phase tool	–	–	Surface reflection twt frame
Band-pass filtering (Low pass–high pass in MHz)	100–700	225–1575	500–3500
Background removal	Whole line/distance range		
Running average (# of traces)	3	3	3

Near-field zone effect should be noted here in the case of ground-coupled antennas. This case happens owing to the powerful EM field within the vicinity of the antenna covering a radius of approximately 1.5 times the wavelength of the central frequency developed by EM energy transmitted from the surface of the antenna. This zone is larger in the case of ground-coupled antennas (since they usually operate with lower frequencies) than in the case of air-coupled antennas often operating with higher central frequencies. Therefore, surface reflection determination is more challenging in the case of ground-coupled antennas than the air-coupled one in this study and the differences in RDP values might stem from the near zone effect in the case of ground-coupled antennas.

A devoted data processing flow was used taking into account the intervals and the width of the sleepers to ensure that RDP computations would be made in the picked traces positioned precisely within the cribs, which would avoid the obstructive impacts of the reinforced concrete sleepers in the GPR signals. For this purpose, a text file was formed including the computed trace spacings corresponding to the start and the end points of each crib. Those trace spacings were identified in compliance with the monitored sleepers from the GPR radargrams, intervals, and the width of the sleepers. Then, this text file was incorporated into the abovementioned processing flow, which enabled to obtain RDP values only in the crib regions. Moreover, 228 traces were assessed in each phase, and RDP values were calculated.

4. Results

According to the preliminary visual inspection in section P2, substandard outcomes were anticipated because of the lacking water drainage utilities particularly in P2U2, hence the substandard status of the ballast layer was assumed. The computed values for the EMWV values were unexpectedly and relatively high, even in the stage before ballast cleaning. At this point, it might be presumed that, although there is a significant rate of finer size material (33.34%) obtained through the laboratory experiments (sample #2), there was not any moisture observed in the trackbed. It is obvious that as a result of the reconstruction/maintenance activities, the overall state of the railway substructure was enhanced, which is supported by the RDP/EMVW values in Figure 9. Particularly between phases 3 and 5, EMWV values raised by 19.3% and 34.4% in the P2U1 and P2U2 sections, respectively.

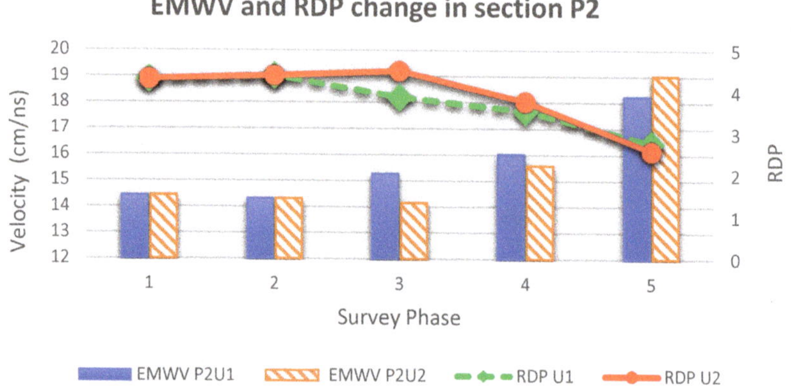

Figure 9. The change of Electromagnetic Wave Velocity (EMWV) and RDP values during the lifecycle of the railway track in section P2.

A generally decent state of the ballast layer was foreseen in the P3 section, which was affirmed both by the SŽDC surveys (phases 1 and 2) and all 2018 surveys (phases 3, 4, and 5). The drainage utilities were working properly. Obviously, as a result of the intervention activities, enhancement can be observed in the substructure also displayed in Figure 10. Specifically, between phases 3 and 5, the EMWV values increased by 24.1% and 21.6% in the P3U1 and P3U2 sections, respectively.

The P6 section was identified by the existence of bothersome vegetation (particularly in P6U1), revealing a high trapped moisture ratio in the railway infrastructure. This situation might stem from a handful of reasons such as substandard draining of the railway substructure, the existence of swampy spots, and/or the water existing beneath the substructure. Despite the percentage of fines obtained by laboratory experiments in this section (14%) is lower than the one found in section P2 (33.34%), water existed in the trackbed apparently, as also verified by the EMWV/RDP values collected by GPR before the ballast cleaning process. This situation was also affirmed by the distinctive reflections at deeper points at P6U1, which may result from the existing aquifer areas. It is evident that as a result of the mechanized ballast cleaning action coupled with other reconstruction/maintenance activities, the ballast layer state is upgraded, which might also be seen in Figure 11. EMWV values raised by 34.4% and 20.3% in the P6U1 and P6U2 sections, respectively, in comparison of the phases between 3 and 5. It is also worth mentioning the abnormal pattern from phases 3 to 5. More explicitly, in phase 4, a higher EMWV value was found than the one in phase 5. It is probable that the combined utilization of the RM 900 VB ballast cleaner/undercutter and Unimat 08-275/3S tamping machine contributed to a hoed ballast layer (thus the bulk density was decreased). It is also significant to consider the precise timing of GPR data collection (as to the point that if the survey occurred just after ballast cleaning or during tamping operation cycles).

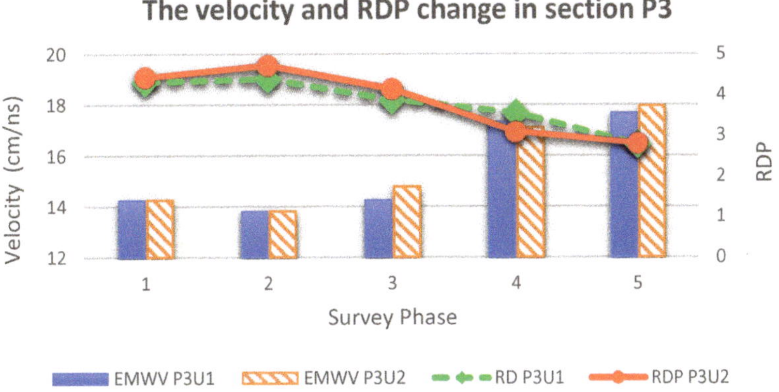

Figure 10. The change of EMWV and RDP values during the lifecycle of the railway track in section P3.

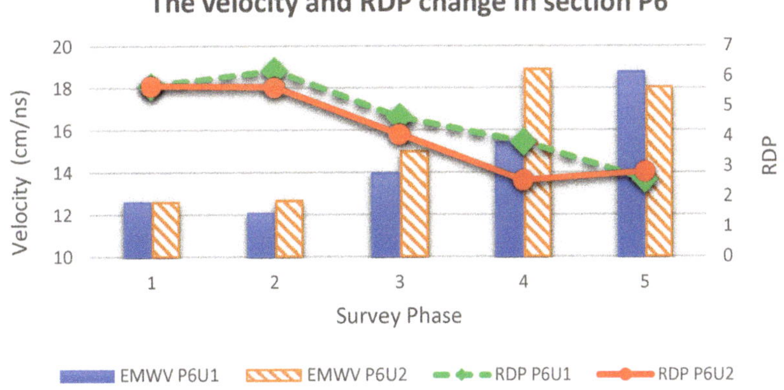

Figure 11. The change of EMWV and RDP values during the lifecycle of the railway track in section P6.

There was also unwanted vegetation in section P8, similar to section P6 (yet, to a lower extent). Obviously, mechanized ballast cleaning and reconstruction activities led to an improvement in the trackbed condition as can be also followed in Figure 12. EMWV values raised by 11.6% and 39.6% in the P8U1 and P8U2 sections, respectively, between survey stages of 3 and 5. Such high improvement in the case of P8U2 section might be associated with the drainage of water achieved by cleaning process.

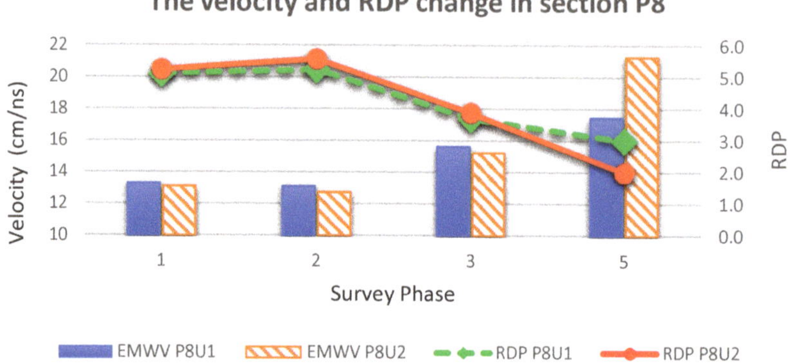

Figure 12. The change of EMWV and RDP values during the lifecycle of the railway track in section P8.

There were no meaningful variations at P2 and P3 between the survey phases of 1, 2, and 3, however, more obvious changes at P6 and P8 were observed. This may occur because of the presumed higher moisture content at P6 and P8—where the different GPR signal attenuation based on the antenna frequency should be considered. Variations noticed between the GPR survey phases 3 (prior to ballast cleaning) and 5 (final state) match with and even surpass the values in the literature, which might occur due to GPR data collection in the final phase immediately after the completion of the mechanized ballast cleaning operation. Data from the survey stage 4 provide complementary values to characterize the ballast cleaning process since all reconstruction activities at this stage include an enormous number of non-measurable cycles in the track infrastructure (recurring vertical alignment variation and supply of new ballast material while tamping process was being undertaken). Due to the ongoing reconstruction/maintenance activities, no entrance was allowed into the P8 section during phase 4.

The condition of the ballast layer was upgraded in all of the sections. The most noteworthy enhancement was found in the P8U2 section, whereas the least improvement was observed in the P8U1 section. This might be because of the existing water before ballast cleaning process in both sections of P8. However, in addition to the standard new ballast material, the leftover/residual material (finer and broken-down material from the bottom of the wagon) was also spilled into the track after cleaning in section P8U1 particularly. This resulted in an unmanageable fraction fill and a higher extent of dust particles, which was clearly monitored visually in situ. Significant improvements were also observed in the P2U2 section, where higher preliminary ballast fouling was visually detected, compared to P2U1. An analogy can be recognized between P6U1 and P6U2 sections, where the existence of water in these sections was recorded (greater in P6U1 than in P6U2).

Differences between the values measured by SŽDC (phase #1 in 2014 and #2 in 2017) and 2018 surveys (from #3 to #5) occurred potentially by dissimilar antenna criteria from variant GPR providers. Any information regarding ballast layer interventions and reconstruction activities was not found between 2014 and 2018 in the surveyed tracks. Obviously, in phases 4 and 5, the EMWV values are significantly different from phase #3, which endorses the eligibility of the mechanized ballast layer cleaning process. In general, the aggregate features of the ballast layer were homogenized and enhanced in the maintained section. The percentage improvement values in EMWV values between phases 3 and 5 for each section are indicated in Table 6 (EMWV values are rounded up to three decimal places in Table 6). An average value of 25.7 % improvement in EMWV values has been achieved using the conventional intervention methods of mechanized ballast cleaning (Table 6).

Table 6. Percentual increase in Electromagnetic Wave Velocity (EMWV) values during the maintenance works between phase 3 and phase 5.

Track Sections	EMWV before Ballast Cleaning Phase #3 (cm/ns)	EMWV in the Final State Phase #5 (cm/ns)	Percentage Improvement
P2U1	0.153	0.182	19.3
P2U2	0.142	0.190	34.4
P3U1	0.142	0.177	24.1
P3U2	0.148	0.180	21.6
P6U1	0.140	0.188	34.4
P6U2	0.150	0.180	20.3
P8U1	0.156	0.175	11.6
P8U2	0.152	0.213	39.6
Average Improvement in all surveyed track sections			25.7

Changes in EMWV values in each section corresponding to GPR survey phases 1, 3, 5 are displayed in Figure 13.

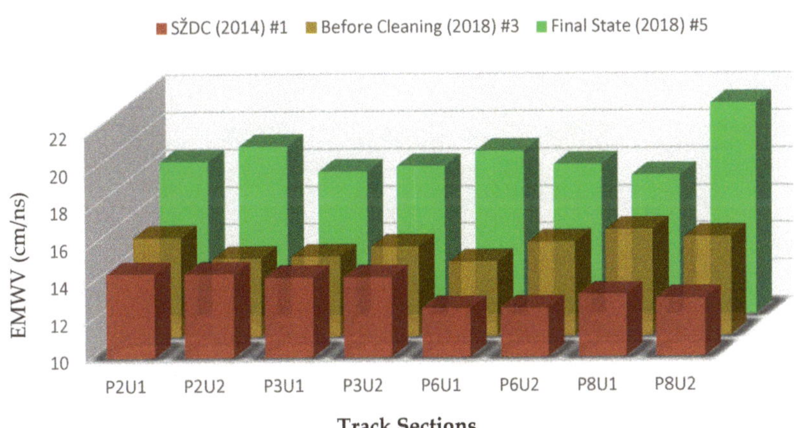

Figure 13. Comparison of each section for phases 1, 3, and 5.

5. Discussion

Based on in situ observations and laboratory tests of the materials taken from the site, and the values published for that particular type of ballast (granite) in various studies [31–34,51,52], authors suggest fouling categories based on RDP/EMWV intervals in Table 7. It should be noted that the suggested values apply solely to the local conditions of railway track sections surveyed in the present study and the parameters (frequency and orientation) of used GPR equipment.

Table 7. Suggested fouling intervals for granite railway ballast.

Fouling Category	Interval for EMWV (cm/ns)	Interval for RDP
Clean [1]	≥16	≤3.52
Slightly fouled [1]	13–16	3.52–5.33
Highly fouled [1]	9–13	5.33–11.11
Unrecommended for traffic [1]	<9	>11

[1] Colors represent the fouling categories based on RDP/EMWV values as they are also used in Table 8.

Table 8. RDP values of granite ballast layers for each section and survey phase (colors correspond to Table 7).

SECTIONS		#1 SZDC 23 September 2014 400 MHz GSSI	#2 SZDC 29 August 2017 400 MHz GSSI	#3 RDP before Cleaning Process 12 October 2018 400 MHz IDS	900 MHz IDS	2 GHz IDS	#4 RDP after Cleaning Process 17 October 2018 400 MHz IDS	900 MHz IDS	2 GHz IDS	#5 RDP after Handing Over the Construction Site 19 October 2018 400 MHz IDS	900 MHz IDS	2 GHz IDS
P2	U1	4.31	4.39	3.65	3.65	4.31	3.09	3.42	4.13	2.20	2.68	3.44
P2	U2	4.31	4.39	4.92	4.34	4.26	3.09	3.79	4.42	1.80	2.88	3.10
P3	U1	4.43	4.72	4.70	4.34	4.29	2.88	2.72	3.45	2.75	2.52	3.50
P3	U2	4.43	4.72	3.85	4.26	4.29	2.93	3.36	2.96	2.72	2.57	3.11
P6	U1	5.68	6.19	4.85	4.68	4.35	3.73	4.29	3.41	2.33	2.57	2.80
P6	U2	6.68	5.63	3.31	4.43	4.43	2.56	2.82	2.23	2.27	3.37	2.82
P8	U1	5.10	5.24	3.35	3.84	3.89	–	–	–	3.63	2.66	2.72
P8	U2	5.24	5.56	3.73	3.98	3.91	–	–	–	1.99	2.04	1.94

There did not exist any critical variations or patterns between the RDP/EMWV values of the longitudinal and the diagonal arrangement of the air-coupled antenna during GPR surveys. However, GPR data obtained from both orientations were utilized for confirming one another and analyzing the locations of the identified interfaces and sleepers, which was the preliminary intention for using various orientations. The comparison and evaluation of the variations in the evaluated indicators were first undertaken in each section independently for each survey stage, and then, all of the sections were analyzed in comparison with one another.

During the GPR surveys, three central frequencies were used (400 MHz, 900 MHz, and 2 GHz) for comparison of collected data. All computed RDP values are tabulated in Table 8 for every section and survey stage, where the variations according to frequencies and antennas can be clearly observed. Even though the RDP values calculated with the GPR data collected with various antennas and frequencies were different, no direct reliance of RDP/EMWV values on central frequency was found. Therefore, the figures (Figure 9, Figure 10, Figure 11, Figure 12, and Figure 13) display the mean values of RDP and EMWV for the frequencies used in the survey stages 3, 4, and 5.

A synopsis might be formulated with regards to the article's aim, i.e., what is required to be figured out consecutively in the evaluation of the railway ballast layer states using time-domain GPR analysis.

1. Determination of GPR survey conditions, equipment criteria, and settings,
2. Determination of a data processing flow on the basis of step 1, (time-domain),
3. Computation of RDP/EMWV values on the basis of steps 1 and 2, (time-domain),
4. Selecting the standard fouling indicator(s) for ballast material(s) and setting its threshold
5. Comparison/correlation of RDP/EMWV values from step 3 and standard fouling indicator from step 4,
6. Decision-making as to when/whether the mechanized ballast cleaning intervention should be initiated.

It should be examined whether the existence of water or fouling material is the reason for the increasing RDP values. When the cause of the rising RDP values is water content, then it is better to investigate that if it is a temporary case (e.g., due to a recent rain) or a permanent one caused by draining incapability, etc. As a rule of thumb, GPR surveys are not suggested to be undertaken in moist circumstances.

6. Conclusions

This paper presented the condition monitoring of the railway granite ballast layer and identifying the variation of the degree of fouling by Ground Penetrating Radar (GPR) technique prior to and following the mechanical ballast cleaning process in a real railway line portion.

GPR surveys were undertaken in eight track sections with a total length of 400 m (each 50 m) within the Čáslav-Kutná Hora railway track, using two types of antennas with three dissimilar central frequencies and two different antenna orientations. GPR surveys were performed thrice for each section at different stages of the lifecycle of the railway track, specifically, prior to and following the mechanized cleaning activity of the ballast layer. The time-domain analyses for recorded data in 2018 (phases #3, #4, and #5) and data obtained from Správa železniční dopravní cesty (phase #1 in 2014, and phase #2 in 2017) were performed. For the calibration, ground truth data were used.

Changes in relative dielectric permittivity (RDP) values between GPR investigation phases 3 (prior to mechanized ballast cleaning) and 5 (final state), which are similar or even greater than the ones in the literature (10–30%), were observed. This occurrence might stem from the fact that the GPR survey was carried out in the final phase immediately after the ballast cleaning process. An average value of 25.7% improvement in EMWV values, which has been measured by GPR surveys, has been achieved using the conventional intervention methods of mechanized ballast cleaning. In the laboratory, percentage of fouling and the gradation tests for fine material distribution were conducted, and the fouling index was calculated. In view of the comparison between laboratory experiments and GPR field surveys, it can be confirmed that the existence of trapped water within the ballast layer has a considerably greater effect in the rise of RDP values (and in the decrease in electromagnetic wave velocity (EMWV) values) compared to the case of ballast fouling by infiltration of fine particles only. However, it should be noted that those two phenomena are closely connected to each other.

On the basis of the GPR track surveys, laboratory experiments, and comparisons with published values of RDP/EMWV for that particular type of ballast (granite), the cleaning process could be, in the authors' opinion, recommended launching approximately at the time when relevant indicators lie in the interval of highly fouled category suggested by authors in Table 7. It should be noted that the suggested values apply solely to the parameters of railway track sections surveyed and the parameters of used GPR equipment in the present study.

As a result, quantification of the influence of the mechanized ballast cleaning process on the condition of the railway ballast layer using GPR technology has been achieved for surveyed track sections. Improvement in the ballast layer status has been observed in all sections after the mechanized ballast cleaning process. Moreover, in situ RDP values of the granite ballast layer obtained in this study provide additional literature data for that particular type of ballast material.

Integration of several NDT methods (e.g., GPR and TDR) for assessment of railway infrastructure layers could task itself as a future work for a better understanding of the ballast fouling and moisture retention mechanisms through data fusion.

This research provides new perspectives into the decision-making process in initiating the mechanized ballast cleaning intervention based on the GPR-measured RDP/EMWV values obtained in different life cycles of a real track segment. The results from this study suggest that quantification of the efficiency of the mechanized ballast cleaning process can be non-destructively obtained through GPR measurements.

Author Contributions: Conceptualization, S.S.A. and V.B.; investigation, A.B., V.B., and F.Š.; data curation, S.S.A. and V.B.; writing—original draft preparation, A.B.; writing—review and editing, S.S.A. All authors have read and agreed to the published version of the manuscript.

Funding: The APC was funded by the project "International mobility of employees of the University of Pardubice II" with project number CZ.02.2.69/0.0/0.0/18_053/0016969.

Institutional Review Board Statement: Not applicable.

Informed Consent Statement: Not applicable.

Data Availability Statement: The data presented in this study are available on request from the corresponding author. The data are not publicly available due to the ongoing processing and examination for further interpretations and results with different analyzing methods.

Acknowledgments: All GPR surveys were conducted using the equipment of the Faculty of Transport Engineering, University of Pardubice. This paper is of interest to COST Action TU1208 Civil Engineering Applications of Ground Penetrating Radar. SŽDC has kindly provided GPR data from surveys held in 2014 and 2017 for the same track sections.

Conflicts of Interest: The authors declare no conflict of interest.

References

1. Qian, Y.; Dersch, M.S.; Gao, Z.; Edwards, J.R. Railroad infrastructure 4.0: Development and application of an automatic ballast support condition assessment system. *Transp. Geotech.* **2019**, *19*, 19–34. [CrossRef]
2. Tamrakar, P.; Azari, H.; Yuan, D.; Nazarian, S. Implementation of spectral analysis of surface waves approach for characterization of railway track substructures. *Transp. Geotech.* **2017**, *12*, 101–111. [CrossRef]
3. Mvelase, G.M.; Gräbe, P.J.; Anochie-Boateng, J.K. The use of laser technology to investigate the effect of railway ballast roundness on shear strength. *Transp. Geotech.* **2017**, *11*, 97–106. [CrossRef]
4. Schmidt, S.; Shah, S.; Moaveni, M.; Landry, B.J.; Tutumluer, E.; Basye, C.; Li, D. Railway Ballast Permeability and Cleaning Considerations. *Transp. Res. Rec.* **2017**, *2607*, 24–32. [CrossRef]

5. Ižvolt, L.; Dobeš, P.; Mečár, M. Calibration of TDR Test Probes for Measuring Moisture Changes in the Construction Layers of the Railway Line. *Procedia Eng.* **2016**, *161*, 1057–1063. [CrossRef]

6. Pieš, J.; Môcová, L. Application of TDR Test Probe for Determination of Moisture Changes of Railway Substructure Materials. *Transp. Res. Proc.* **2019**, *40*, 74–81. [CrossRef]

7. Bianchini Ciampoli, L.; Calvi, A.; D'Amico, F. Railway ballast monitoring by GPR: A test site investigation. *Rem Sens.* **2019**, *11*, 2381. [CrossRef]

8. Tosti, F.; Bianchini Ciampoli, L.; Calvi, A.; Alani, A.M.; Benedetto, A. An investigation into the railway ballast dielectric properties using different GPR antennas and frequency systems. *NDT E Int.* **2018**, *93*, 131–140. [CrossRef]

9. Benedetto, A.; Pajewski, L. *Civil Engineering Applications of Ground Penetrating Radar*; Springer International Publishing: Cham, Germany, 2015. [CrossRef]

10. Artagan, S.S.; Bianchini Ciampoli, L.; D'Amico, F.; Calvi, A.; Tosti, F. Non-destructive Assessment and Health Monitoring of Railway Infrastructures. *Surv. Geophys.* **2020**, *41*, 447–483. [CrossRef]

11. Jack, R.; Jackson, P. Imaging attributes of railway track formation and ballast using ground probing radar. *NDT E Int.* **1999**, *32*, 457–462. [CrossRef]

12. Gallagher, G.P.; Leiper, Q.; Williamson, R.; Clark, M.R.; Forde, M.C. The application of time domain ground penetrating radar to evaluate railway track ballast. *NDT E Int.* **1999**, *32*, 463–468. [CrossRef]

13. Hugenschmidt, J. Railway track inspection using GPR. *J. Appl. Geophys.* **2000**, *43*, 147–155. [CrossRef]

14. Hyslip, J.P.; Smith, S.S.; Olhoeft, G.R.; Selig, E.T. Assessment of railway track substructure condition using ground penetrating radar. In Proceedings of the 2003 annual conference of AREMA, Chicago, IL, USA, 5–8 October 2003.

15. Alemu, A.Y. Survey of Railway Ballast Selection and Aspects of Modelling Techniques. Master's Thesis, Royal Institute of Technology, Stockholm, Sweden, 2011.

16. Anbazhagan, P.; Dixit, P.S.N.; Bharatha, T.P. Identification of Type and Degree of Railway Ballast Fouling Using Ground Coupled GPR Antennas. *J. Appl. Geophys.* **2016**, *126*, 183–190. [CrossRef]

17. Misar, H. Criteria for cost-effective ballast cleaning: Machine design considerations. *Rail. Eng. Int.* **2005**, 4.

18. Schilling, R. Ballast cleaning of single-track railway lines: A strategic analysis. *Rail. Eng. Int.* **2005**, *1*, 8–12.

19. Korolev, V.; Loktev, A.; Shishkina, I.; Zapolnova, E.; Kuskov, V.; Basovsky, D.; Aktisova, O. Technology of crushed stone ballast cleaning. *IOP Conf. Ser. Earth Environ. Sci.* **2019**, *403*, 012194. [CrossRef]

20. Tzanakakis, K. *The Railway Track and Its Long Term Behavior: A Handbook for a Railway Track of High Quality. Springer Tracts on Transportation and Traffic*; Springer: Berlin/Heidelberg, Germany, 2013.

21. Annan, A.P.; Davis, J.L. Ground penetrating radar—Coming of age at las!! In *Geophysics and Geochemistry at the Millennium*; GEO F/X: Johnsburg, IL, USA, 1997; pp. 515–522.

22. Jol, H. *Ground Penetrating Radar Theory and Applications*, 1st ed.; Elsevier: Amsterdam, The Netherlands, 2018.

23. Grote, K.; Hubbard, S.; Harvey, J.; Rubin, Y. Evaluation of infiltration in layered pavements using surface GPR reflection techniques. *J. Appl. Geophys.* **2005**, *57*, 125–129. [CrossRef]

24. Daniels, D.J. *Ground penetrating Radar*, 2nd ed.; The Institution of Electrical Engineers: London, UK, 2004.

25. Artagan, S.S. Use of Ground Penetrating Radar in Condition Assessment of Railway Ballast. Ph.D. Thesis, Univesity of Pardubice, Pardubice, Czechia, 2018.

26. Selig, E.T.; Waters, J.M. *Track Geotechnology and Substructure Management*; Thomas Telford Publications: London, UK, 1994.

27. Foun, D. Planning ballast cleaning using ballast fouling levels determined with ground-penetrating radar online. In Proceedings of the CORE 2012: Global Perspectives; Conference on railway engineering, Brisbane, Australia, 10–12 September 2012; pp. 697–705.

28. Anbazhagan, P. *Characterization of Rail Track Ballast Fouling Using Ground Penetration Radar and Field Sampling. Project Report*; Indian Institute of Science: Bangalore, India, 2013.

29. Fontul, S.; Fortunato, E.; Chiara, F.D. Evaluation of ballast fouling using GPR. In Proceedings of the 15th International Conference on Ground Penetrating Radar, Brussels, Belgium, 30 June–4 July 2014; ISBN 978-14-799-6789-6. [CrossRef]

30. Lalagüe, A. Use of Ground Penetrating Radar for Transportation Infrastructure Maintenance. Ph.D. Thesis, Norwegian University of Science and Technology, Trondheim, Norway, 2015.

31. Clark, M.R. Electromagnetic properties of railway ballast. *NDT E Int.* **2001**, *34*, 305–311. [CrossRef]

32. Sussmann, T.R. Application of Ground Penetrating Radar to Railway Track Substructure Maintenance Management. Ph.D. Thesis, University of Massachusett–Amherst, Amherts, MA, USA, 1999.

33. Leng, Z.; Al-Qadi, I. Railroad Ballast Evaluation Using Ground-Penetrating Radar: Laboratory Investigation and Field Validation. *Transp. Res. Rec. J. Transp. Res. Board.* **2010**. [CrossRef]

34. Artagan, S.S.; Borecký, V. Advances in the nondestructive condition assessment of railway ballast: A focus on GPR. *NDT E Int.* **2020**, *115*, 102290. [CrossRef]

35. Maturana, R.M.; Bautista, B.D.; Aguacil, Á.A.; Plaza, M.R.; Castaño, S.S. Preventive Maintenance of Railway Infrastructures using GPR–Ground Penetrating Radar. In Proceedings of the 9th World Congress on Railway Research, Lille, France, 22–26 May 2011.

36. Roberts, R.; Al-Qadi, I.L.; Tutumluer, E.; Kathage, A. Ballast fouling assessment using 2 GHz horn antennas-GPR and ground truth comparison from 238 km of track. In Proceedings of the 9th International Railway Engineering Conference, London, UK, 1 June 2007.

37. Roberts, R.; Rudy, J.; GSSI, S. Railroad ballast fouling detection using ground penetrating radar. A new approach based on scattering from voids. In Proceedings of the 9th European Conference on NDT, Berlin, Germany, 25–29 September 2006. ECNDT 2006–Th. 4.5 1..

38. Al-Qadi, I.L.; Xie, W.; Roberts, R. Scattering analysis of ground-penetrating radar data to quantify railroad ballast contamination. *NDT E Int.* **2008**, *41*, 441–447. [CrossRef]

39. Su, L.-J.; Rujikiatkamjorn, C.; Indraratna, B. An Evaluation of Fouled Ballast in a Laboratory Model Track Using Ground Penetrating Radar. *Geotech. Test. J.* **2010**, *33*, 103045. [CrossRef]

40. Al-Qadi, I.L.; Xie, W.; Roberts, R.; Leng, Z. Data analysis techniques for GPR used for assessing railroad ballast in high radio-frequency environment. *J. Transp. Eng.* **2010**, *136*, 392–399. [CrossRef]

41. De Bold, R.P. Non-Destructive Evaluation of Railway Trackbed Ballast. Ph.D. Thesis, University of Edinburgh, School of Engineering, Glasgow, Scotland, 2011.

42. Zhang, Q.; Gascoyne, J.; Eriksen, A. Characterisation of ballast materials in trackbed using ground penetrating radar: Part 1. In Proceedings of the Railway Condition Monitoring and Non-Destructive Testing, 2011, (RCM 2011), 5th IET Conference On. IET, Kuala Lumpur, Malaysia, 5–6 September 2018; pp. 1–8.

43. Anbazhagan, P.; Lijun, S.; Buddhima, I.; Cholachat, R. Model track studies on fouled ballast using ground penetrating radar and multichannel analysis of surface wave. *J. Appl. Geophys.* **2011**, *74*, 175–184. [CrossRef]

44. De Chiara, F.; Fontul, S.; Fortunato, E. GPR Laboratory Tests for Railways Materials Dielectric Properties Assessment. *Remote Sens.* **2014**, *6*, 9712–9728. [CrossRef]

45. Kashani, H.; Ho, C.L.; Oden, C.P.; Smith, S.S. Model Track Studies by Ground Penetrating Radar (GPR) on Ballast with Different Fouling and Geotechnical Properties. In Proceedings of the Joint Rail Conference ASME, San Jose, CA, USA, 23–26 March 2015. [CrossRef]

46. Benedetto, A.; Tosti, F.; Bianchini Ciampoli, L.; Calvi, A.; Brancadoro, M.G. Alani, A.M. Railway ballast condition assessment using ground-penetrating radar—An experimental, numerical simulation and modelling development. *Constr. Build. Mater.* **2017**, *140*, 508–520. [CrossRef]

47. Kashani, H. Evaluating the Influence of Breakdown Fouling and Moisture Content on Mechanical and Electromagnetic Properties of Ballasted Railroad Track. Ph.D. Thesis, University of Massachusett–Amherst, Amherst, MA, USA, 2017.

48. Fontul, S.; Paixão, A.; Solla, M.; Pajewski, L. Railway Track Condition Assessment at Network Level by Frequency Domain Analysis of GPR Data. *Remote Sens.* **2018**, *10*, 599. [CrossRef]

49. Barrett, B.E.; Day, H.; Gascoyne, J.; Eriksen, A. Understanding the capabilities of GPR for the measurement of ballast fouling conditions. *J. Appl. Geophys.* **2019**, *169*, 183–198. [CrossRef]

50. Keogh, T.; Mesher, D.E.; Keegan, T.R. *An Integrated System for Accurate Tie and Ballast Condition Assessment*; Rail Radar ™ Inc; Canadian National Railway: Alberta, Canada, 2006.

51. Sussmann, T.R.; O'Hara, K.R.; Selig, E.T. Development of material properties for railway application of ground-penetrating radar. In Proceedings of the Resented at the Ninth International Conference on Ground Penetrating Radar (GPR2002), Santa Barbara, CA, USA, 29 April–2 May 2002; pp. 42–47. [CrossRef]

52. Göbel, C.; Hellmann, R.; Petzold, H. Georadar-model and in-situ investigations for inspection of railway tracks. In Proceedings of the Fifth International Conferention on Ground Penetrating Radar, Kitchener, Canada, 12–16 June 1994.

53. Artagan, S.S.; Borecký, V.; Bartoš, J.; Kurel, R. Moisture Influence on the GPR-Measured RDP Values of Granite Ballast under Clean and Fouled Conditions. In Proceedings of the 7th International Scientific Conference of the Faculty of Transport Engineering, Pardubice, Czech Republic, 6–7 September 2018.

54. Benedetto, F.; Tosti, F.; Alani, A.M. An entropy-based analysis of GPR data for the assessment of railway ballast conditions. *IEEE Trans. Geosci. Rem. Sens.* **2017**, *55*, 3900–3908. [CrossRef]

55. Bianchini Ciampoli, L.; Tosti, F.; Brancadoro, M.G.; D'Amico, F.; Alani, A.M.; Benedetto, A. A spectral analysis of ground-penetrating radar data for the assessment of the railway ballast geometric properties. *NDT E Int.* **2017**, *90*, 39–47. [CrossRef]

56. Benedetto, A.; Bianchini Ciampoli, L.; Brancadoro, M.G.; Alani, A.M.; Tosti, F. A computer-aided model for the simulation of railway ballast by random sequential adsorption process. *Comput. Aided. Civ. Infrastruct. Eng.* **2018**, *33*, 243–257. [CrossRef]

57. Fontul, S.; Mínguez, R.; Solla, M.; Santos-Assunção, S. The use of geophysics for the condition assessment of railway infrastructure. In *Non-Destructive Techniques for the Evaluation of Structures and Infrastructure*; Riveiro, B., Solla, M., Eds.; CRC Press: Boca Raton, FL, USA, 2016.

58. Solla, M.; Pérez-Gracia, V.; Fontul, S. A Review of GPR Application on Transport Infrastructures: Troubleshooting and Best Practices. *Remote Sens.* **2021**, *13*, 672. [CrossRef]

59. Ferrante, C.; Bianchini Ciampoli, L.; Benedetto, A.; Alani, A.M.; Tosti, F. Non-destructive Technologies for a Sustainable Assessment and Monitoring of Railway Infrastructures: A Focus on GPR and InSAR Methods. 2021. Available online: https://assets.researchsquare.com/files/rs-216510/v1_stamped.pdf (accessed on 1 March 2021).

60. Khakiev, Z.; Shapovalov, V.; Kruglikov, A.; Yavna, V. GPR determination of physical parameters of railway structural layers. *J. Appl. Geophys.* **2014**, *106*, 139–145. [CrossRef]

61. Khakiev, Z.; Shapovalov, V.; Kruglikov, A.; Morozov, A.; Yavna, V. Investigation of long term moisture changes in trackbeds using GPR. *J. Appl. Geophys.* **2014**, *110*, 1–4. [CrossRef]

62. Bianchini Ciampoli, L.; Tosti, F.; Economou, N.; Benedetto, F. Signal Processing of GPR Data for Road Surveys. *Geoscience* **2019**, *9*, 96. [CrossRef]
63. Bianchini Ciampoli, L.; Artagan, S.S.; Tosti, F.; Gagliardi, V.; Alani, A.M.; Benedetto, A. A comparative investigation of the effects of concrete sleepers on the GPR signal for the assessment of railway ballast. In Proceedings of the 17th international conference on ground penetrating radar (GPR2018), Rapperswil, Switzerland, 18–21 June 2018. [CrossRef]

remote sensing

Article

Identifying Spatial and Temporal Variations in Concrete Bridges with Ground Penetrating Radar Attributes

Vivek Kumar [1,*], Isabel M. Morris [2], Santiago A. Lopez [2] and Branko Glisic [1]

[1] Department of Civil and Environmental Engineering, Princeton University, Princeton, NJ 08540, USA; bglisic@princeton.edu

[2] Department of Civil and Environmental Engineering, New Mexico Institute of Mining and Technology, Socorro, NM 87801, USA; isabel.morris@nmt.edu (I.M.M.); santiago.lopez@student.nmt.edu (S.A.L.)

* Correspondence: vivekkr1809@gmail.com

Abstract: Estimating variations in material properties over space and time is essential for the purposes of structural health monitoring (SHM), mandated inspection, and insurance of civil infrastructure. Properties such as compressive strength evolve over time and are reflective of the overall condition of the aging infrastructure. Concrete structures pose an additional challenge due to the inherent spatial variability of material properties over large length scales. In recent years, nondestructive approaches such as rebound hammer and ultrasonic velocity have been used to determine the in situ material properties of concrete with a focus on the compressive strength. However, these methods require personnel expertise, careful data collection, and high investment. This paper presents a novel approach using ground penetrating radar (GPR) to estimate the variability of in situ material properties over time and space for assessment of concrete bridges. The results show that attributes (or features) of the GPR data such as raw average amplitudes can be used to identify differences in compressive strength across the deck of a concrete bridge. Attributes such as instantaneous amplitudes and intensity of reflected waves are useful in predicting the material properties such as compressive strength, porosity, and density. For compressive strength, one alternative approach of the Maturity Index (MI) was used to estimate the present values and compare with GPR estimated values. The results show that GPR attributes could be successfully used for identifying spatial and temporal variation of concrete properties. Finally, discussions are presented regarding their suitability and limitations for field applications.

Keywords: structural health monitoring; ground penetrating radar; attribute analysis; in situ material property; machine learning; maturity method

Citation: Kumar, V.; Morris, I.M.; Lopez, S.A.; Glisic, B. Identifying Spatial and Temporal Variations in Concrete Bridges with Ground Penetrating Radar Attributes. *Remote Sens.* **2021**, *13*, 1846. https://doi.org/10.3390/rs13091846

Academic Editor: Mercedes Solla, Vega Perez-Gracia and Simona Fontul

Received: 30 March 2021
Accepted: 30 April 2021
Published: 9 May 2021

Publisher's Note: MDPI stays neutral with regard to jurisdictional claims in published maps and institutional affiliations.

1. Introduction

A significant proportion of the United States' infrastructure has exceeded its intended service life [1]. ASCE's 2021 report-card on the state of US infrastructure grades bridges at C with approximately 7.5% of the assessed bridges classified as structurally deficient, with 42% of the total bridge stock being over 50 years old [2,3]. To extend the remaining useful life (RUL) of these structures, structural health monitoring (SHM) and nondestructive evaluation (NDE) techniques are used by engineers. SHM can be considered as a set of processes that are aimed at providing actionable information to the stakeholders using various monitoring strategies [4]. The monitoring could be permanent or periodic to evaluate mechanical, physical, and chemical parameters over multiple scales (material or structural). Various monitoring strategies based on strain, vibration, deflection, electrical impedance, and pulse velocity in a medium have been commonly employed for SHM of transport infrastructure. Their use for regular condition assessment of transport infrastructure [5–10] has increased to reduce the total life-cycle cost of the structure [11–14].

Estimation of in situ material properties is key to these infrastructure assessment steps. These properties provide key data which form the basis for structural computations, including those regarding safety, remaining capacity, and actual maintenance needs. Between SHM and NDE, NDE is favored for evaluation of existing structures as these techniques can be quickly deployed and often do not require information about the structure's prior history, which can only be available if SHM systems were installed from the time of construction. SHM systems can often act as a complementary source of information to the NDEs if installed later.

NDE for transport infrastructure involves techniques to examine and qualify the structural condition of bridges, pavements, and roads without modifying the structure. Development of reliable non-destructive techniques has been the focus of research in the last few decades [15,16] and some methods have been standardized by technical committees through organizations such as International Union of Laboratories and Experts in Construction Materials, Systems, and Structures (RILEM), The American Concrete Institute (ACI), and American Society for Testing and Materials (ASTM). Methods such as acoustic emissions [17–19], thermal methods [20,21], ultrasonic [22–27], magnetic methods [28,29], vibrational analysis [30,31], rebound hammer [32], combined methods such as SonReb [33,34], and ground penetrating radar (GPR) [35–38] have been popular. These approaches focus on estimating the current strength levels of structural members, estimating the moisture content, predicting damage due to corrosion, and detection of cracks [39,40]. Often a single NDE method does not provide reliable estimates of these parameters and data fusion techniques are employed to converge towards a better estimate [40–44]. In this paper, GPR is used as a nondestructive evaluation technique for comprehensive assessment of concrete structures. GPR enables the inspectors to rapidly cover large surfaces without disrupting the traffic, saving on time and expenses required for performing such evaluations.

GPR-based condition assessment has gained great traction over the last two decades [7,45] and has become one of the most successful NDE approaches for investigation of transport infrastructure such as bridges and pavement [6]. In the GPR approach, low-frequency short electromagnetic signal pulses (≤ 1 ns) are emitted from the transmission antenna on the surface and the subsequent reflections are recorded at virtually the same location on the surface by the receiving antenna. The recorded signal is called an A-scan (or trace) and is a time-series of the reflected amplitude. Most surveys consist of a series of adjacent A-scans collected along a transect, forming a B-scan or profile. The high spatial resolution and sensitivity of the electromagnetic signal to factors such as electrical conductivity and moisture content make GPR particularly useful for feature detection and visual inspection of concrete structures [46]. They have been used for location of reinforcements and tendon ducts [46–48], voids [49], determination of concrete cover thickness [45], delamination, and crack detection [50]. Concrete moisture content [51,52] and pavement density [53] are other application areas of GPR-based condition assessment using amplitude of the signal as the quantifying parameter to estimate material properties.

Even though the use of GPR has gained prevalence, GPR attributes are rarely used for infrastructure assessment. In previous work by the authors, GPR attributes were successfully employed for estimating the material properties (compressive strength, porosity, density) of concrete [54]. Previous works by the authors' research group focused on feasibility of identifying material features in concrete deck using GPR attributes and quantitative estimation of concrete properties in laboratory conditions. The current work extends the previous studies to estimate relative variations in material properties using GPR attributes in real-life conditions. Attribute analysis is chosen in this work too as it is computationally inexpensive and does not require knowledge of the physical structure. As such it can efficiently and effectively be applied to buildings, bridges, pavements, and highways. Commonly used attributes in civil engineering applications include transient and time lapse attributes, imaging analysis such as coherence, and amplitude- and frequency-based attributes such as energy and frequency spectra [47,55].

The objective of the paper was to use GPR attributes to identify spatial and temporal variations in concrete properties. This would enable both qualitative and quantitative assessment of concrete bridges for both monitoring and inspection needs. To achieve this aim, GPR scans on a concrete pedestrian bridge are collected. These scans are used to obtain attributes which provide qualitative differences and indicate the spatial variability of concrete strength of the bridge deck. Using the machine learning models that predict physical concrete properties from GPR attributes developed by the authors in a previous work [54], quantitative assessment of the bridge deck is carried out. These quantitative assessments involve using GPR attributes to estimate in situ material properties of concrete, such as density, porosity, and compressive strength and establishing the differences over both spatial and temporal domain. One of the challenges of using approaches such as GPR is the lack of calibrated data about these properties to verify the results; in this work, a variety of other data sources and analyses are used to provide some validation. Compressive strength values estimated by GPR attributes were compared with the strengths of the reserved core samples to validate the proposed approach. Since the bridge was instrumented with a fiber-optic SHM system, available temperature measurements are used to determine the present compressive strength using maturity index and compared with the compressive strength estimated using GPR attributes for control.

2. Data Description

In this section, the data collection and processing applied for the GPR attribute-based inspection of concrete bridges are described. For the purposes of establishing the use of GPR as an approach to identify the spatial and temporal variations in concrete, four different types of data sets were used. Three of these data sets have been published in previous works by SHM*lab* and are being used as is [47,54,56] and the fourth is the GPR scan collected on a real-life bridge for this study. Interested readers can refer to the cited works for more details. These three data sets are:

1. Laboratory data: Concrete cylinders of 3in diameter and 6in height as per ASTM testing standards, and beams (15×15 cm $\times$ 90 cm) made of various industrial mixes and cured under different conditions were fabricated and tested. Direct tests were used to obtain values for density, porosity, and compressive strength and the corresponding beams were scanned using a 900-MHz GSSI antenna. The direct test values and calculated attributes from scans were used to develop machine learning models to predict the material properties on Streicker bridge in Section 4. The details of data collection can be found in the previous work by the authors [54]; the lab data and connection with the modeling pipeline (Section 3) are summarized in Figure 1.
2. Streicker bridge: Streicker bridge at Princeton University is the real-life application structure for this paper. The pedestrian bridge provides strategic connection between the east and west ends of the campus. It is a post-tensioned prestressed bridge in the shape of a chromosome. It is 105 m long and consists of a 35 m deck-stiffened arch (the main span) and four approach legs. The approach legs are continuous curved concrete girders. The front view of the bridge is shown in Figure 2. The main span and all legs except the southeast leg were constructed in August 2009. The southeast leg was constructed in October of that year using the same specified concrete mix (Class A HPC with design strength of 41 MPa at 28 days).
    Streicker bridge was instrumented with long-gage fiber optic sensors at the time of construction for research and teaching purposes. A schematic of Fiber Bragg Grating (FBG) sensor used for temperature monitoring is shown in Figure 3. A brief description of the working is as follows: When a light is sent from the reading unit, specific wavelengths of the light are reflected back by the Bragg gratings in the fiber. Depending on the strain in the fiber, the reflected wavelength changes. The fiber optic sensors installed in Streicker bridge measure both the temperature and strain.
    The sensor locations along the main span and southeast leg are shown in Figure 4 by black boxes. The sensors have been continuously monitoring strain and temperature with periodic interruptions for maintenance since their installation in 2009. These sensors are installed at various locations on the bridge which provide the temperature

data that are required for the maturity index calculations. The temperature readings have an uncertainty of 0.14 °C. These sensors are used for compressive strength calculations based on the maturity method. Two typical locations are also highlighted in Figure 4. The GPR attributes collected on Streicker bridge in 2016 and 2020 are used to identify the spatial and temporal variation in concrete in a structure. There are two types of data sets used from previous studies for validation of those predicted values:

(a)   Temperature measurements: Streicker bridge is instrumented with Fiber Bragg Grating (FBG) fiber-optic sensors which have been regularly collecting temperature and strain data from the time of construction in 2009. The sensor locations along the main span and southeast leg are shown in Figure 4 by black boxes. The sensors have been continuously monitoring strain and temperature with periodic interruptions for maintenance since their installation in 2009. These sensors are installed at various locations on the bridge which provide the temperature data that are used for the maturity index estimates of compressive strength. The temperature readings have an uncertainty of 0.14 °C. The location of the typical sensors in the main span (P8h9) and southeast leg (P10h11) are highlighted in Figure 4.

(b)   Core reserves: Class A HPC concrete with a design strength of 41 MPa (at 28 days) was used for the construction of the bridge. The bridge was constructed in two phases; one in August 2009 (main span and northeast leg included) and the other in October 2009 (southeast leg). Even though the design strengths were the same for the two construction phases, the measured compressive strength on the reserved cores indicate a nominal compressive strength of 51 MPa for the main span (MS) and northeast (NE) leg and 59 MPa for southeast (SE) leg at 28 days [56]. Figure 5 shows the compressive strengths of the reserved cores based on the strength tests performed at US Laboratories Inc. (Broomall, PA, USA). Six samples were tested for the concrete poured in August (MS and NE leg) at 2, 3, 7, and 28 days and four samples were tested for the October pour (SE leg) at 3, 4, 7, 14, and 28 days. In this work, we try to identify the spatial variation using GPR attributes.

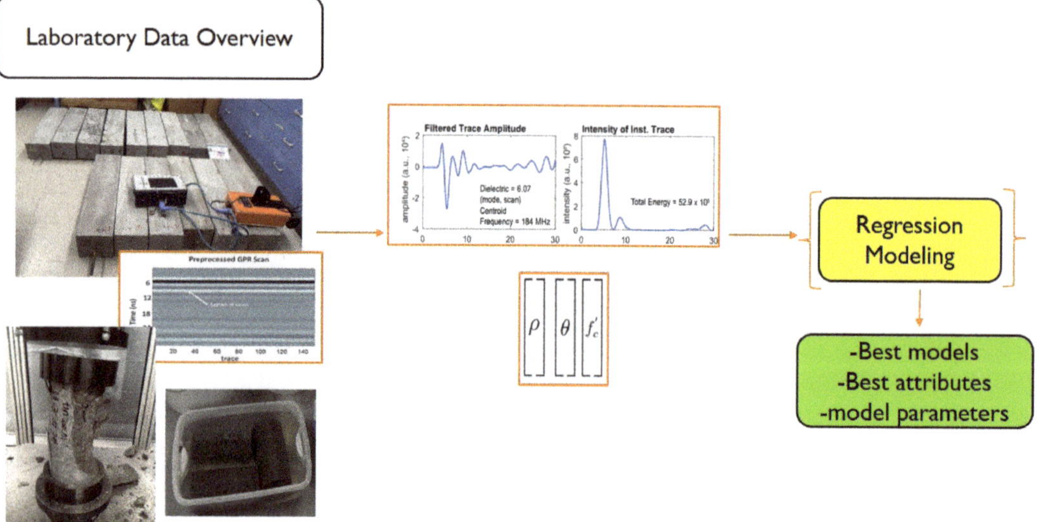

Figure 1. Overview of the experimental approach and data collected in the laboratory study.

Figure 2. Front view of the Streicker bridge at Princeton University.

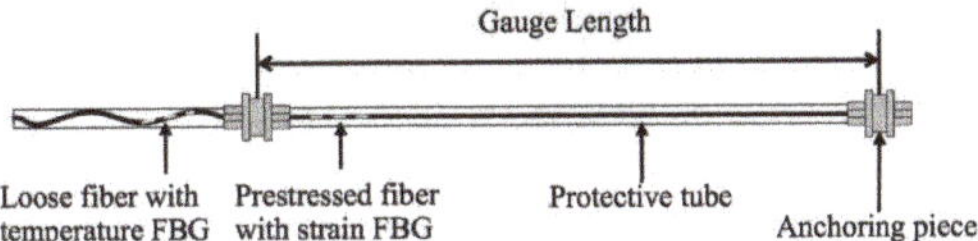

Figure 3. A single fiber-optics sensor present as part of the SHM system in the Streicker bridge (Adapted from [4]).

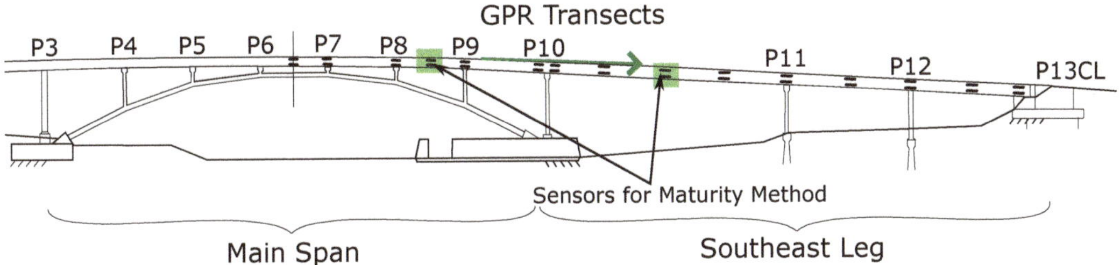

Figure 4. Elevation view of the Streicker bridge with sensor locations and direction of GPR transects.

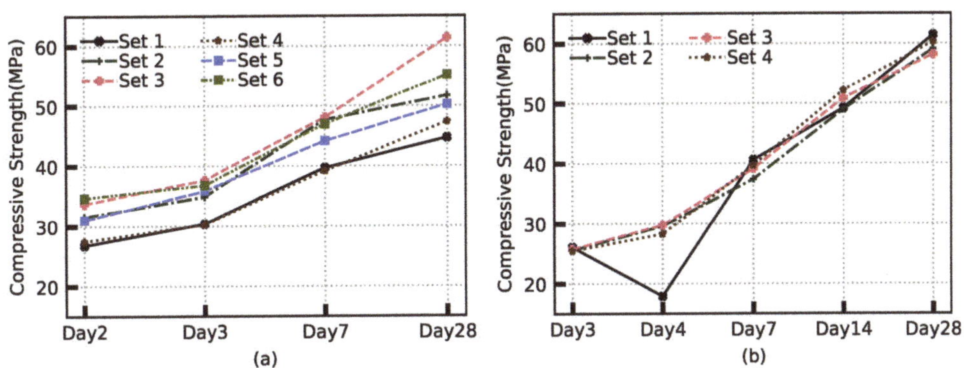

Figure 5. (**a**) Compressive strength for the August pours (NE and main span.) (**b**) Compressive strength for the October pour (SE leg).

2.1. GPR Survey of Streicker Bridge

For identifying the temporal and spatial variations in concrete, the GPR data collection on Streicker bridge was performed using a 900 MHz antenna (from GSSI) on 16 March 2020 (Note: the same antenna was used for laboratory testing and model development). Leading up to the day of the scan, the weather conditions were dry with no precipitation or surface moisture due to accumulated snow. The GPR unit is 15 cm × 23 cm × 18 cm and was rolled on the surface while being connected to a data logger (Figure 6). Longitudinal transects were collected along the northeast and southeast legs to assess the differences in the two construction phases of the concrete structure.

Figure 6. GPR data collection on Streicker bridge by the authors.

These longitudinal scans were collected with an average spacing of 20 cm. The location of the transects with respect to the internal structure of the cross-section is shown in Figure 7. Only longitudinal scans were collected as the purpose of this survey was to differentiate between the legs; from SHM of the structure, no damage to the bridge was expected. The longitudinal scans were straight but their length varies due to the curved plan of the bridge. Constant lengths were obtained and used for this study as part of the data processing.

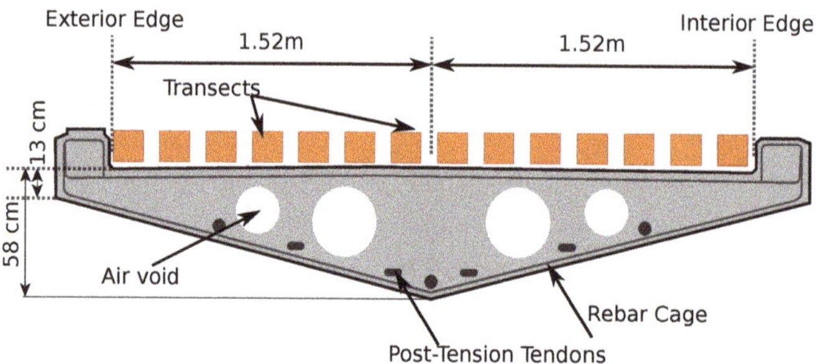

Figure 7. Southeast leg section showing location of GPR scans. Transects run longitudinally from main span down the SE and NE legs (adapted from [47]).

In addition to the 900 MHz data described, a previous survey of the bridge from 2016 was also included for some comparisons. This survey was collected and processed in the same manner, but a 2.6 GHz antenna was used. With this high resolution antenna, the survey focused primarily on feature detection, as described in [57].

2.2. Data Processing

The raw data obtained from the GPR scans were processed to calculate the primary and derived attributes. The raw amplitudes of the GPR data are highly variable with respect to the electromagnetic properties of the material and to antenna ground coupling, surface roughness, and other factors [58]. A standard set of basic filters were applied while preserving the relative amplitudes of the traces. The initial reflection or first break is trimmed using the STALTA algorithm, the traces are dewowed, and the scans are truncated to include only the upper 6 cm of the deck (Figure 8) [59–61]. Truncating the scans ensures that there are reflections from rebar present in the traces, but the spatially variable reflections at deeper depths (such as air voids, the bottom surface of the deck, and other reinforcing) are removed (Figure 8). The length of the scans are also trimmed to the same length. The trimmed and filtered traces are then converted into continuous complex traces using the Hilbert transform before attributes are computed.

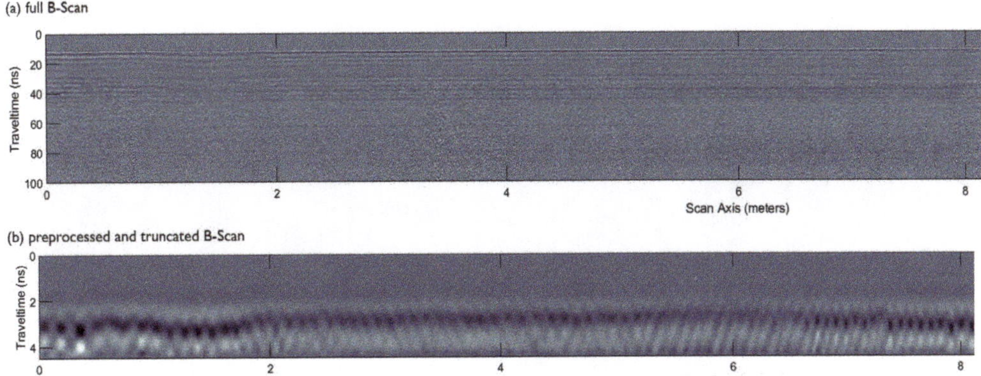

Figure 8. (**a**) Raw B-scan along the deck of the bridge collected with 900 MHz antenna and (**b**) preprocessed B-scan truncated to remove all internal reflections except the upper rebar layer.

The attributes used in this paper for the assessment of concrete bridges are summarized in Table 1. The attributes are calculated from the instantaneous trace that results from the Hilbert transform, which has the form $A(t) = a(t) + \phi(t)i$, where $A(t)$ is the instantaneous amplitude and ϕ is the phase of the signal. This selection of attributes is chosen from the best performing attributes in [54], which tested the attribute's correlation with the porosity, strength, and density of a range of lab concrete samples. Therefore, the selected attributes are expected to have correlation with the physical properties of the concrete, particularly the subtle differences between the two legs of Streicker bridge. Most of the attributes are related to the cumulative behavior of the traces and capture differences between the attenuation in different materials.

Table 1. Attribute calculation for GPR scans. The attributes are based on a Hilbert transform. Here, c = speed of light, d = depth of concrete, t_{top} = travel time top reflection, t_{bottom} = travel time bottom reflection, α_r = attenuation ratio, α_c = attenuation constant, DW = direct wave, RW = reflected wave, TR = top reflected surface.

Main Attribute	Equation	Derived Attribute	Equation
Instantaneous Amplitude	$A(t)$	Total energy	$\sum A(t)^2$
Intensity	$A(t)^2$	Raw average amplitude	$\text{Average}(A)$
Phase	$\phi(t)$	Signal-to-Noise (SNR)	$\frac{P_{\text{clean}}}{P_{\text{avg}}}$
Dielectric Constant (ϵ)	$\left(\frac{c(t_{\text{top}} - t_{\text{bottom}})}{2d}\right)^2$	$\alpha_{r,DW}$	$\frac{A_{DW}}{A_{TR}}$
		$\alpha_{r,RW}$	$\frac{A_{TR}}{A_{RW}}$
		$\alpha_{c,DW}$	$\frac{20}{d\log(\alpha_{r,DW})}$
		$\alpha_{c,RW}$	$\frac{20}{d\log(\alpha_{r,RW})}$

The data sets used for comparison with the GPR data (attributes and predicted properties) have been previously published and validated, the errors, outliers, and gaps in data are well understood. The GPR data collection and processing scheme includes a number of standard and additional steps which reduce random and measurement errors. Standard GPR data recording includes stacking, which records one compiled trace at a location that has been created from multiple recordings, thereby removing random errors and improving resolution. The data are prepared for use by the described preprocessing, including attribute calculation. The attributes used in this study represent the "average" B-scan from each leg (averaged across the width of the deck in each leg), thereby reducing the effects of any measurement errors and lateral variation in the already truncated scans. In addition to the truncation, this reduces the effect of variations from the internal structure of the bridge such as voids, tendons, and reinforcement locations that are oriented longitudinally.

3. Modeling

In this section, we describe the machine learning model used for quantitative prediction of compressive strength (f_c'), density (ρ), and porosity (θ), and the maturity method used to estimate the in situ strength using the temperature measurements from the SHM system collected over the 8 year period from 2009 to 2017.

3.1. Machine Learning Pipeline

The laboratory samples described in Section 2 were used to train the machine learning models for prediction of compressive strength, density, and porosity using GPR scans collected from Streicker bridge. The machine learning model development process is shown in Figure 9.

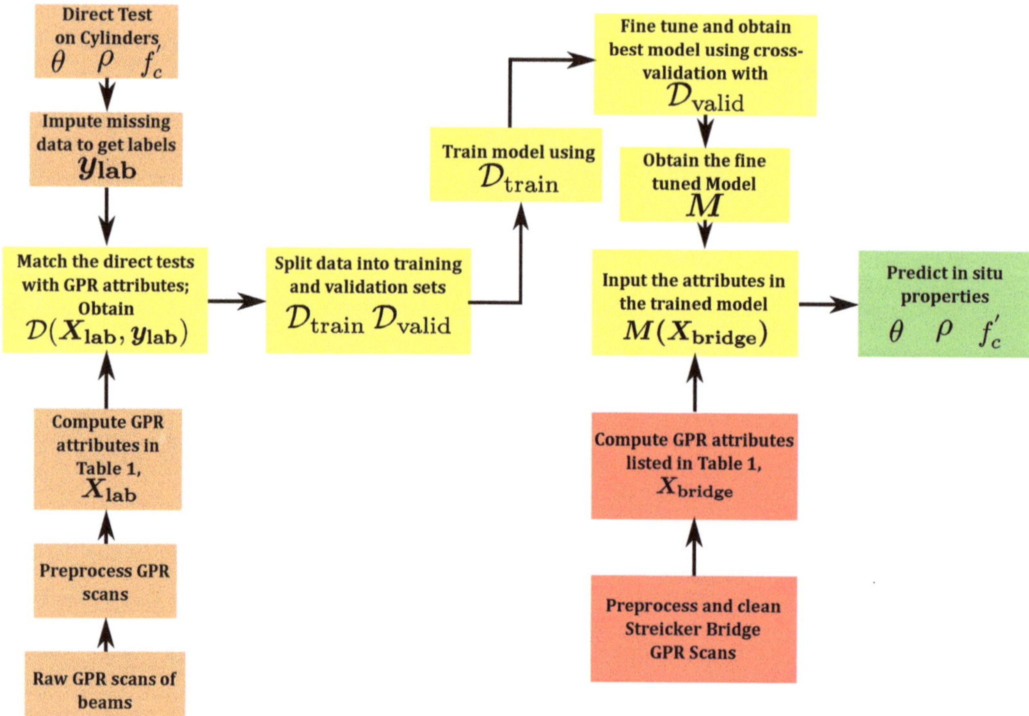

Figure 9. Flowchart depicting the steps involved in estimating the material properties using a machine learning model.

To describe the machine learning pipeline, let the set of GPR trace attributes and their corresponding imputed labels be represented by $\mathcal{D}(X_{\text{lab}}, y_{\text{lab}})$. Here, X_{lab} are the calculated GPR attributes which act as the "features" for the machine learning models and y_{lab} represent the imputed labels corresponding to the direct test values of material properties of the cylinders (see Figure 1). This set was separated into training and validation sets, $\mathcal{D}_{\text{train}}$ and $\mathcal{D}_{\text{validation}}$, respectively. Using Random Forest [62] as the supervised learning model, the cross-validation [63] scores are used to determine the best performing features and fine tune the model. Random Forest was chosen as it had shown the best performance based on authors previous work [54]. These best performing attributes are then calculated from the GPR scans collected from the Streicker bridge (X_{bridge}). These attributes are used for the final estimation of the material properties using the Random Forest model, y_{bridge}.

The key features for the current machine learning pipeline are as follows (details can be found in [54]):

1. Data imputation: The total samples tested directly in the laboratory for density, compressive strength, and porosity were 219, 146, and 73, respectively. Since the number of samples tested for compressive strength and porosity were small, a data imputation was performed. Mean substitution was chosen as it is a standard practice in data science even if it sometimes results in statistically correlated samples [64].
2. Stratified split of train-validation data: For a small sample set such as ours, stratified splitting of training and validation sets avoids overfitting. The stratified approach further guarantees that a sample in the validation set would have the same mix of concrete in the training set while preserving the distribution of properties [63].
3. Feature selection: Since many of the attributes had more than 500 features, feature selection was adopted to improve the computational efficiency of the machine learning models. The feature selection was performed using the f_1 score. The top "n" features were chosen heuristically based on the f_1 scores.
4. Model tuning approach: All the machine learning models were first trained using a baseline set of hyper-parameters. These hyper-parameters were then fine-tuned to improve the predictions using randomized search and grid search [65]. The cross-validation score was used to determine the best parameters in all these cases.

3.2. Maturity Index Model

The maturity method is a technique that estimates the strength of concrete from either the equivalent age (at certain temperature) or the temperature–time factor based on the reserved cores of the concrete mix [66]. The temperature–time factor (or the Nurse–Saul maturity index) is given by Equation (1)

$$M = \sum_{0}^{t}(T - T_0)\Delta t \tag{1}$$

where M is the maturity or temperature–time factor; t is time; T is the average temperature during the time interval Δt; and T_0 is the datum temperature. Based on experimental studies conducted by previous researchers to develop the relationship between concrete maturity and compressive strength, a straight line best fit was found between the logarithmic value of maturity and compressive strength [67,68]. After combining Equation (1) with the logarithmic best fit line for the strength of the cylinders, a relationship between compressive strength ($f'c$) and Nurse–Saul maturity (M) can be found as per Equation (2)

$$f'c = C_1 ln(M(t)) + C_2 \tag{2}$$

where C_1 and C_2 are constants obtained from the cylinder test data. The compressive strength and maturity index of the cylinder tests (Figure 5) was fit to the logarithmic curve as shown in Figure 10.

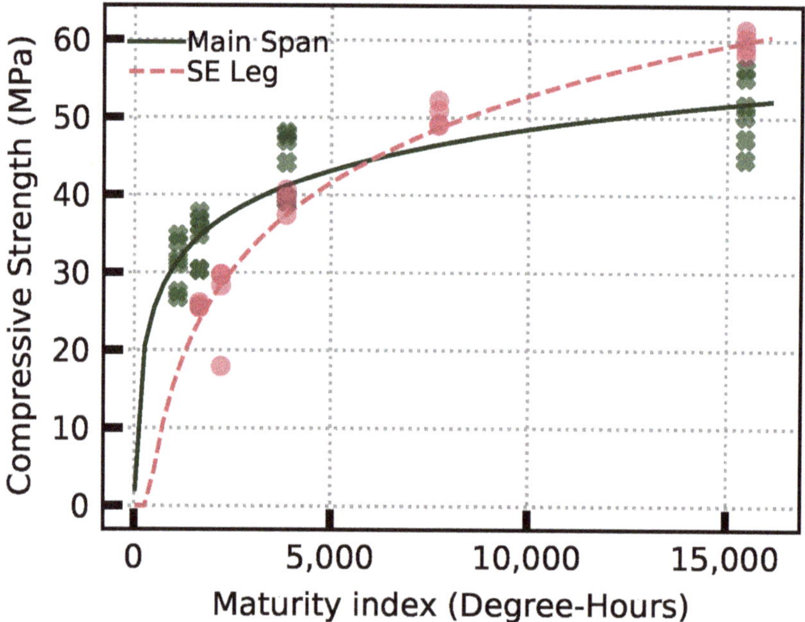

Figure 10. Compressive strength obtained as a function of maturity index based on the initial concrete core sample tests.

Based on the logarithmic fit for the cylinder data for Streicker bridge, the constants, C_1 and C_2, are obtained for the main span/northeast leg (MS/NE) and southeast leg (SE) and shown in Table 2.

Table 2. Constants of the maturity index and compressive strength logarithmic fit.

Span	C_1	C_2
Main Span	1116.3	−3198.7
South-east Leg	2249.7	−12,951.9

The temperature–time factor is calculated using the temperature data recorded by the SHM system installed in the bridge. The data used in this paper consist of the time-stamped temperature measurements from each sensor located in the main span and southeast leg through 2017.

The maturity and strength functions used in this paper take the value of temperature–time factor to be zero for time intervals when the average concrete temperature is less than the datum temperature. As recommended by ASTM C1074-19, a datum temperature of 0 °C was used. This ensures that the analytical model does not predict loss of concrete strength when exposed to sub-zero temperature. The choice of datum temperature is important to avoid overestimation or underestimation of concrete strengths. Previous studies have shown that concrete can potentially gain strength at temperatures as low as −10° while most common values for different mixes were above freezing point, roughly 2.2 °C [68]. Estimating the actual datum temperature is not possible after construction. Since over the 8-year period only 10% had sub-zero temperature, the choice of datum is acceptable for the present work. A slight overestimation of strength is expected which would not affect the main findings. For the periods of time when the system is not active, the average temperature for the inactive period was used. The temperature–time factors for each sensor are used in the corresponding maturity–strength relationship found from the cylinder data to obtain a strength development curve for each sensor.

4. Results

In this section, the results of identifying the spatial and temporal variations in concrete of the bridge are provided. First, the results from GPR scans are provided to differentiate qualitatively and quantitatively between the northeast and southeast legs of the Streicker bridge. Then, the maturity method results are presented as an upper limit for the GPR's compressive strength prediction.

4.1. GPR Attributes: Qualitative Spatial and Temporal Variation

In this subsection, we highlight the key attributes that identify differences between the two scanned segments, northeast and southeast, of Streicker bridge. Recall that the main span and northeast legs were constructed at the same time. First, the spatial variation between the northeast and southeast legs is demonstrated using GPR attributes collected by the authors (See Section 2) followed by the temporal variation using the attributes computed from an earlier GPR scan of the bridge.

4.1.1. Spatial Variation

Figure 11 shows that raw means (raw average amplitude), total energy, attenuation constant, and attenuation ratio (relative to the direct wave) are attributes which qualitatively identified the differences in the two legs. Figure 12 shows that attributes such as total phase, SNR, attenuation constant, and centroid of instantaneous frequency do not visually capture differences between the two legs.

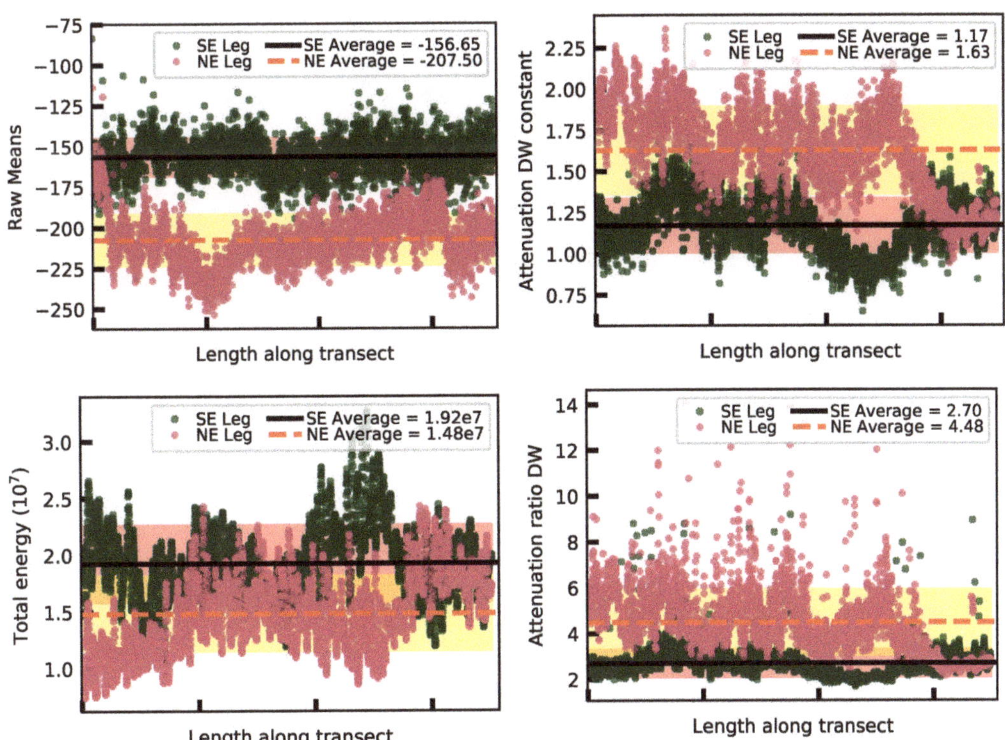

Figure 11. Attributes capturing qualitative differences in the two construction phases.

The red and yellow bands around the average value of attributes represent one standard deviation range. For the first set of attributes in Figure 11, there is little or no overlap and hence a new measurement is likely to indicate whether the concrete was from the southeast leg or the northeast leg with 85% confidence. However, for the attributes shown in Figure 12, there is considerable overlap between the 1−standard deviation bands and hence these attributes could not be used to determine which concrete phase the GPR scan is from.

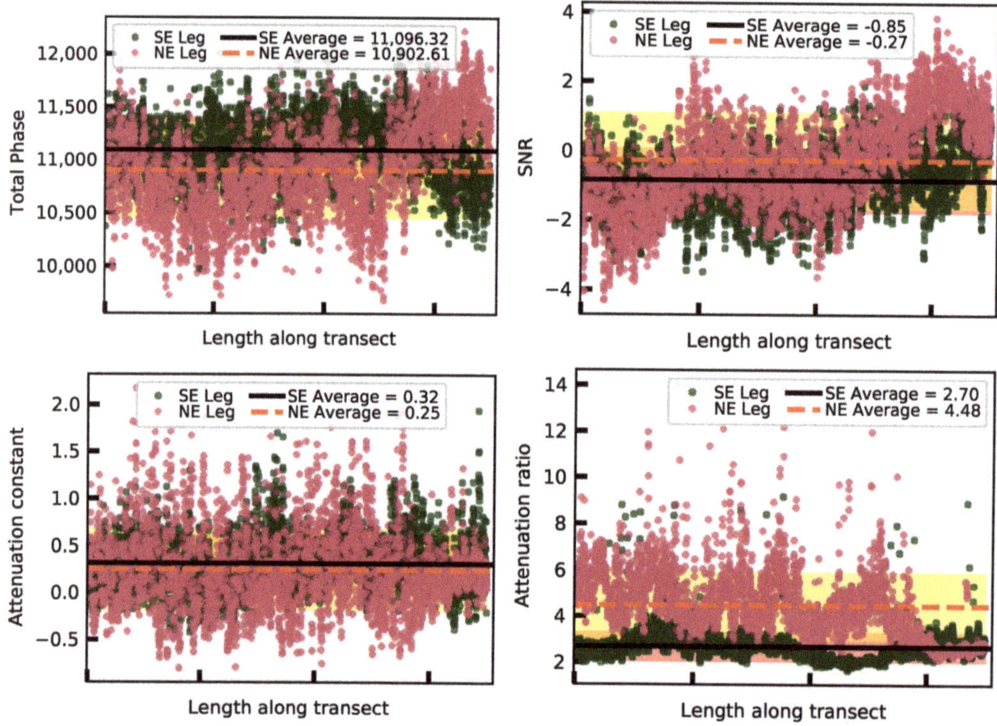

Figure 12. Attributes unsuccessful in capturing differences between the construction phases.

For a more quantitative comparison, the range of measured values for the two different legs are shown in Table 3.

Table 3. Range of values for various measured attributes in the southeast and northeast legs.

Attribute	Range SE Leg	Range NE Leg
Raw Means	$[-542.13, 196.38]$	$[-464.25, 60.93]$
Attenuation DW constant	$[-0.73, 6.16]$	$[-0.47, 6.23]$
Denergy	$[5.64 \times 10^5, 1.11 \times 10^8]$	$[1.8 \times 10^5, 9.37 \times 10^7]$
Attenuation DW ratio	$[0.58, 101.32]$	$[0.70, 107.22]$
Total Phase	$[0.58 \times 10^4, 1.70 \times 10^4]$	$[0.48 \times 10^4, 1.65 \times 10^4]$
SNR	$[-16.797, 12.224]$	$[-16.899, 14.009]$
Attenuation constant	$[-6.51, 15.00]$	$[-6.18, 20.17]$
Dielectric constant	$[3.01, 78.74]$	$[3.10, 93.71]$

Two key statistical measures used for determining whether the means of two different test samples can be considered different are Welch's t-test and Mann–Whitney U-Test [69]. These tests are used to further characterize the spatial variation in concrete. Welch's t-test tests the hypothesis that two populations have the same mean. A low p-value in the test indicates that the means are different. The test is most commonly employed to differentiate samples which have many overlapping values. Mann–Whitney U-test determines whether the two samples themselves are independent, with low scores indicating stronger independence. The results are shown in Table 4. The absolute value of the scores are useful for relative comparison. Hence, based on the combination of results from the two tests, it can be inferred that the mean values of GPR scan attributes for the two legs are different and that they are obtained from different populations.

Table 4. Statistical comparison (with *p*-values) of the attributes for the northeast and southeast legs.

Attribute	Welch's t-Score	*p*-Value	Mann–Whitney U-Score	*p*-Value
Raw Means	154.52	0.0	12.41×10^6	0.0
Attenuation DW constant	86.57	0.0	1.03×10^6	0.0
Denergy	55.60	0.0	1.01×10^7	0.0
Attenuation DW ratio	−67.88	0.0	1.20×10^6	0.0
Total Phase	21.19	1.11×10^{-96}	7.88×10^6	1.14×10^{-95}
SNR	21.06	1.48×10^{-95}	4.52×10^6	1.12×10^{-80}
Attenuation constant	7.73	1.26×10^{-14}	7.06×10^6	1.96×10^{-28}
Dielectric constant	−22.41	1.62×10^{-107}	4.54×10^6	3.6×10^{-79}
Attenuation ratio	−11.42	6.38×10^{-30}	5.50×10^6	7.58×10^{-14}

The low *p*-value of both the Welch's t-score and Mann–Whitney U-score as shown in Table 4 indicates that (i) the mean values of both the legs are different in a statistically significant way and (ii) the samples themselves are taken from different populations. Those are independent and interesting findings in their own regard which indicate the value in qualitative attribute based comparisons of GPR attributes in concrete inspections. These statistical measures confirm the capability of GPR attributes to qualitatively distinguish between different concrete mixes used in the same structure.

4.1.2. Temporal Variation

To identify the temporal variation in concrete, the above attributes from the 900 MHz antenna (2020) are compared with GPR data collected in Nov, 2016 with a 2.6 GHz antenna for a previous study more focused on high resolution feature mapping [47]. A special deicing agent is used on the bridge to avoid corroding the exposed rebar on the surface, so seasonal accumulation of chlorides over the winter should be negligible in the November 2016 to March 2020 comparison. The temperature, humidity, and very low precipitation [70] for the week preceding GPR measurements in both surveys were similar, so the effect of moisture related seasonal changes in the dielectric properties and other attributes are minimal. The results are presented in Table 5. Since two different antennas were used in 2016 and 2020 (2.6 GHz vs. 900 MHz), attributes that consider amplitude values directly, such as raw means and total energy were normalized for 2016 to match 2020 values and minimize the effect of using two different antennas. Recall that the 900 MHz antenna used primarily in this work matches the laboratory data and predictive modeling; the 2.6 GHz antenna was not available and not used in 2020.

Table 5. Comparison of attribute values for southeast and northeast legs between 2016 and 2020 measurements. Note, * represents normalized value to account for different antennas.

Attribute	Southeast Leg		Northeast Leg		Ratio SE/NE	
	2016	2020	2016	2020	2016	2020
Raw Means	−160.4 *	−156.65	−169.5 *	−207.50	0.95	0.75
Attenuation constant (DW)	−0.60	1.17	−1.29	1.63	0.46	0.72
Total energy	3.49×10^7 *	1.92×10^7	3.50×10^7 *	1.48×10^7	1.0	1.30
Attenuation ratio (DW)	3.41	2.70	1.84	4.48	1.85	0.60
Total Phase	0.419×10^4	1.11×10^4	0.450×10^4	1.09×10^4	0.93	1.01
Attenuation constant	−1.07	0.32	−0.36	0.25	2.97	1.28
Attenuation ratio	21.72	3.19	25.64	5.39	0.85	0.59

The qualitative comparison of the attributes over a gap of 4 years using two antennas provides some interesting insights. The ratio of attribute values for the two legs increased for attenuation constant (with respect to direct wave), total energy, and total phase, while the ratio decreased for raw means, attenuation ratio (with respect to both the direct wave and the reflected wave), and attenuation constant (with respect to reflected wave). These could be explained by a higher loss in concrete cover in the southeast leg compared to the northeast leg due to higher pedestrian and golf cart traffic. As the concrete cover decreases, the reflected signals have thinner concrete medium to travel before they penetrate into the voids resulting in proportionally lower travel time in the concrete than the second medium (grout, air, etc.). As a result of the change in the medium, the computed attributes could have higher negative values. This is reflected in the higher negative raw means value and the decrease in the ratio over the 4 years period. However, an increase in negative values of the attribute would still result in higher total energy as the the total energy only takes the square of the values into account. For a reduced cover, the values of α_{DW} decrease while the α_{TR} remains constant (see Table 1) which would explain the change over time in attributes such as attenuation constant and ratio. The total phase comparison does not change significantly over time, as it is largely determined by the antenna frequency characteristics and therefore should not change significantly between legs. This increased loss of concrete cover can be seen in Figure 13. The aggregates are visible on the southeast leg while it is not the case with the northeast leg.

Figure 13. Figure showing increased loss of cover on the southeast leg compared to northeast leg.

4.2. Quantitative Differences between Legs

In this section, quantitative analyses of the GPR attributes of the northeast and southeast legs are presented to identify the spatial differences. The quantitative analysis is based on the machine learning framework and maturity method described in Section 3.

4.2.1. In Situ Property Estimation Using GPR Attributes

Different material properties required different attributes for the best prediction. The most successful attributes were instantaneous amplitude for porosity and compressive strength, while it was intensity for the density prediction. Using these GPR attributes and fine tuned Random Forest models from [54], the compressive strength, density, and porosity were predicted using all the GPR transects collected. These predicted values and the average for each leg is shown in Figure 14 (Note: outlier datum points were removed).

Figure 14 shows the predicted values for each of these properties are fairly scattered. This is attributed to the macro-scale variation along the length of the scans from the irregular presence of plastic tubes (voids in Figure 7) and reinforcing along each transect. Hence, in the discussion of the results, we only focus on the average values of these material properties over all the transects. Though not an accurate representation, it is in line with the practice of using a single number to quantify material properties for civil engineering structures.

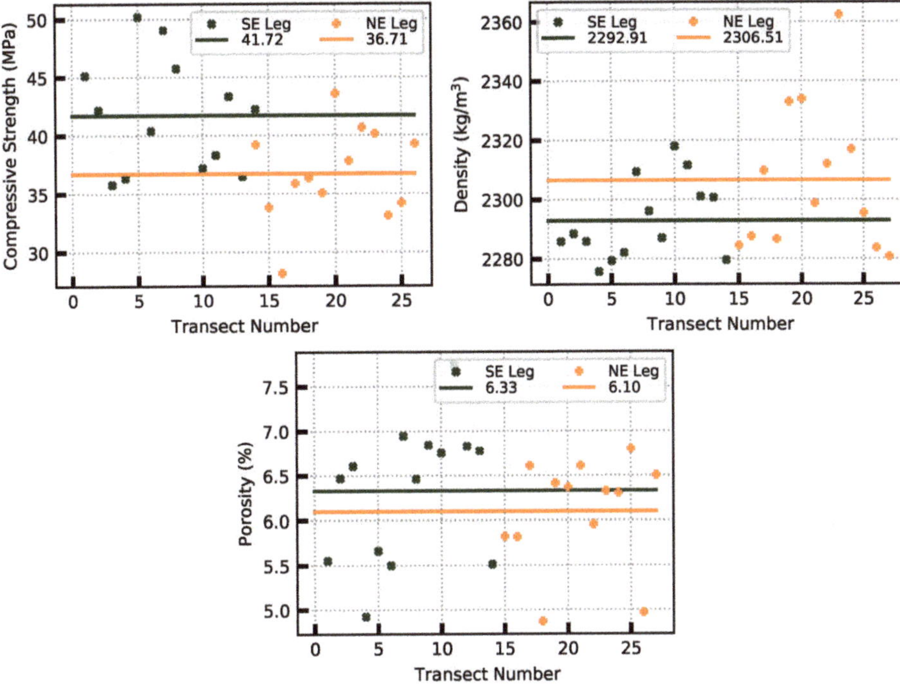

Figure 14. Spatial differences in material properties between the southeast and northeast legs.

The compressive strength predicted by the machine learning model is less than the measured compressive strength from the reserve cores. The compressive strength is not actually expected to have reduced over time (it is expected to have increased) and the lower predicted values are a result of using small training sets of lab data to create the models. The predicted values still identify a 12% difference between the two legs, while the initial difference between the reserved cores was 13.6%. This shows the potential of using GPR attributes to identify relative difference over space.

The density predicted by the machine learning model is close to those expected in real-life settings for this type of concrete. The predicted density is ≈2300 kg/m^3 for both legs while the concrete mix used for the construction of the bridge had a density of 2400 kg/m^3. It is a 4% lower prediction than the expected value but it is acceptable. This variation could be attributed to embedded air voids in the structure lowering the effective density compared to the training set based on laboratory fabricated samples with only one piece of reinforcing steel embedded.

Finally, the predicted porosity values for both the legs are around 6%. This porosity includes the connected pores and not include dead-end pores and unconnected air voids. Saturated pores would not be included in this porosity. Although there is no direct way to verify the porosity values, it is possible to compare them with those found in literature. The in situ porosity values reported for concrete structures is between 5 and 10%. The reported value falls comfortably in this range. The relative values do not follow the expected trend of higher porosity-lower strength, though this could be due to other factors in the mix design, such as the high rate of strength development required in the SE leg.

4.2.2. Compressive Strength Calculation Using Maturity Method

Based on the Equation (2) and the constants obtained using the reserved concrete cores (Table 2), the mean 8-year compressive strength for all sensors in main span is 57.12 MPa and 74.8 MPa in the southeast leg (see Figure 15). The higher compressive strength of the southeast leg compared to the main-span is attributed to the higher initial strength of the concrete mix and rapid early strength gain of that particular concrete mix. Hence, the method predicts a gain of 6.2 MPa for main span and 15.1 MPa for the southeast leg for the same period of 8-years. These estimates should be considered as upper bounds on the compressive strengths as the maturity method overestimates the strength at later stages. This results from the logarithmic function used to predict the strength gain for the entire life of the concrete. The issue is aggravated when the method is applied to concrete mixes designed for early strength gain [68]. For strength development at specific sensor locations, readers can refer to [71].

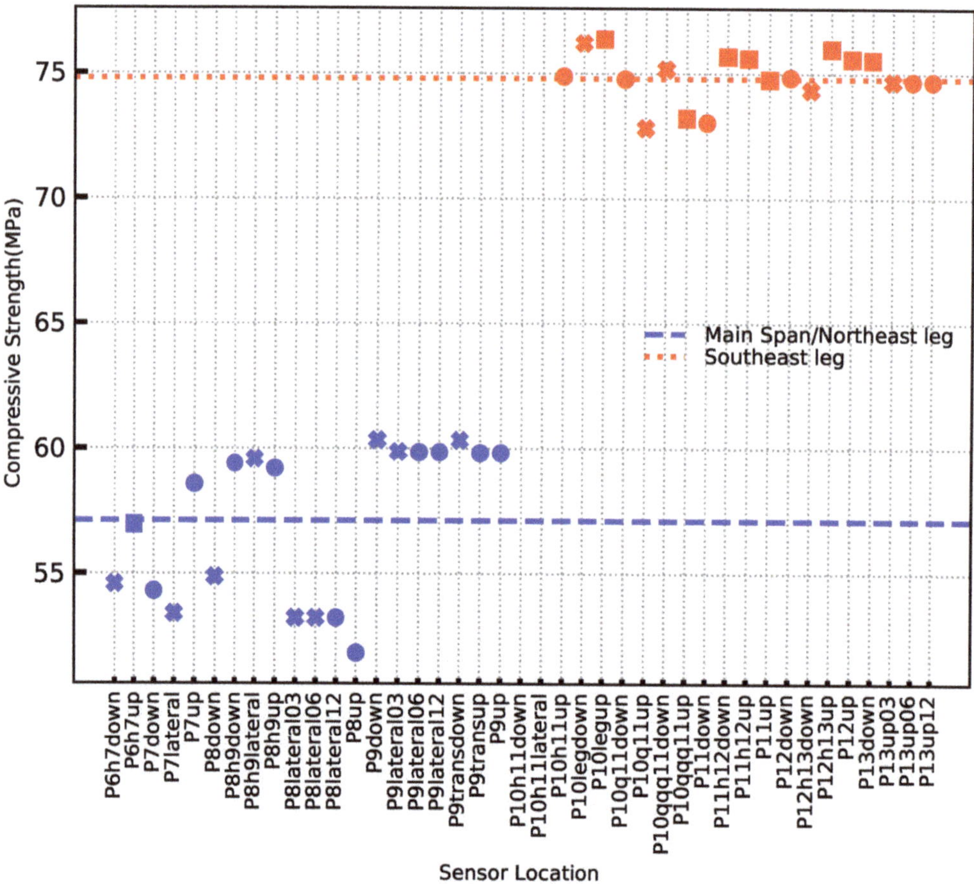

Figure 15. Strength calculation at different sensor locations using maturity method.

5. Conclusions and Discussion

In this paper, GPR attributes were used to estimate temporal and spatial variation of in situ material properties of mature concrete. Variation in in situ material properties was evaluated on a concrete pedestrian bridge at Princeton University campus using GPR attributes. The novelty of the present work is three-fold: (1) GPR attributes are used to qualitatively differentiate between different mixes of concrete in place, (2) machine learning algorithms are developed for quantitative differentiation, and (3) maturity index is demonstrated as a method for establishing bounds on the predictions of mature concrete strength. The conclusions are summarized below:

1. Instantaneous amplitude and summary attributes can statistically distinguish between the concrete in Streicker bridge on the basis of relative material properties.
2. Spatial variation in the physical properties of the two phases of concrete is identified with amplitude-based attributes such as raw means, total energy, and two different measures of attenuation.
3. Temporal variation in the physical properties over a four-year period is difficult to determine due to the use of different antennas and seasonal differences, but the comparison does identify how sensitive attributes are to the antenna relative to the different concrete.
4. The GPR attributes predict a 5.01 MPa difference in the mean compressive strength, a 13.6 kg/m^3 difference in density, and a 0.23% difference in porosity between the southeast and northeast legs of the bridge.
5. The quantitative strength predictions from the GPR attributes are reasonable and fall between the lower bound of the 28-day reserved concrete core strength and the upper bound from the maturity method and temperature history of the concrete.

The quantitative estimates for density and porosity cannot be verified through alternative measurements but the values are within reasonable ranges for the concrete mix used in Streicker bridge. While the current work is limited due to the availability of samples to train models on, the availability of temperature and GPR data from the bridge enable a range of techniques to be applied. For GPR attributes to become an accepted industry practice for establishing spatial variability in properties like rebound hammer technique, further research into the development of calibration charts for attribute values and material properties are required. In future works, these studies would be carried out to identify point, local, global relationships between material properties and GPR and more fully describe the effects of data collection settings like antenna frequency, seasonal variability, and variability in internal reflectors. Some immediate directions towards these would be a one-on-one comparison between traces with consistent profiles to understand and calibrate the GPR attributes with concrete properties. Long term projects using total phase could present a solution to evaluate material property evolution as these attributes are only dependent on frequency of the internal reflected waves and not affected by concrete cover. Finally, other NDT techniques like rebound hammer or ultrasonic testing could be conducted to provide additional data about spatial and material variability. The present work paves the way for future research in expanding the use of GPR attributes for assessments and laying the groundwork for developing codes of practice for using GPR for maintenance of transport infrastructure.

Author Contributions: Conceptualization, V.K., I.M.M., B.G.; methodology, I.M.M., V.K.; software, V.K.; validation, I.M.M.,V.K.; formal analysis, V.K., I.M.M., S.A.L.; data curation, I.M.M.; writing—original draft preparation, V.K.; writing—review and editing, I.M.M.; visualization V.K., I.M.M., S.A.L.; supervision, I.M.M., B.G.; funding acquisition, I.M.M., B.G. All authors have read and agreed to the published version of the manuscript.

Funding: I.M.M. was supported by NSF GRFP #1148900 and S.A.L. was supported by the NM AMP Program (NSF HRD-1305011).

Institutional Review Board Statement: Not Applicable.

Informed Consent Statement: Not Applicable.

Data Availability Statement: The data supporting the current study can be provided at a reasonable request to the corresponding author.

Acknowledgments: The authors would like to acknowledge and thank Joseph Vocaturo for the help in preparation of laboratory samples; the past and present members of SHM*lab* for maintaining the SHM system on Streicker Bridge; Mauricio Periera for GPR scan collection on Streicker Bridge.

Conflicts of Interest: The authors declare no conflict of interest. The funders had no role in the design of the study; in the collection, analyses, or interpretation of data; in the writing of the manuscript, or in the decision to publish the results.

References

1. Betti, R. *Aging Infrastructure: Issues, Research, and Technology. Buildings and Infrastructure Protection Series. Infrastructure Protection and Disaster Management Division*; Science & Technology Directorate, US Department of Homeland Security: Wahsington, DC, USA, 2010.
2. ASCE. *2021 Infrastructure Report Card*; ASCE: Reston, VA, USA, 2021.
3. ASCE. *Failure to Act: Economic Impacts of Status Quo Investment Across Infrastructure Systems*; Technical Report; ASCE: Reston, VA, USA, 2021. Available online: https://www.asce.org/failuretoact/ (accessed on 2 February 2021).
4. Glisic, B.; Inaudi, D. *Fibre Optic Methods for Structural Health Monitoring*; John Wiley & Sons: Hoboken, NJ, USA, 2008.
5. Scott, M.; Rezaizadeh, A.; Delahaza, A.; Santos, C.; Moore, M.; Graybeal, B.; Washer, G. A comparison of nondestructive evaluation methods for bridge deck assessment. *NDT E Int.* **2003**, *36*, 245–255. [CrossRef]
6. Rehman, S.K.U.; Ibrahim, Z.; Memon, S.A.; Jameel, M. Nondestructive test methods for concrete bridges: A review. *Constr. Build. Mater.* **2016**, *107*, 58–86. [CrossRef]
7. Alani, A.M.; Aboutalebi, M.; Kilic, G. Applications of ground penetrating radar (GPR) in bridge deck monitoring and assessment. *J. Appl. Geophys.* **2013**, *97*, 45–54. [CrossRef]
8. An, Y.; Chatzi, E.; Sim, S.H.; Laflamme, S.; Blachowski, B.; Ou, J. Recent progress and future trends on damage identification methods for bridge structures. *Struct. Control. Health Monit.* **2019**, *26*, e2416. [CrossRef]
9. Chen, Z.; Zhou, X.; Wang, X.; Dong, L.; Qian, Y. Deployment of a smart structural health monitoring system for long-span arch bridges: A review and a case study. *Sensors* **2017**, *17*, 2151. [CrossRef] [PubMed]
10. Ko, J.; Ni, Y.Q. Technology developments in structural health monitoring of large-scale bridges. *Eng. Struct.* **2005**, *27*, 1715–1725. [CrossRef]
11. Cao, W.; Liu, W.; Koh, C.G.; Smith, I.F. Exploring potential benefits of bridge condition assessment in highway operations. In Proceedings of the IABSE Congress 2019, New York, NY, USA, 4–6 September 2019; IABSE: Zurich, Switzerland, 2019.
12. Fu, G.; Devaraj, D. Bridge Management Using Pontis and Improved Concepts. In *Bridge Engineering Handbook*, 2nd ed.; CRC Press: Boca Raton, FL, USA, 2014; pp. 233–246.
13. Kim, J.; Ahn, Y.; Yeo, H. A comparative study of time-based maintenance and condition-based maintenance for optimal choice of maintenance policy. *Struct. Infrastruct. Eng.* **2016**, *12*, 1525–1536. [CrossRef]
14. Ni, Y.; Wong, K. Integrating bridge structural health monitoring and condition-based maintenance management. In Proceedings of the 4th International Workshop on Civil Structural Health Monitoring, Berlin, Germany, 6–8 November 2012; pp. 6–8.
15. Rens, K.L.; Wipf, T.J.; Klaiber, F.W. Review of nondestructive evaluation techniques of civil infrastructure. *J. Perform. Constr. Facil.* **1997**, *11*, 152–160. [CrossRef]
16. Giannini, R.; Sguerri, L.; Paolacci, F.; Alessandri, S. Assessment of concrete strength combining direct and NDT measures via Bayesian inference. *Eng. Struct.* **2014**, *64*, 68–77. [CrossRef]
17. Li, Z.; Li, F.; Zdunek, A.; Landis, E.; Shah, S.P. Application of acoustic emission technique to detection of rebar corrosion in concrete. *Mater. J.* **1998**, *95*, 68–81.
18. Rhazi, J.; Kharrat, Y.; Ballivy, G.; Rivest, M. Application of acoustical imaging to the evaluation of concrete in operating structures. *Spec. Publ.* **1997**, *168*, 221–232.
19. Holford, K.M.; Davies, A.; Pullin, R.; Carter, D. Damage location in steel bridges by acoustic emission. *J. Intell. Mater. Syst. Struct.* **2001**, *12*, 567–576. [CrossRef]
20. Maser, K.R.; Roddis, W.K. Principles of thermography and radar for bridge deck assessment. *J. Transp. Eng.* **1990**, *116*, 583–601. [CrossRef]
21. Clark, M.; McCann, D.; Forde, M. Application of infrared thermography to the non-destructive testing of concrete and masonry bridges. *NDT E Int.* **2003**, *36*, 265–275. [CrossRef]
22. Chang, Y.F.; Wang, C.Y. A 3-D image detection method of a surface opening crack in concrete using ultrasonic transducer arrays. *J. Nondestruct. Eval.* **1997**, *16*, 193–203. [CrossRef]
23. Mohamed, O.A.; Rens, K.L. Ultrasonic testing of properties of 50 year old concrete. *Mater. Eval.* **2001**, *59*, 1426–1430.
24. Shiotani, T.; Aggelis, D.G.; Makishima, O. Global monitoring of large concrete structures using acoustic emission and ultrasonic techniques: Case study. *J. Bridge Eng.* **2009**, *14*, 188–192. [CrossRef]
25. Rens, K.L.; Greimann, L.F. Ultrasonic approach for nondestructive testing of civil infrastructure. *J. Perform. Constr. Facil.* **1997**, *11*, 97–104. [CrossRef]
26. Davis, A.; Ansari, F.; Gaynor, R.; Lozen, K.; Rowe, T.; Caratin, H.; Heidbrink, F.; Malhotra, V.; Simons, B.; Carino, N. *Nondestructive Test Methods for Evaluation of Concrete in Structures*; American Concrete Institute, ACI: Farmington Hills, MI, USA, 1998; Volume 228.
27. Ervin, B.L.; Kuchma, D.A.; Bernhard, J.T.; Reis, H. Monitoring corrosion of rebar embedded in mortar using high-frequency guided ultrasonic waves. *J. Eng. Mech.* **2009**, *135*, 9–19. [CrossRef]

28. Azari, H.; Nazarian, S.; Yuan, D. Assessing sensitivity of impact echo and ultrasonic surface waves methods for nondestructive evaluation of concrete structures. *Constr. Build. Mater.* **2014**, *71*, 384–391. [CrossRef]

29. Krause, M.; Milmann, B.; Mielentz, F.; Streicher, D.; Redmer, B.; Mayer, K.; Langenberg, K.J.; Schickert, M. Ultrasonic imaging methods for investigation of post-tensioned concrete structures: A study of interfaces at artificial grouting faults and its verification. *J. Nondestruct. Eval.* **2008**, *27*, 67–82. [CrossRef]

30. Hsieh, K.H.; Halling, M.W.; Barr, P.J. Overview of vibrational structural health monitoring with representative case studies. *J. Bridge Eng.* **2006**, *11*, 707–715. [CrossRef]

31. Samman, M.M.; Biswas, M. Vibration testing for nondestructive evaluation of bridges. I: Theory. *J. Struct. Eng.* **1994**, *120*, 269–289. [CrossRef]

32. Kazemi, M.; Madandoust, R.; de Brito, J. Compressive strength assessment of recycled aggregate concrete using Schmidt rebound hammer and core testing. *Constr. Build. Mater.* **2019**, *224*, 630–638. [CrossRef]

33. Breccolotti, M.; Bonfigli, M.F.; Materazzi, A.L. Influence of carbonation depth on concrete strength evaluation carried out using the SonReb method. *NDT E Int.* **2013**, *59*, 96–104. [CrossRef]

34. Cristofaro, M.; Viti, S.; Tanganelli, M. New predictive models to evaluate concrete compressive strength using the SonReb method. *J. Build. Eng.* **2020**, *27*, 100962. [CrossRef]

35. Diamanti, N.; Annan, A.P.; Redman, J.D. Concrete bridge deck deterioration assessment using ground penetrating radar (GPR). *J. Environ. Eng. Geophys.* **2017**, *22*, 121–132. [CrossRef]

36. Chen, D.H.; Wimsatt, A. Inspection and condition assessment using ground penetrating radar. *J. Geotech. Geoenviron. Eng.* **2010**, *136*, 207–214. [CrossRef]

37. Maser, K.R. Condition assessment of transportation infrastructure using ground-penetrating radar. *J. Infrastruct. Syst.* **1996**, *2*, 94–101. [CrossRef]

38. Barnes, C.L.; Trottier, J.F. Ground-penetrating radar for network-level concrete deck repair management. *J. Transp. Eng.* **2000**, *126*, 257–262. [CrossRef]

39. Sbartaï, Z.; Laurens, S.; Viriyametanont, K.; Balayssac, J.; Arliguie, G. Non-destructive evaluation of concrete physical condition using radar and artificial neural networks. *Constr. Build. Mater.* **2009**, *23*, 837–845. [CrossRef]

40. Sbartaï, Z.M.; Laurens, S.; Elachachi, S.M.; Payan, C. Concrete properties evaluation by statistical fusion of NDT techniques. *Constr. Build. Mater.* **2012**, *37*, 943–950. [CrossRef]

41. Ploix, M.A.; Garnier, V.; Breysse, D.; Moysan, J. NDE data fusion to improve the evaluation of concrete structures. *NDT E Int.* **2011**, *44*, 442–448. [CrossRef]

42. Villain, G.; Sbartaï, Z.M.; Dérobert, X.; Garnier, V.; Balayssac, J.P. Durability diagnosis of a concrete structure in a tidal zone by combining NDT methods: Laboratory tests and case study. *Constr. Build. Mater.* **2012**, *37*, 893–903. [CrossRef]

43. Völker, C.; Shokouhi, P. Multi sensor data fusion approach for automatic honeycomb detection in concrete. *NDT E Int.* **2015**, *71*, 54–60. [CrossRef]

44. Dérobert, X.; Lataste, J.F.; Balayssac, J.P.; Laurens, S. Evaluation of chloride contamination in concrete using electromagnetic non-destructive testing methods. *NDT E Int.* **2017**, *89*, 19–29. [CrossRef]

45. Hugenschmidt, J.; Mastrangelo, R. GPR inspection of concrete bridges. *Cem. Concr. Compos.* **2006**, *28*, 384–392. [CrossRef]

46. Maierhofer, C. Nondestructive evaluation of concrete infrastructure with ground penetrating radar. *J. Mater. Civ. Eng.* **2003**, *15*, 287–297. [CrossRef]

47. Morris, I.; Abdel-Jaber, H.; Glisic, B. Quantitative attribute analyses with ground penetrating radar for infrastructure assessments and structural health monitoring. *Sensors* **2019**, *19*, 1637. [CrossRef]

48. Jazayeri, S.; Kruse, S.; Hasan, I.; Yazdani, N. Reinforced concrete mapping using full-waveform inversion of GPR data. *Constr. Build. Mater.* **2019**, *229*, 117102. [CrossRef]

49. Clemena, G.G.; Sprinkel, M.M.; Long, R.R., Jr. *Use of Ground-Penetrating Radar for Detecting Voids under a Jointed Concrete Pavement*; Transportation Research Record: Washington, DC, USA, 1987.

50. Anderson, N.L.; Ismael, A.M.; Thitimakorn, T. Ground-penetrating radar: A tool for monitoring bridge scour. *Environ. Eng. Geosci.* **2007**, *13*, 1–10. [CrossRef]

51. Laurens, S.; Balayssac, J.; Rhazi, J.; Klysz, G.; Arliguie, G. Non-destructive evaluation of concrete moisture by GPR: Experimental study and direct modeling. *Mater. Struct.* **2005**, *38*, 827–832. [CrossRef]

52. Dérobert, X.; Iaquinta, J.; Klysz, G.; Balayssac, J.P. Use of capacitive and GPR techniques for the non-destructive evaluation of cover concrete. *NDT E Int.* **2008**, *41*, 44–52. [CrossRef]

53. Plati, C.; Loizos, A. Estimation of in-situ density and moisture content in HMA pavements based on GPR trace reflection amplitude using different frequencies. *J. Appl. Geophys.* **2013**, *97*, 3–10. [CrossRef]

54. Morris, I.M.; Kumar, V.; Glisic, B. Predicting material properties of concrete from ground-penetrating radar attributes. *Struct. Health Monit.* **2020**, 1475921720976999.

55. Rodés, J.P.; Pérez-Gracia, V.; Martínez-Reguero, A. Evaluation of the GPR frequency spectra in asphalt pavement assessment. *Constr. Build. Mater.* **2015**, *96*, 181–188. [CrossRef]

56. Sigurdardottir, D.H. Strain-Based Monitoring Methods for Beam-Like Structures. Ph.D. Thesis, Princeton University, Princeton, NJ, USA, 2015.

57. Morris, I.M.; Abdel-Jaber, H.; Glisic, B. Ground penetrating radar as a monitoring tool: Mapping internal features with attribute analysis. In Proceedings of the SHMII 8, Brisbane, Australia, 5–8 December 2017.
58. Allroggen, N.; Tronicke, J. Attribute-based analysis of time-lapse ground-penetrating radar data. *Geophysics* **2016**, *81*, H1–H8. [CrossRef]
59. Annan, A. *Ground Penetrating Radar Principles, Procedures, and Applications*; Sensors and Software Inc.: Mississauga, ON, Canada, 2003; p. 285.
60. Allen, R. Automatic earthquake recognition and timing from single traces. *Bull. Seismol. Soc. Am.* **1978**, *68*, 1521–1531.
61. Wong, P.T.; Lai, W.W.; Sham, J.F.; sun Poon, C. Hybrid non-destructive evaluation methods for characterizing chloride-induced corrosion in concrete. *NDT E Int.* **2019**, *107*. [CrossRef]
62. Breiman, L. Random forests. *Mach. Learn.* **2001**, *45*, 5–32. [CrossRef]
63. Kohavi, R. A study of cross-validation and bootstrap for accuracy estimation and model selection. In Proceedings of the IJCAI, Montreal, QC, Canada, 20–25 August 1995; Volume 14, pp. 1137–1145.
64. Rubin, L.; Witkiewitz, K.; Andre, J.; Reilly, S. Methods for handling missing data in the behavioral neurosciences: Don't throw the baby rat out with the bath water. *J. Undergrad. Neurosci. Educ.* **2007**, *5*, A71. [PubMed]
65. Bergstra, J.; Bengio, Y. Random search for hyper-parameter optimization. *J. Mach. Learn. Res.* **2012**, *13*, 281–305.
66. C1074-19e1. *Standard Practice for Estimating Concrete Strength by the Maturity Method*; ASTM International: West Conshohocken, PA, USA, 2019.
67. Plowman, J.M. Maturity and the strength of concrete. *Mag. Concr. Res.* **1956**, *8*, 13–22. [CrossRef]
68. Carino, N.J. The Maturity Method. In *Handbook on Nondestructive Testing of Concrete*; Malhotra, V.M., Carino, N.J., Eds.; CRC Press: West Conshohocken, PA, USA, 2003; pp. 5.1–5.47.
69. Ruxton, G.D. The unequal variance t-test is an underused alternative to Student's t-test and the Mann–Whitney U test. *Behav. Ecol.* **2006**, *17*, 688–690. [CrossRef]
70. Past Weather in Princeton, NJ, USA. Available online: https://www.timeanddate.com/weather/@5102922/historic (accessed on 24 April 2021).
71. Morris, I.M.; Kumar, V.; Lopez, S.A.; Glisic, B. Comparing predicted in situ 8-year concrete strength by ground penetrating radar attributes and maturity method. In *Nondestructive Characterization and Monitoring of Advanced Materials, Aerospace, Civil Infrastructure, and Transportation XV*; Yu, T.Y., Gyekenyesi, A.L., Eds.; International Society for Optics and Photonics, SPIE: Bellingham, WA, USA, 2021; Volume 11592, pp. 38–47. [CrossRef]

Article

Bridge Foundation River Scour and Infill Characterisation Using Water-Penetrating Radar

Kris E. J. Campbell [1,2,*], Alastair Ruffell [2], Jamie Pringle [3], David Hughes [2], Su Taylor [2] and Brian Devlin [4]

[1] Department for Infrastructure (Northern Ireland), Clarence Court, Belfast BT2 8GB, UK
[2] School of the Natural Built Environment, Queens University, Belfast BT7 1NN, UK; a.ruffell@qub.ac.uk (A.R.); d.hughes@qub.ac.uk (D.H.); s.e.taylor@qub.ac.uk (S.T.)
[3] School of Geography, Geology & Environment, Keele University, Keele, Staffordshire ST5 5BG, UK; j.k.pringle@keele.ac.uk
[4] Bedford House, 16-22 Bedford Street, Belfast BT2 7FD, UK; Brian.Devlin@arup.com
* Correspondence: kristopher.campbell@infrastructure-ni.gov.uk

Abstract: Inspections of engineered structures below water level are essential to ensure the long-term serviceability of bridge infrastructure and to avoid major damage or failure. This research aimed to investigate integrated geophysical technologies for the underwater inspection of bridge foundation-related scour and erodible scour-based infill. Survey methods focused on Water-Penetrating Radar (WPR), supplemented by sonar. Whilst the survey benefits of the sonar imaging water–sediment interface and structures are well known, those of WPR are not. However, it is ideally suited to the survey of the water base and sub-sediment in shallow (>10 m) freshwater, especially where suspended sediment, weed infestation or methane impede sonar results. Our work produced good WPR imagery acquired from small, manoeuvrable boats that allowed bathymetric profiles to be plotted, as well as the likely locations of soft-sediment scour in future high-water flow events. This study provides clear benefits for integrated sonar and WPR surveys in the quantitative assessment of engineered structures within freshwater.

Keywords: bridge monitoring; sonar; water-penetrating radar; river scour

Citation: Campbell, K.E.J.; Ruffell, A.; Pringle, J.; Hughes, D.; Taylor, S.; Devlin, B. Bridge Foundation River Scour and Infill Characterisation Using Water-Penetrating Radar. *Remote Sens.* **2021**, *13*, 2542. https://doi.org/10.3390/rs13132542

Academic Editors: Mercedes Solla, Vega Perez-Gracia and Simona Fontul

Received: 27 April 2021
Accepted: 15 June 2021
Published: 29 June 2021

Publisher's Note: MDPI stays neutral with regard to jurisdictional claims in published maps and institutional affiliations.

1. Introduction

The failure of bridges may be due to incorrect geotechnical evaluation, over-loading, concrete/metal corrosion or lack of inspection [1] and, rarely, poor design. Water-scour of foundations is a significant and often unpredictable cause of problems in bridge supports [2,3]. Deng and Cai (2010: [4]) considered it to be one of the main causes of failure, presenting over 1000 cases from 1961 to 1991 in the United States alone, with an attendant annual maintenance cost of over USD 30 million. Whilst modern design practices take increased flood events into account, future climate change is not currently reliably predicted. Older bridges, which would benefit from closer monitoring, typically did not consider these factors at the design stage and so rarely benefit from the mitigatory construction of modern structures, such as that seen in flexi-arch bridges [5].

Deng and Cai (2010, [4] p. 125) stated that "Scour is the result of the erosive action of flowing water, which excavates and carries away materials from the bed and banks of streams, and from around the piers and abutments of bridges". They observed (also p. 125) that "there are generally three types of scour that affect the performance and safety of bridges, namely, local scour, contraction scour, and degradational scour". These are common in the gyratory currents of the marine realm (see [6] on the collapse between Belfast and Dublin, Ireland, for instance) but also occur in rivers and, to some extent, lakes; in this work, we focused on a fluvial system with two different bridge types.

Sonar surveys have been typically applied to quantify water depths and water–sediment interfaces, and to identify/locate objects [7]. Sonar does, however, have limitations, including: the inability to penetrate beneath the sediment–water interface, rocky substrates, weed infestation and gas bubbles [8]. Ground-penetrating radar (GPR) is commonly used to investigate terrestrial engineering structures and the surrounding environment, but has rarely been used when surveying in fresh water.

The aim of this paper was to first evaluate previous studies where water-penetrating radar (WPR from hereon) was used for the inspection of sub-bottom channel profiles around structures such as bridge piers in freshwater; secondly, to detail a case study where we have used both WPR and sonar to investigate bridge scour and infill; and thirdly, to combine the results to create a 3D model of the study site as an exemplar.

Ground-Penetrating Radar (GPR) Applied to Water-Penetrating Radar (WPR)

It is beyond the scope of this paper to review in detail all GPR and electrical geophysical survey methodologies; only a brief introduction to the relevant concepts is given here and the interested reader is referred to geophysical textbooks [9–11]. GPR uses the transmission and reflection of electromagnetic waves (typically 25 to 1000 MHz depending on the desired resolution and depth penetration) in soil, rock, water/ice and sediment. A GPR system requires source and receiving antennas (measuring the same central frequency). The transmitter–receiver array is typically moved along a survey profile and radar traces are collected against a pre-set time or distance interval to produce a time–distance cross-section (called a radargram in some studies in the literature); these may have been collated into plan views typical of other geophysical surveys. Alternatively, one antenna may be moved away from a static antenna to create a moveout profile. Radargrams possess two properties that are of particular interest here: (a) reflections at interfaces between two geological media of differing dielectric permittivity; and (b) radar wave attenuation (or signal loss) as a function of fluid electrical conductivity and wavelength [9]. A number of key features of GPR antennas relevant to this study are summarised. Antennas may be built to transmit and receive a range of frequencies, from low (e.g., 25 MHz) to very high (e.g., 2–3 GHz). Most common for subsurface surveying (including in freshwater) would be antennae in the 50 to 500 MHz range.

This study tested 50, 100, 200 and 250 MHz antennas, the results of which are outlined below. Antennas may be left unshielded, making the results of GPR surveying prone to nontarget artefacts (e.g., metal poles, power lines and out-of-plane objects) but leaving a lightweight and flexible design. Shielding of the antennae partly removes this problem of such interference but creates a relatively inflexible design. Antenna orientation can influence results, with different orientations achieving varied subsurface resolutions, balanced with being prone to out-of-plane reflections [12]. For boat-borne GPR surveying, the dimensions of the vessel often determines antenna size and orientation for lower-frequency antennas such as 100 MHz or lower [13]. GPR data can be extensively processed to remove artefacts and improve clarity; a full description of what processing steps are available and what they do can be found in: [9,11]. Freshwater GPR surveys are generally successful [14,15], although, as with all geophysical surveys, there is the choice of improved resolution at shallow depths (in GPR with higher-frequency antennas) vs. poorer resolution but with a greater depth range (in GPR with lower-frequency antennas). Most surveys of sediment thickness/type deploy lower-frequency antennas, attempting to image subsurface sediments [16–19]. The primary difference between terrestrial GPR and aqueous WPR is antenna deployment; in GPR, these are used in assessing ground conditions, with the antenna moved across the surface of the area to be investigated. In WPR, such antennas are positioned in contact with fresh water, be it within a non-metallic, single-skin boat, or on/within the water (water-proofed as necessary). Some work (see above) has trialled suspending antennas above water, with a resultant airgap that may diminish results. Variations in water conductivity (e.g., salt content in brackish/coastal waters) and type(s) of suspended matter affect radar wave propagation and reflection [20] such that in some brackish lakes and lagoons, WPR will not work well. This is because fresh- and saltwater have similar dielectric properties (~80 each) and radar velocities (fresh is 0.033 m/ns; saltwater is 0.01 m/ns) but very different conductivities (freshwater is 0.5 mS/m; saltwater is 30,000 mS/m). This results in a low radar wave attenuation in freshwater and high deterioration in saltwater, which simply absorbs radar waves at the frequencies needed. A summary of water-penetrating radar or WPR can be found in [20], who suggest that because water is relatively homogeneous, radar waves penetrate easily but slowly. Radar wave transmission is facilitated along the water–air interface, causing out-of-plane anomalies when floating objects are present. Conversely, excellent cross-sections of water depth, with suspended objects, as well as sediment subsurface, are obtained using WPR in fresh water. Two WPR data outputs are possible: most common are 2D radargrams (vertical soundings of water and sediment) and, more rarely, 3D plan views at various depths. These map-like outputs comprise digital data and can be subjected to manipulation in a Geographic Information System such as ARC-GIS, but require accurate geolocation on a grid or by GPS. Although WPR has been used to successfully image scour around bridge supports (e.g., Gorin and Haeni, 1989, and see below), the current work showed 2D radargrams, digitised 3D plan views, selected processed data of bridge scours and made an evaluation of each data output type. A critical factor in WPR data acquisition is the mode of antenna deployment (in water/out of water; within boat/type of vessel and base; antenna orientation): [21] reviewed previous methods of WPR surveying for a range of applications, and they commented that many authors do not state this, something we (here) do explicitly.

2. Previous Work

A pioneering study in [13] used low-frequency 80 MHz WPR data, compared to three seismic (or acoustic) methods (black and white fathometer; colour fathometer; tuned transducer), in their study of sediment scour around road bridge supports along the Connecticut River (USA). In their study, WPR suffered less sidescan anomalies and penetrated the sub-bottom sediments more effectively than the acoustic methods, but few details on WPR data acquisition were provided (e.g., antenna orientation). Refs. [22,23] had similar intentions to [13] in assessing scour around bridge supports in rivers; their work compared both their own 100 and 450 MHz antenna profiles to those from the literature and the water conductivities/water depths in each. These both clearly showed the interplay between low frequency and better depth penetration, vs. higher antenna frequencies with improved stratigraphic detail but less penetration—both limited by increasing water conductivities. Ref. [23] also showed how an initial review of the location, water conductivity, engineering, river flow, previous work and local geology are essential in planning a WPR survey. They also elucidated some of the issues in WPR surveying in that objects/surfaces not directly under the boat may be imaged, due to the high transmissivity of radar waves at the air–water interface (surface out-of-plane reflections such as jetties, piers and buoys); water-bottom multiples may dominate the sub-bottom reflections without processing (see below) and the limits of water depth.

Ref. [24] used a 300 MHz radar antenna array to gather 22 lines around 16 bridges in New Hampshire (USA), documenting pre- and post-flood scour holes plus infilled scour at seven locations. Where possible, the results were verified by insertion of a steel rod, for both water depth and sub-bottom sediment thickness. A similar deployment method was used as in this study, by placing a commercially available GPR system in the base of an inflatable boat; water levels (and, thus, changing depths) were elucidated. WPR data processing included migration and signal amplification (gain) on profiles on deeper water, to enhance sediment reflections.

Refs. [25,26] followed the work of [27] with evaluations of the use of WPR in assessing bridge-support-related scour. Ref. [27] acquired WPR data from ten bridge sites over shallow water in Missouri (USA), using boat-borne methods or lowering the radar antenna from bridge decks. Their 2D profiles successfully imaged the base of each water body, some at peak flood times. Both works considered the use of 2D profiling before and during floods as a predictive tool in risk management, as well as WPR surveys over time, to build up a model of changing fluvial processes that they consider superior to reflection seismic 2D profiling, echo sounding and electrical conductivity profiling. Ref. [25] made the contentious comment that the radar system does not need to be in contact with water, which is correct (see [17] although, naturally, optimal coupling between the antenna and water is desirable.

Following abundant publications on the use of WPR in assessing scour through the 1980s and 1990s, there appears fewer mainstream publications until 2020, when [28] considered the complicating issue of wood debris accumulating at bridge supports, which may accelerate circulatory scour by expanding the effective footprint of the bridge pier, as well as setting up internal vortices between wooden objects, especially tree trunks and branches. [21] summarised many of the above articles and commented that advances in radar technology (e.g., the wide use of multiplexer systems); the use of borehole antennas in deep (over 20–30 m) water; a conjunctive approach with sonar (especially remotely deployed sidescan systems such as the Codaoctpus), and CHIRPS; and data processing and visualisation may still further advance the use of WPR in assessing bridge scour.

3. Site Description

This site was used for two reasons. First, because, within a relatively short distance of 40 m, there is both a road and rail bridge, crossing a navigable section of the River Bann, on the southern outskirts of Portadown, County Armagh (N. Ireland)—Figure 1 shows a satellite image of the two bridges.

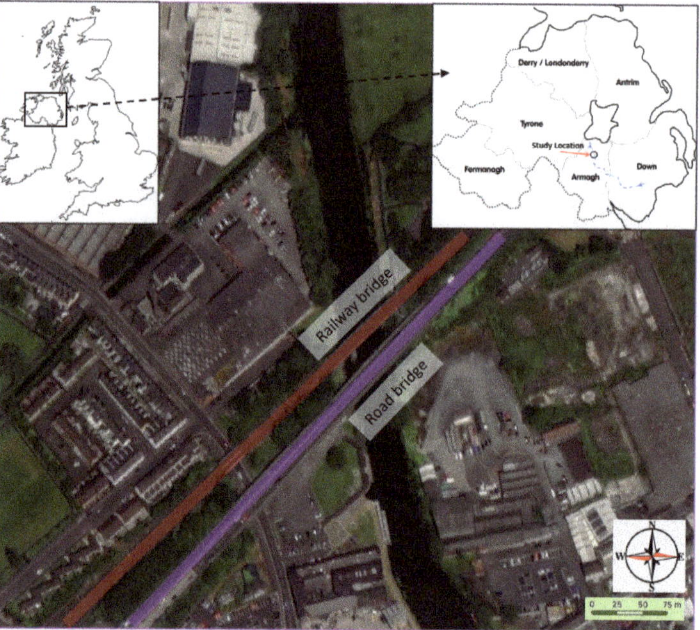

Figure 1. Location of the study site in the British Isles (**top left**) and Northern Ireland (with location and route of River Bann marked, **top right**); modified GeoEye satellite image of the two railway and road bridges and river studied (**bottom**).

The road bridge is a reinforced concrete deck bridge supported over the river on four circular reinforced concrete piers (in the foreground of Figure 2). The railway bridge is seen in the background of Figure 2 and consists of a metal truss deck on circular metal supports on top of two piled foundation-reinforced concrete piers.

Figure 2. Annotated oblique view of the River Bann and the railway and road bridges studied, taken from the southwest bank (see Figure 1 for plan view map).

Secondly, the River Bann is one of the longest rivers in Ireland, with this section being adjacent to a slipway facility; this section of river experiences a significant volume of water, as well as wash effects from powerboats and other river activities. At the location of these two bridges, the River Bann narrows and becomes slightly shallower, which leads to pressurized flow at the supports which, in conjunction with motorized boat movements, may encourage scouring action at these structures [29,30].

The A3 Northway Dual Carriageway Bridge (road bridge built in two phases through the early 1970s) has four circular reinforced concrete vertical supports (Figure 2) whilst the Belfast–Dublin railway bridge (current construction 1958–1960, replacing an 1848 wooden structure) has two, elongate vertical reinforced piled concrete support (lozenge-shaped, or rectangular, with acute ends) piers (Figure 2).

This location presented a good opportunity to consider the scouring effects around two separate bridge pier types of two separate bridges adjacent to each other in the same short stretch of river. The main caveat to this is that the relative proximity of each, implies the possibility of alterations in current flow from the upstream road bridge supports, may influence scour around the downstream railway bridge structures (Figure 2). Conversely, for infrastructural and topographic reasons, road, rail and other (e.g., pipeline) bridges are often constructed side-by-each, making this study relevant. The road bridge comprises a multi-span-reinforced concrete bridge which carries 3 lanes of arterial route traffic. There are four circular piers located in the water: two are near the edge of the riverbank (depending on flood conditions) each consisting of circular concrete columns of 1.5 m in diameter. These columns are spaced ~10 m apart and the span between the four piers is ~20 m (approximate distances as the sides of the river are not linear and are sometimes arcuate). The railway bridge has two tracks (North-East Bound and South-West Bound) with two 15 m long and 2 m wide piers, founded on piled foundations in the riverbed, but closer (1–2 m) to the riverbank than the central roadway piers, leaving a clear navigable channel down the centre of the watercourse.

The River Bann is typically 30–60 m wide at the study location and is popular for both fishing and water-sports, with a 30% "good", 55% "moderate", 10% "poor" and 5% annual water quality classification by the Department of Agriculture and Economic Affairs/Northern Ireland Environment Agency [31]. Conductivities were measured for both survey times (March and September) at between 0.02 and 0.04 µS/m; pH was consistently measured at 6.5 and eH at +20, reflecting the iron-rich Palaeogene basaltic (Antrim Lava

Group) river hinterlands, the underlying Palaeogene Lough Neagh Clay Group, and the red, oxidised glacial till of the river and its catchment in this area.

There is no marine influence, as the site is inland, with the river draining north to Lough Neagh, Europe's largest freshwater lake, and being fed by rainfall riverine catchments upstream. Depth-soundings using a plumb bob indicated that the river has variable depth in and around the bridges, from 0 m at the banks, to an average of 3 m, with isolated locations up to 5 m (see Methods, below). This variability makes hydrographic surveying particularly relevant, as physical soundings (such as rods or plum-bob) are specific to discrete locations and may not be representative of the entire area.

4. Methods

Visual examination of the riverbed from the bank and dinghy, plus indications from plumb-bob soundings, suggest the riverbed to comprise both rocks and soft silt/mud. A carbon-fibre peat probe (Van Walt Ltd., 75 cm extension lengths, Surrey, UK) was also deployed and, as with the plumb-bob, no obvious firm substrate could be felt in the river centre by the user: rocks were detected adjacent to the riverbanks. The use of WPR is very relevant under such mixed subsurface conditions as sediment type may be indicated, buried objects located and sediment depth (or indeed lack of sediment, say, rock at surface) determined.

Three phases are considered in WPR surveys: air/gas, water and sediment/rock. Of the three phases present, air/gas bubbles were only present when sediment was disturbed, and they have a negligible effect on radar waves; freshwater was considered here and the boundary between it and the underlying rock/sediment was always imaged. In order to study the efficacy of the two main data outputs in WPR, (i) 2D profiles and (ii) 3D fence diagram views were generated (see below). A Sidescan Sonar survey using a Lowrance Elite 12Ti Totalscan was also conducted in order to further calibrate/compare to WPR depths and to identify any submerged hazards for health and safety considerations.

Initial tests used 50 MHz and 100 MHz rough terrain (unshielded) antennas, housed in thick plastic sheathing and towed behind the electric motor-powered (MinKota Ltd., Racine, WA, USA, 12v Trolling engine) inflatable rib (Figure 3A). These antennas proved hard to manoeuvre in the 30–60 m-wide river, showed limited detail in the 0–6 m of water depth and suffered from numerous out-of-plane reflections from the bridge piers, metal riverbank signage, subsurface debris and trees.

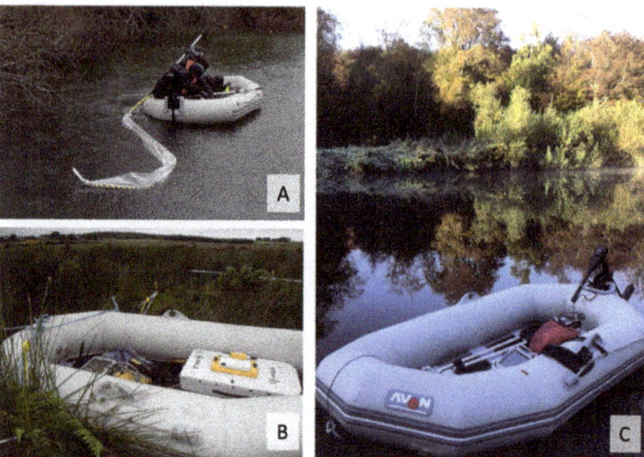

Figure 3. Methods of WPR deployment. (**A**) the 100 MHz rough terrain antenna (straightens out on boat movement), housed in plastic sheathing; (**B**) 250 MHz antenna, positioned in the dinghy (wooden foot-slats removed for optimal antenna to water contact); (**C**) dinghy with electric outboard engine, in preparation for deployment with radar and GPS aboard.

Then, 100 and 250 MHz shielded antennas in parallel and broadside endfire mode were used for the bulk of the surveys, with an unshielded 200 MHz antenna (also in both parallel endfire and parallel broadside mode) deployed to compare with data from the 250 MHz shielded antenna. A grid of survey data around all six piers was gathered and positioned using a Leica Icon Total Station, checked by dead-reckoning of landmarks such as bridge piers: differential GPS was only accurate for the start and end points of the river-parallel survey lines, when not under either bridge, a common problem also encountered when surveying under buildings and tree cover.

Each radar antenna was placed in the base of a 4-person, 6 m-long, 3 m-wide (Avon Boats Ltd., Stratford-upon-Avon, UK) inflatable plastic dinghy with foot-slats removed (Figure 3), propelled by an electric outboard engine. Running a survey with the engine off (drifting in the wind or current) and then running one with the motor on had negligible effect on the radar data (noted on figures, where present). The radar antennas were set directly onto the rubber floor of the dinghy, allowing for the contact with the water through a minimal thickness of 5 mm of rubber. We considered mounting unshielded antennas outside of the dinghy, but this was precluded by the decreased manoeuvrability around bridge piers and the presence of 5–10 cm-amplitude waves from wind, passing vessels and our own survey.

Problems encountered included (Table 1): loss of positioning information around both bridge piers (the total station could only be placed on the west bank footpath; the east bank was inaccessible due to industrial wire fence and thick vegetation); drift in the dinghy during windy conditions; the need to stop or re-run cross-river survey lines when other water craft approached; and the presence of debris in shallow water, adjacent to riverbanks such as tree branches, shopping trollies (carts), discarded metal work (railings) and dumped bicycles. If a wind-generated drift of more than the width of the dinghy (and, thus, for all antennas used, roughly the radar-wave footprint at the boat-deck to water interface) was recorded, the survey line was deleted and re-recorded.

Table 1. Summary of the problems encountered in this WPR survey: similar issues can be anticipated by others doing similar work, depending on local environment (for instance, no significant water flow was encountered here but, under turbulent conditions, will require a higher-power vessel, or suspension of radar antennas above the water).

Problems Encountered	Effect
Loss of positioning information around both bridge piers	Total station could only be placed on the west bank footpath; the east bank was inaccessible due to industrial wire fence and thick vegetation.
Drift in the dinghy during windy conditions	If a wind-generated drift of more than the width of the dinghy (and, thus, for all antennas used, roughly the radar-wave footprint at the boat-deck to water interface) was recorded, the survey line was deleted and re-recorded.
The need to stop or re-run cross-river survey lines when other watercraft approached	Resultant issue of accurately re-locating survey lines under bridges with no line of sight to total station or shore markers.
Presence of debris in shallow water, adjacent to riverbanks	Tree branches, shopping trollies [carts], discarded metal work [railings] and dumped bicycles.

The WPR data were initially viewed onsite and, later, in Mala Geoscience MV Ground-vision (Version 1.4) software in which clear images of both the riverbed and sub-bottom sediment profiles were obvious, even when air-water multiples, causing ringing in the data, were prevalent. Data were also viewed in Sandmeier's ReflexW (Version 7.2.4) software for comparison. Both Gain and Background Removal filters were applied to all 2D profile data (for consistency), with no further requirement for other data processing steps. Velocity could only be calculated to the water depths (as normal moveout requires two dinghies in water, and hyperbola-fitting needs accurate distance measurement), calibrated against plumb-bob and probe estimates. Depths to water-bottom and sediment reflectors were interpreted and digitised for import to ARC-GIS and AutoCAD visualisation software. The latter outputs allowed secondary import to Google SketchUp to visualise the bridge support scour and sub-scour sediment geometries (see Table 2 below, and Results). A complimentary sidescan sonar survey of the study site was also collected using a Lowrance Elite 12Ti Totalscan, operating at 10 Hz in continuous data collection mode, with no data processing.

Table 2. Typical WPR reflection characteristics seen in 2D profiles, used to interpret and digitise the datasets collected in this study.

WPR Profile	WPR Characteristic
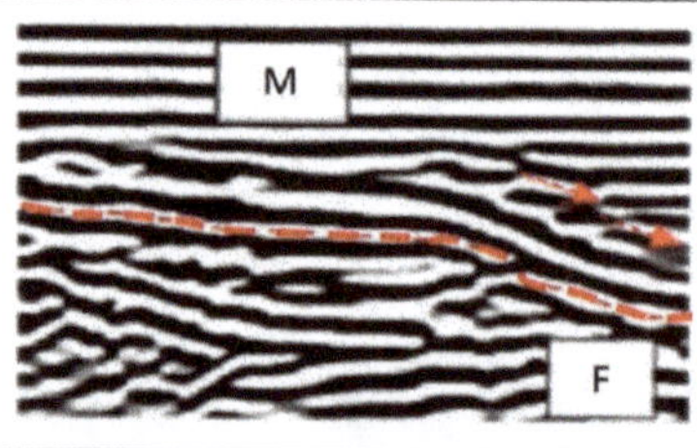	Three hyperbolas (H), buried in draped sediment, overlying glacial till. Typical digitized subsurface reflection (red dash).
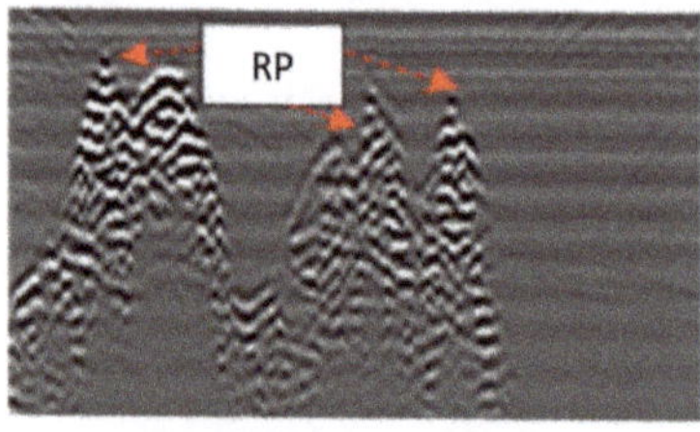	Horizontal water-bottom multiples (M), two layers (red arrows) of dipping sediments. Unconformably draped over folded (F) sediments.
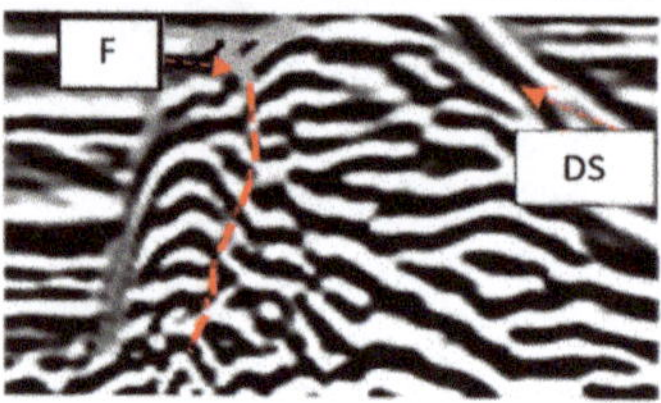	Submerged rock pinnacles (RP) with no internal layering (speckled appearance) and no substantial overlying sediment drape.
	Submerged and faulted (red dashed line—F) sediment mound, with draped sediment (DS)

5. Results

The sidescan sonar data were successfully acquired, detailing any potential submerged hazards present in the water for the subsequent WPR surveys, showed the sides of the bridge piers effectively (an example in Figure 4), and imaged some riverbed objects, but they suffered from gas bubble interference and from sonar scattering from rocks that are typically observed at such shallow water depths.

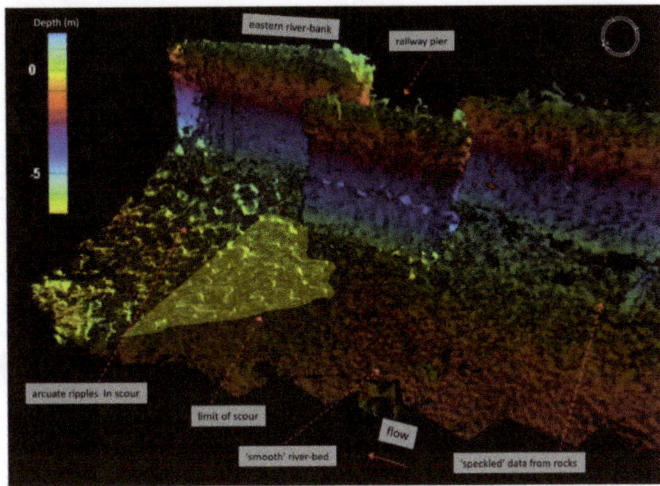

Figure 4. Annotated sidescan sonar image (Lowrance Elite 12Ti Totalscan) of the eastern river Bann margin, with elongated railway bridge pier support also being imaged.

Notably, the 100 MHz data showed the best balance between resolution and depth, so these were deployed for the bulk of the survey area, in conjunction with the 250 MHz data, especially in shallow waters on the riverbank sides of the bridge piers where deeper water penetration was not needed (Figure 5).

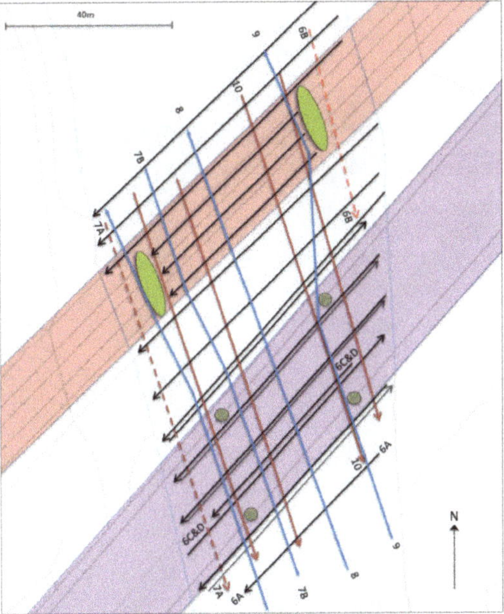

Figure 5. Survey grid line map of the WPR 2D profiles gathered within the River Bann (shaded blue). Numbers shown refer to figures in this paper. Profile solid lines = 100 MHz data; profile dashed lines = 250 MHz data. Green lozenges are railway bridge supports; green circles are the road bridge supports. River flow is ~south to ~north; coloured bridge decks included to indicate Road (purple) and Railway (red).

The short cross-lines (all original data are available, on request) shown in Figure 6 show changes in bathymetry across the river, as well as some detail below the water-sediment surface. Processing of the data, as shown in the examples of Figures 6 and 7, improves the data quality, allowing rocky outcrops, hollows (from scour) around bridge piers, subsurface objects and an indication of sediment thickness to be observed.

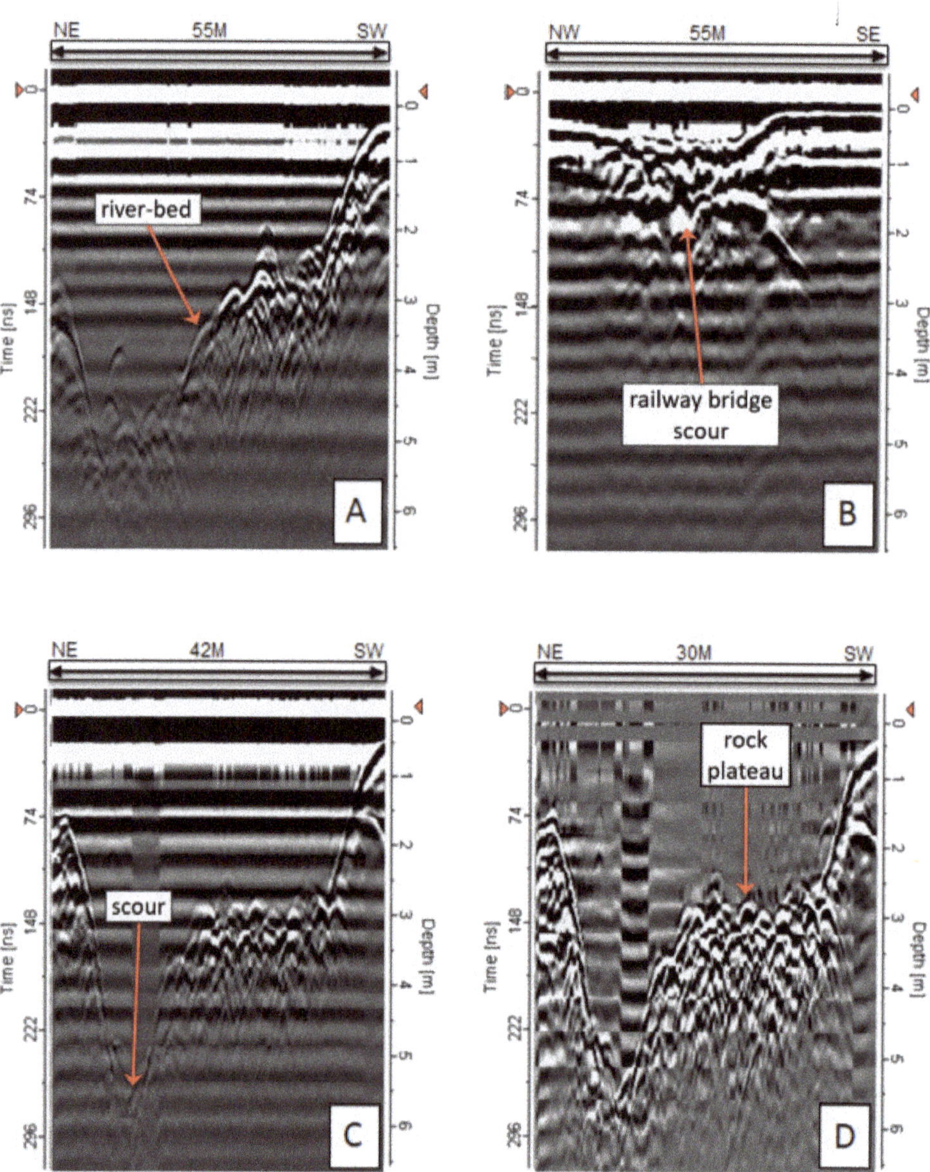

Figure 6. Examples of WPR 2D profile data collected. (**A**) 100 MHz (raw); (**B**) 250 MHz (raw); (**C,D**) 100 MHz raw and processed (gain and background removal) 2D profile, respectively (see Figure 5 for respective locations).

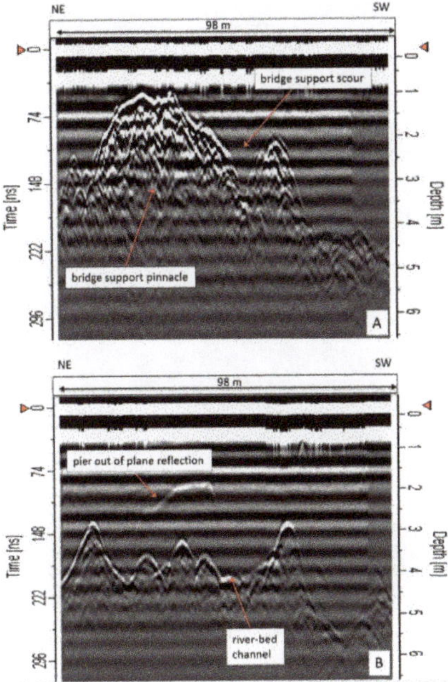

Figure 7. 100 MHz WPR 2D profile data, gathered along the side of the River Bann, adjacent to the western bridge pier. (**A**) adjacent to the pier; (**B**) 5 m from the southwestern pier on Figure 5.

The longer river-parallel lines (Figures 7–9) show more subtle changes in riverbed topography, with similar improvement in sub-riverbed imaging on the processed data (Figures 7–9).

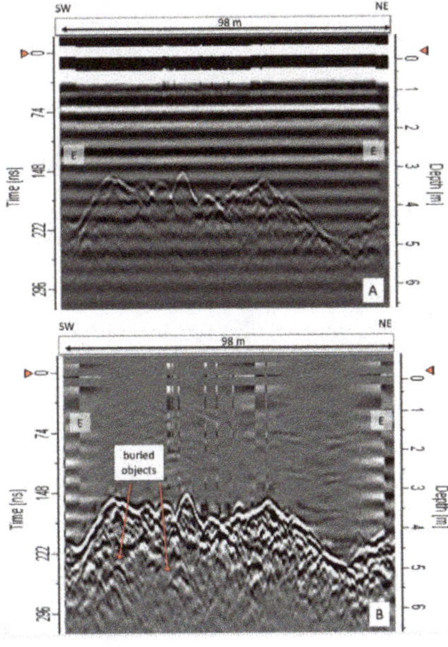

Figure 8. Raw (**A**) and processed (gain and background removal); (**B**) WPR 100 MHz 2D profile data. E = change in engine speed at start and end of survey line. See Figure 4 for location.

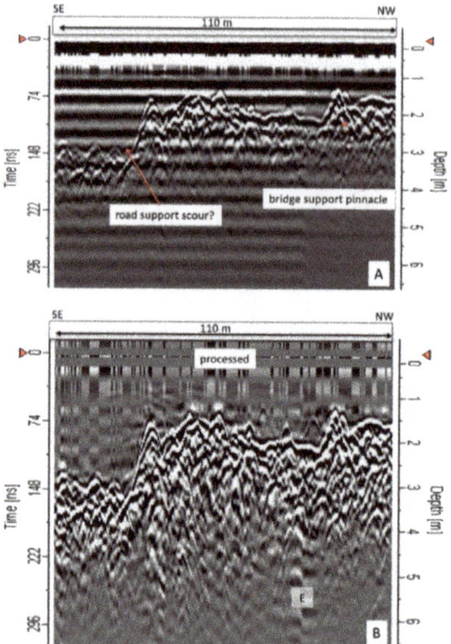

Figure 9. Raw (**A**) and processed (gain and background removal); (**B**) WPR 100 MHz 2D profile data. E = change in engine pitch. See Figure 4 for location.

Data interpretation followed the methods of [11] in picking strong, continuous reflections for digitising: these surfaces separate distinct radar facies (see Table 2). When interpreted and digitised, these 2D profiles were then combined (Figure 10), which allowed a crude view of the area under both the road bridge and its four supports, as well as the railway bridge with the two elongated piers plus deeper areas of water and a central rock outcrop (Figure 11) to be determined. The plan view of water depth in the same survey area (Figure 11) shows greater detail, with the deep-water area now resolved as two ~5 m deep hollows with a third area not detected from looking at the 2D WPR profiles alone.

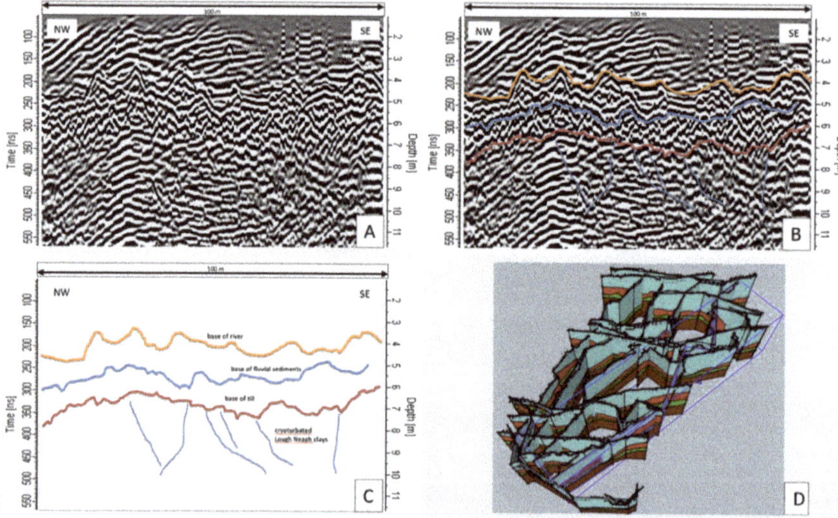

Figure 10. Method diagrams to show: (**A**) processed 2D profile data; (**B**) interpretation of 2D profile data; (**C**) digitised 2D profile data output; (**D**) fence diagram of digitised 2D profiles, for subsequent export and 3D visualisation (see Figure 11).

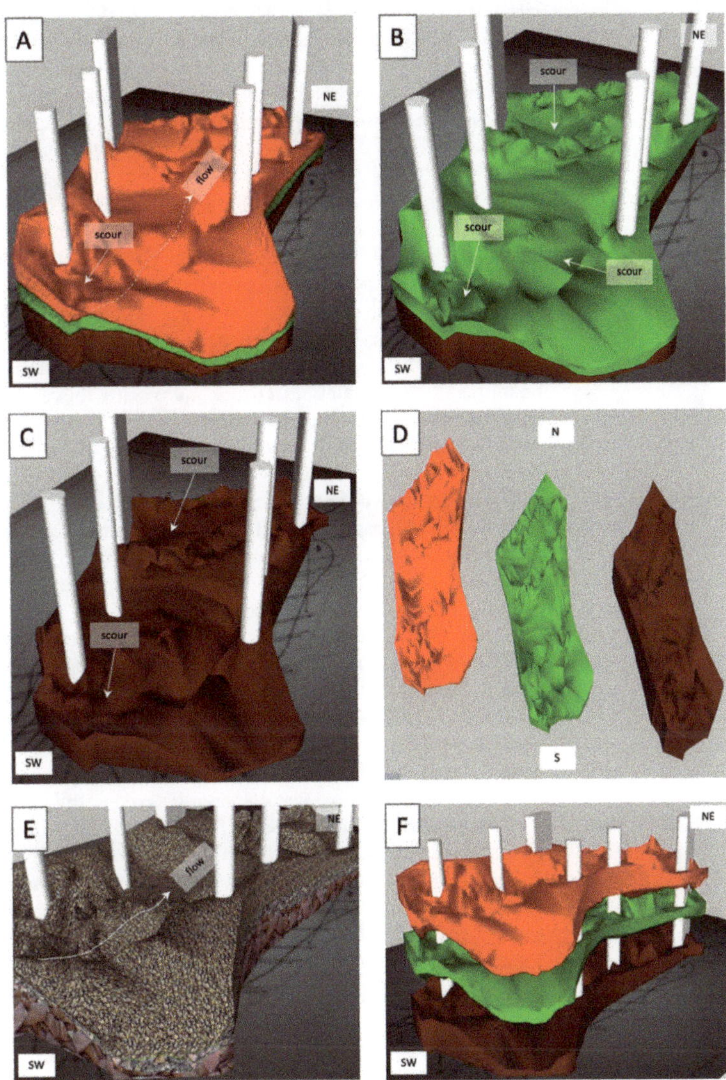

Figure 11. Visualisation of WPR data around the 6 bridge piers surveyed (white vertical bridge piers), plotted by colour to correspond to the interpreted WPR reflection in Figure 10. (**A**) 3D schematic with all surfaces marked (red = river floor, fluvial sediments; green = fluvial sediments; brown = Lough Neagh Clays). (**B**) top surface of the second layer of fluvial sediments, with water and top layer removed. (**C**) top surface of the Lough Neagh Clays, with river water and fluvial sediments removed. (**D**) all three layers, visualised separately. (**E**) colours replaced by textures for easy understanding and translation to geotechnical properties if further work is to be undertaken. (**F**) image capture of an interactive display in Google SketchUp, showing the three surface layers, artificially separated.

These depressions in the river floor are adjacent to the circular bridge supports. In order to gain some insight into the nature of these hollows, each was examined in detail and their sediment thickness mapped; all depocentres were aligned SW-NE in parallel to the main flow direction of the river in this location. Significant scour was detected adjacent to, and upstream of, the elongated railway piers. Scour was also observed upstream of the circular roadway supports. This was anticipated—contributory factors may be that the railway piers are positioned in shallower water, with an attendant higher current flow. The road piers are positioned in similar conditions to the railway piers; however, it is located adjacent to the slipway used by motorised river craft, and the railway bridge constricts flow. No obstructions are present for another 100 m upstream.

6. Discussion

This paper first reviewed the typical equipment setup and previous use of Water-Penetrating Radar (WPR) for bridge engineering studies, with different antenna configurations, shielded/unshielded antenna types and different boat deployments all having significant effects on the datasets collected by earlier authors. By far, the dominant collection strategy is the collection of multiple 2D profile datasets, using low-frequency radar antennas, which may or may not need to be combined into 3D datasets. For simple rural surveys with little geophysical cultural noise, an unshielded antenna on nonconductive boats may be optimal, but in urban areas with above-ground conductive items present, shielded antennas are thus recommended.

This paper secondly detailed a case study of surveying a navigable section of river around two bridges (a railway and road bridge) with different-shaped vertical supports. WPR results showed that the two bridges had correspondingly different river sediment scour geometries associated with them, which could be both detected and characterised by relatively closely spaced 2D profiles being collected, processed (with simple gain filters) and combined into a 3D dataset with interpreted riverbed scours, below the water–sediment interface boundaries and bedrock all being imaged.

One area of deep water within the survey site, with an adjacent platform of rock outcrop, was identified from the 2D data—this was resolved using the 3D fence diagram data as more than three areas of deep water, each with 4–5 m of sediment infill. Areas of riverbed infill were found to be adjacent to both bridge supports, and oriented along the axis of river flow. They could be due to scour-fills from the bridge supports, or possibly locations of fossil scours where the bridge piles were originally sunk when bridge construction took place. Results largely agree with [24,27] who performed similar WPR surveys around bridges, with data from 100 MHz shielded antennas providing optimal 2D profile data. Side-scan sonar data were also collected to compare with WPR data, which was favourable, as well as to identify any submerged hazards to the data acquisition stage. Limited intrusive 1D investigations, using a plumb bob/metal probe, can also give some ground truth of results, but are relatively slow and spatially unrepresentative; side scan sonar was evidenced in this study to be more effective for bathymetry.

7. Conclusions

This paper briefly reviewed Water-Penetrating Radar (WPR) surveys, equipment configurations and relevant case studies to evidence what is currently known and the potential major variables associated with data collection in such challenging environments.

Whilst physical depth sounding with a plumb bob or probe may be quick and inexpensive, it might not provide the comprehensive coverage of a hydrographic survey using echo sounding or WPR. For estimates of sediment thickness, buried objects and rock outcrops in confined locations such as our study location, WPR appeared to be advantageous when compared to other methods such as side-scan sonar.

The WPR 2D datasets collected in the case study allowed the interpretation of the survey site water bathymetry, the presence of rock vs. sediment on the riverbed floor, and some estimate of the likely distribution of water and sediment depths. Combining 2D profiles into 3D datasets were deemed less useful in assessing likely sub-bottom conditions, but did give an accurate image of the distribution of both water depths and sediment thicknesses. The unshielded antennae were not suitable for such an urban small site with above-ground conductive objects present, so 100 MHz shielded antennas were determined to be optimal and were used for the main survey.

Remote Sens. **2021**, *13*, 2542

The nature of interpreting radar traces is open for debate and the accuracy difficult to ascertain without physical excavations to validate assumptions on site. The analysis largely relies on the experience of the user and the quality of the radar data to determine interfaces between distinct geological/sedimentary features. The challenges faced with undertaking an accurately positioned grid of radar traces from a small inflatable boat were considered and more work would be required in improving this element of the data retrieval. As such, the accuracy of the positioning and interpretation of these WPR surfaces are somewhat unknown without further investigation. However, the principle of identifying bed level and potential scour features subsurface was demonstrated in this work and warrants further consideration.

This work demonstrates the applicability of the WPR technique using existing non-adapted technology (terrestrial GPR) to give good indicative results useful for bridge managers in order to decide on a suitable maintenance plan. It can be seen that WPR can be used for the identification of sub bottom profiling around bridges. Although this concept of using GPR on water or around bridges is not new in itself, there is still a clear opportunity to exploit this technology and develop it further as a useful tool for bridge inspections and 3D channel profiling.

Author Contributions: K.E.J.C.—data acquisition, processing, visualisation and writing; A.R.—project coordination and writing; J.P.—writing; D.H.—bridge foundations and scour; S.T.—bridge design and maintenance; B.D.—scour measurement. All authors have read and agreed to the published version of the manuscript.

Funding: Knowledge Transfer Partnership Grant 090519 between Collins CEI Engineering and Queen's University Belfast.

Data Availability Statement: Can be found at: http://doi.org/10.21252/ybse-4815.

Acknowledgments: Mary Hart of the Queen's University Belfast KTP; Conor Graham (QUB) for use of GPS; Mike Langton (GuidelineGeo/Mala) for loan of GPR equipment; Keith Bennett (University of St. Andrews) and Lisa Coyle-McLung (Queen's, Belfast) for the use of the dinghy. Karl Sandmeier is thanked for providing Queen's University, Belfast (under licence) ReflexW data processing software.

Conflicts of Interest: The authors declare no conflict of interest.

References

1. Biezma, M.V.; Schanack, F. Collapse of steel bridges. *J. Perf. Cons. Fac.* **2007**, *21*, 398–405. [CrossRef]
2. Ettema, R.; Melville, B.W.; Constantinescu, G. *Evaluation of Bridge Scour Research: Pier Scour Processes and Predictions*; The National Academies Press: Washington, DC, USA, 2011.
3. Sathurusinghe, P.; Hamill, G.; Taylor, S.; Robinson, D. Hydraulics of scour in the vicinity of a FlexiArch bridge. *Proc. Civ. Eng. Res. Ire.* **2020**, *5*, 83–86.
4. Deng, L.; Cai, C.S. Bridge scour: Prediction, modeling, monitoring and countermeasures—Review. *Prac. Period. Struct. Des. Cons.* **2010**, *15*, 125–134. [CrossRef]
5. Sathurusinghe, P.; Hamill, G.; Taylor, S.; Robinson, D. Foundation scour and its effect on the performance of FlexiArch™ modern arch bridge system. 2021; in press.
6. RAIU (Railway Accident Investigation Unit). *Malahide Viaduct Collapse on the Dublin to Belfast Line, on the 21st August 2009*; RAIU: Blackrock, Ireland, 2010.
7. Topczewski, L.; Ciesla, J.; Mikolajeski, P.; Adamski, P.; Markowski, Z. Monitoring of scour around bridge piers and abutments. *Trans. Res. Proc.* **2016**, *14*, 3963–3971. [CrossRef]
8. Ruffell, A.; Pringle, J.K.; Cassella, J.P.; Morgan, R.; Ferguson, M.; Heaton, V.G.; Hope, C. The use of geoscience methods for aquatic forensic searches. *Earth-Sci. Rev.* **2017**, *171*, 323–337. [CrossRef]
9. Reynolds, J.M. *An Introduction to Applied and Environmental Geophysics*, 2nd ed.; John Wiley and Sons Inc.: Chichester, UK, 2011; 696p.
10. Sharma, P.V. *Environmental and Engineering Geophysics*; Cambridge University Press: Cambridge, UK, 1997; 349p.
11. Daniels, D.J. *Ground Penetrating Radar*, 2nd ed.; The Institution of Engineering and Technology: Hong Kong, 2004; 725p.
12. Kruk, J.; van der Slob, E.C. Reduction of reflections from above surface objects in GPR data. *J. Appl. Geophys.* **2004**, *55*, 271–278. [CrossRef]
13. Gorin, S.R.; Haeni, F.P. *Use of Surface-Geological Methods to Assess Riverbed Scour at Bridge Piers*; Report 88-4212; Department of the Interior, US Geological Survey: Reston, VR, USA, 1989.

14. Beres, M.; Haeni, F.P. Application of ground-penetrating radar methods in hydrological studies. *Ground Water* **1991**, *29*, 375–386. [CrossRef]
15. Nobes, D.C.; Rother, H.; Kruk, J.; Jol, H.M. Radar 'lensing' by a small river: Can a layer of surface water improve the signal? *Near Surf. Geophys.* **2006**, *4*, 69–74. [CrossRef]
16. Haeni, F.P. Use of ground-penetrating radar and continuous seismic reflection profiling on surface water bodies in environmental and engineering studies. *J. Environ. Eng. Geophy.* **1996**, *1*, 27. [CrossRef]
17. Haeni, F.P.; McKeegan, D.K.; Capron, D.R. *Ground-Penetrating Radar Study of the Thickness and Extent of Sediments beneath Silver Lake, Berlin & Meriden, Connecticut*; Report 85-4108; US Department of the Interior, Geological Survey: Reston, VR, USA, 1987; 19p.
18. Kovacs, A. *Impulse Radar Bathymetric Profiling in Weed-Infested Fresh Water. U.S.A*; Cold Reg. Res Eng Lab. (CRREL): Hanover, NH, USA, 1991.
19. Sellmann, P.V.; Delaney, A.J.; Arcone, S.A. *Sub-Bottom Surveying in Lakes with Ground-Penetrating Radar*; US Army Cold Regions Research and Engineering Laboratory: Hanover, NH, USA, 1992; pp. 92–98.
20. Parker, R.; Ruffell, A.; Hughes, D.; Pringle, J. Geophysics and the search of freshwater bodies: A review. *Sci. Just* **2010**, *50*, 141–149. [CrossRef] [PubMed]
21. Ruffell, A.; Parker, R. Water penetrating radar. *J. Hydrol.* **2021**, *597*, 126300. [CrossRef]
22. Davidson, N.C.; Hardy, M.S.A.; Forde, M.C. Bridge scour assessment by impulse radar. Institution of Electrical Engineers (IEE) Colloquium on Radar and Microwave. *Tech. Non-Destr. Eval.* **1995**, *8*, 1–8.
23. Forde, M.C.; McCann, D.M.; Clark, M.R.; Broughton, K.J.; Fenning, P.J.; Brown, A. Radar Measurement of Bridge Scour. *NDT E Int.* **1999**, *32*, 481–492.
24. Olimpio, J.R. *Use of a Ground-Penetrating Radar System to Detect Pre and Post-Flood Scour at Selected Bridge Sites in New Hampshire, 1996–1998*; US Department of the Interior, US Geological Survey: Reston, VR, USA, 2000.
25. Anderson, N.L.; Ismael, A.M.; Thitimakorn, T. Ground penetrating radar: A tool for monitoring bridge scour. *Environ. Eng. Geosci.* **2007**, *13*, 1–10. [CrossRef]
26. Park, I.; Lee, J.; Cho, W. Assessment of bridge scour and riverbed variation by a ground penetrating radar. In Proceedings of the Tenth International Conference on Grounds Penetrating Radar, Delft, The Netherlands, 21–24 June 2004; pp. 411–414.
27. Webb, D.J.; Anderson, N.L.; Newton, T.; Cardimona, S. Bridge scour: Application of ground penetrating radar. In Proceedings of the First International Conference on the Application of Geophysical Methodologies and NDT to Transportation Facilities and Infrastructure, Los Angeles, CA, USA, 15–19 May 2002.
28. Ebrahimi, M.; Djordjevic, S.; Panici, D.; Tabor, G.; Kripakaran, P. A method for evaluating local scour depth at bridge piers due to debris accumulation. *Proc. Inst. Civ. Eng.* **2020**, *173*, 86–99. [CrossRef]
29. Cumberlidge, J. *The Inland Waterways of Ireland*; Imray, Laurie, Norie & Wilson Publishers Limited: Cambridge, UK, 2002; p. 135.
30. Delany, R. *Ireland's Inland Waterways—Celebrating 300 Years*; Appletree Press: Mankato, MN, USA, 2004; p. 216.
31. DAERA (Department of Agriculture, Environment and Rural Affairs); NIEA (Northern Ireland Environment Agency). *Planning for the Third Cycle River Basin Management Plan 2021–2027, Appendix 2: Overall Water Quality Status Interim Results 2018 and Pressure Analysis*; Northern Ireland Environment Agency: Belfast, UK, 2019; 22p.

 *remote sensing*

Communication

Diagnostics of Reinforcement Conditions in Concrete Structures by GPR, Impact-Echo Method and Metal Magnetic Memory Method

Karel Pospisil [1,2], Monika Manychova [3], Josef Stryk [1,*], Marta Korenska [1], Radek Matula [4] and Vaclav Svoboda [5]

[1] CDV—Transport Research Centre, Lisenska 33a, 636 00 Brno, Czech Republic; karel.pospisil@cdv.cz (K.P.); korenska.m@email.cz (M.K.)
[2] Institute of Forensic Engineering, Brno University of Technology, Purkynova 464/118, 612 00 Brno, Czech Republic
[3] Institute of Building Structures, Faculty of Civil Engineering, Brno University of Technology, Veveri 95, 602 00 Brno, Czech Republic; manychova.m@fce.vutbr.cz
[4] SQZ Ltd., U Mistni Drahy 939/5, 779 00 Olomouc, Czech Republic; matula@sqz.cz
[5] Preditest Ltd., Pod Višňovkou 23, 140 00 Praha 4, Czech Republic; svoboda@preditest.cz
* Correspondence: josef.stryk@cdv.cz

Citation: Pospisil, K.; Manychova, M.; Stryk, J.; Korenska, M.; Matula, R.; Svoboda, V. Diagnostics of Reinforcement Conditions in Concrete Structures by GPR, Impact-Echo Method and Metal Magnetic Memory Method. *Remote Sens.* **2021**, *13*, 952. https://doi.org/10.3390/rs13050952

Academic Editors: Roberto Orosei and Massimiliano Pieraccini

Received: 31 January 2021
Accepted: 25 February 2021
Published: 3 March 2021

Publisher's Note: MDPI stays neutral with regard to jurisdictional claims in published maps and institutional affiliations.

Abstract: It is important to use adequately reliable non-destructive methods that would be capable of determining the reinforcement conditions in concrete structures. Three different methods: ground penetrating radar, impact-echo method, and metal magnetic memory method were used for testing laboratory-prepared reinforced concrete beams (with a reinforcing bar of the same diameter along its whole length, reinforcing bar locally impaired, and reinforcing bar interrupted). The ground-penetrating radar proved the correlation of signal parameters with the reinforcing bar condition. An impairment/interruption reinforcing bar appeared in the record from measurements in the transversal and longitudinal direction by changes of the observed depth of the reinforcing bar from the concrete surface and direct wave attenuation. The impact-echo method proved that the shifts of the dominant frequencies from the response signal correspond with the impairment/interruption of the reinforcing bar. Results of diagnostics by the metal magnetic memory method were presented by a magnetogram of the magnetic field strength and field gradient on the measured distance. The changes in the magnetic field strength proved different stress concentration zones due to the reinforcing bar condition. The used non-destructive methods showed that they are capable of indicating the different reinforcement conditions in reinforced concrete beams. This paper indicates in which cases and for what reason it is appropriate to use these three methods and in what way they differ from each other.

Keywords: reinforced concrete beam; steel reinforcing bar; ground penetrating radar; impact-echo method; metal magnetic memory method

1. Introduction

It is evident that the determination of the current condition of reinforced concrete structures and the verification of their safety is of utmost importance. This is why non-destructive testing (NDT) and combinations of different diagnostic methods are favored for evaluations of structure conditions.

The basic method is a visual inspection and evaluation of the occurrence of changes and defects visible on the surface. Preference is given to methods working in 3D, enabling automatic evaluation, such as photogrammetry, interferometry, laser scanning, etc. [1–3].

Diagnostic test methods can be divided into destructive, partially destructive, and non-destructive methods. Ground penetrating radar (GPR) has an irreplaceable place among other diagnostic methods (sonic, ultrasonic, microwaves, radiography, infrared thermography) that are used for testing reinforced concrete materials and structures [4].

259

GPR is commonly used as a diagnostic method for a variety of applications [5]. There are more or less sophisticated applications that focus on determining the rebar position and concrete cover thickness [6–8], some are trying to determine rebar diameter [8,9], or identify rebar corrosion in concrete structures [10]. Combinations of GPRs with other methods are also used to refine the measurement results, e.g., with electromagnetic induction [8], electrical resistivity [11], ultrasonic techniques [12] or to compare the possibilities of individual methods with each other [13].

The presented work aims to investigate the potential of three non-destructive methods to detect different reinforcement conditions in reinforced concrete beams. It is a single-channel ground penetrating radar, with a common commercial way of evaluating the recording of this measurement, compared to one commonly used method, i.e., impact-echo, which allows measurements outside the area (not directly above it) and the metal magnetic memory method, which is not yet commonly used for the purpose of evaluating built-in reinforcement in concrete.

The measurements described below were performed as the first step in a project aimed at evaluating the condition of prestressed concrete beams.

In the case of GPR, it can be stated in this context that it allows the measurement of the position of the reinforcement and the thickness of the concrete cover of the reinforcement. It is a fast measurement that can be preliminarily evaluated directly from the recorded radargram in situ and indicate the position of the reinforcement directly on the surface of measured specimens/structures. In the case of dense reinforcement in rows one above the other, it has limitations, and if the reinforcement were placed in a steel duct (e.g., prestressing reinforcement), it is not possible to assess the condition of this reinforcement. This method does not record the state of reinforcement in terms of stress/load to which the specimen/structure has been subjected.

In the case of the impact-echo method, the method is suitable for the detection, localization, and evaluation of discontinuities inside concrete [14–16], including evaluation of nonlinear effects [17–19]. Its advantage is that it allows measuring even in cases where it is not possible to evaluate the condition of the reinforcement directly in the monitored area (directly above it). It is possible to measure at different distances between the exciter and the sensor. This method is less user-friendly than GPR and, like GPR, cannot detect the state of reinforcement in terms of stress/load to which the specimen/structure has been subjected.

The metal magnetic memory method can record the stress/load condition of steel elements, even when they are no longer exposed to this stress [20–23]. The condition of the concrete cover does not affect the results of this measurement. At present, the method is used mainly for evaluating the condition of iron and steel structures, or underground pipelines and there is not enough experience with its application on reinforced concrete structures.

These three methods were applied in a laboratory environment where the different reinforcement conditions were simulated by the use of reduced diameter of reinforcing bars in concrete beams. The comparison of the measured parameters obtained for different specimens was performed.

2. Description of Specimens

For this experiment, three variations of concrete beams with dimensions of 100 mm × 100 mm × 400 mm were made. They were reinforced with one steel reinforcing bar with a 10 mm diameter and a length of 400 mm, passing through the center of the specimen, see Figure 1.

The following reinforcement variations were used: an undamaged reinforcing bar with the same diameter along the whole beam length (denoted N), reinforcing bar impaired to a diameter of 5 mm in the length of 50 mm (denoted Z), and reinforcing bar interrupted in the length of 50 mm (denoted P). These variations cover the basic three possibilities with the biggest difference between each other.

Figure 1. Three reinforcement variations used for the preparation of concrete beams: N—undamaged reinforcing bar; Z—reinforcing bar locally impaired; P—reinforcing bar interrupted in the central part [16].

3. Ground Penetrating Radar

Ground-penetrating radar (GPR) is equipment that uses high-frequency electromagnetic waves. In combination with the corresponding software, it provides the location and an evaluation of the electrical and magnetic features of the studied environment in which these waves radiate.

A transmitting antenna (Tx) sends a wave into the structure that is examined. Because of discontinuities in the structure, a part of the energy is reflected back while a part continues through the structure. The reflected signal is recorded by the receiving antenna (Rx) and analyzed in a central unit (Figure 2a). A discontinuity could be a boundary or interface between two materials/layers with different dielectric properties, their debonding or delamination, or the presence of inbuilt objects, such as reinforcement. The amplitudes of the detected echoes (Figure 2b) and the corresponding arrival times can then be used to locate the discontinuity. Low-frequency antennas allow performing measurements at greater depths, but with lower resolutions. On the other hand, high-frequency antennas provide higher resolutions, but in a smaller depth. Concrete structures are commonly inspected with the use of antennas with a central frequency above 1 GHz.

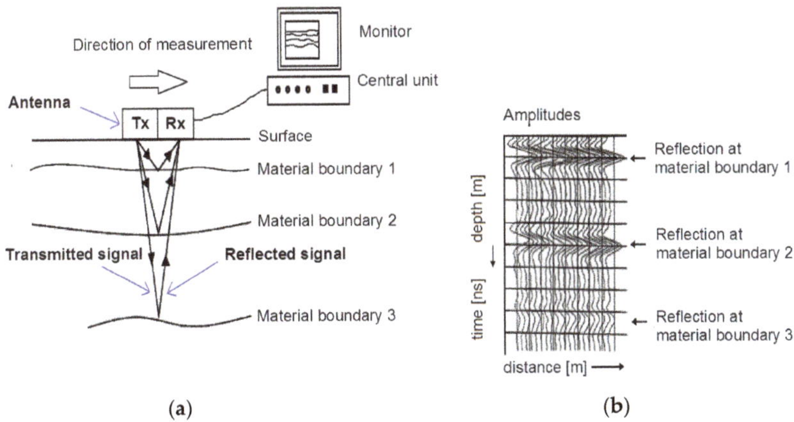

Figure 2. (a) Principle of ground penetrating radar (GPR) measurement with one antenna; (b) radargram record from continual GPR measurement in one line [13].

3.1. Data Acquisition

For this experiment the measurement system SIR-20 of the American company Geophysical Survey Systems, Inc. was used (Figure 3). An antenna 2600 MHz was attached to a trolley in order to measure in a single line and allow recording of the traveled distance.

Figure 3. Experimental device set-up of GPR during the test.

Measurement device adjustment was as follows:

- Vertical high pass filtering HP F = 400 MHz;
- Vertical low pass filtering LP F = 5070 MHz;
- Range Gain: in order to highlight reinforcement bar surroundings;
- Position measurement by trolley: 2000 scans/m, 153 scans/s;
- Resolution: 512 samples/scan.

Timber ramps were used for the trolley movement. The measurement was repeated three times in laboratory conditions at a temperature of 20 °C. Measurement orientations were as follows, see Figure 4:

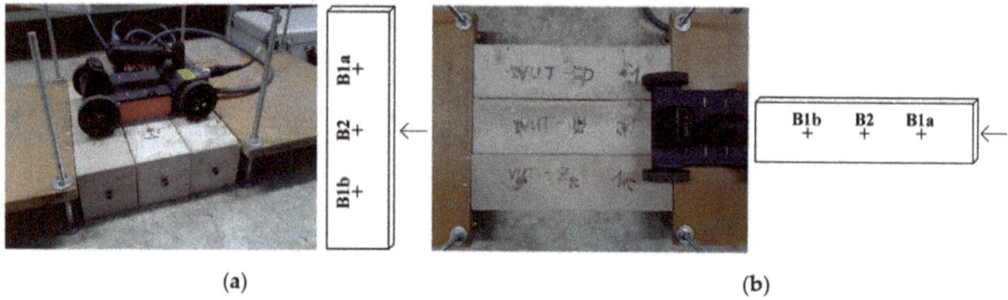

(a)   (b)

Figure 4. GPR measurement orientation: (**a**) transversal direction; (**b**) longitudinal direction.

Transversal direction:

(a) crossing above the right part of specimens (N-B1a; P-B1a; Z-B1a);
(b) crossing above the center of specimens (N-B2; P-B2; Z-B2);
(c) crossing above the left part of specimens (N-B1b; P-B1b; Z-B1b).

Longitudinal direction:

(a) crossing above specimen N (line N-B1a; N-B2; N-B1b);
(b) crossing above specimen Z (line Z-B1a; Z-B2; Z-B1b);
(c) crossing above specimen P (line P-B1a; P-B2; P-B1b).

3.2. Results and Discussion

Records from GPR measurements performed on specimens with reinforcing bar type (N, P, Z) in transversal direction are shown in Figure 5.

Three identical hyperboles were evident in GPR records across the right parts and the left parts of the specimens (Figure 5a,c) representing the reinforcing bar upper surface in all specimens (N, P, Z). The peaks of hyperboles with the highest amplitude were cut through by a reference horizontal (white) line. It confirmed that the reinforcing bar surfaces were at the same depth. Figure 5b) illustrates the results of measurements across the center of the specimens. In this case, we could see that the plus amplitude value of the central hyperbole, corresponding to the interrupted reinforcing bar denoted as P, was much lower and its peak was located under the reference horizontal line. This corresponded to the fact that during the crossing, the end of the interrupted reinforcement 2.5 cm from the measurement axis was captured in the record. The hyperbole in Figure 5b on the right corresponded to specimen Z with an impaired reinforcing bar and it was visible that the hyperbole peak was located under the horizontal reference line.

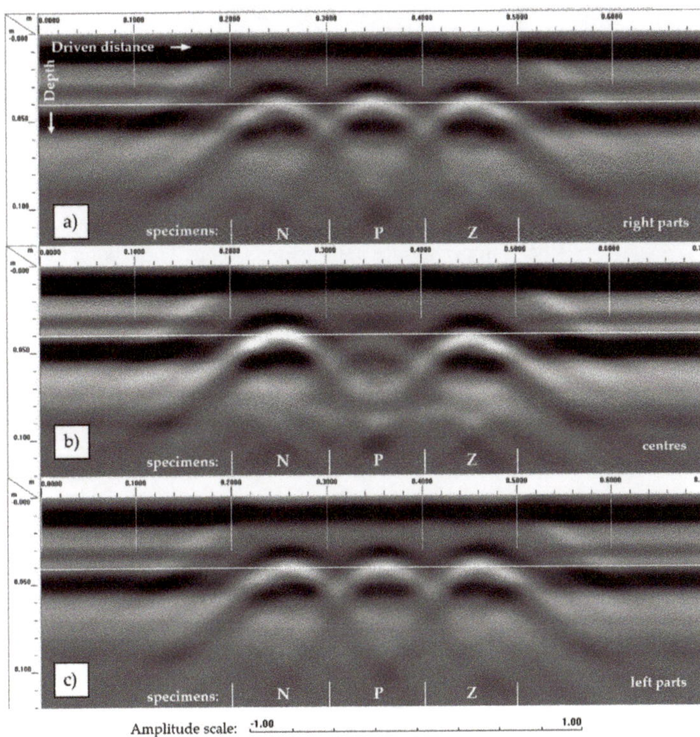

Figure 5. Radargrams from GPR measurements performed in transversal direction: (**a**) across the right parts of specimens crossing *N*-B1a; *P*-B1a; *Z*-B1a; (**b**) across the centers of specimens crossing *N*-B2; *P*-B2; *Z*-B2; (**c**) across the left parts of specimens crossing *N*-B1b; *P*-B1b; *Z*-B1b.

The measured data were evaluated using software RADAN 7.4 in the Interactive Interpretation module which allowed the determination of the depth of the reinforcing bars from the concrete specimen's surface on the basis of the known velocity of electromagnetic signal propagation. The propagation velocity ($v = 0.113$ m/ns) was determined by the use of the known depth of the reinforcing bar at the beam's ends. The measurements were repeated above the measurement positions three times. Calculated mean values of direct wave attenuation (expressing the ratio of the signal amplitude corresponding to the surface of the reinforcement and the surface of the concrete) [11] (p. 365) and reinforcing bar location depth are summarized in Table 1.

Table 1. Mean values of GPR measurement results in a transversal direction.

Measurement Position	Direct Wave Attenuation [dB]	Reinforcing Bar Depth h [mm]
N–B1a	4.5	45
N–B2	4.2	44
N–B1b	3.7	44
Z–B1a	4.0	44
Z–B2	**5.6**	47
Z–B1b	4.0	44
P–B1a	4.0	46
P–B2	**17.4** [1]	**49** [1]
P–B1b	4.1	44

[1] The end of the interrupted reinforcement 2.5 cm from the measurement axis was captured in the record.

The bold numbers from Table 1 show different measurement results. In the case of the impaired reinforcing bar—specimen Z—there was a moderate increase of direct wave attenuation and partial increase of reinforcing bar depth. In the case of the interrupted reinforcing bar 2.5 cm from the measurement axis—specimen P—there was a striking increase of direct wave attenuation and increase of reinforcing bar depth.

The evaluation of records from GPR measurements performed on specimens in the longitudinal direction is shown in Figure 6. The analyzed boundaries represent the reinforcing bar's upper surface in measured beams. These boundaries were visible at a depth of about 0.04 m at a distance of approximately 0.2 to 0.6 m in a longitudinal direction. The shape and length of reinforcing bar surface boundaries are evaluated and sketched in the lower right corners of Figure 6. A change in reinforcing bar surface depth was not observed in the case of specimen N with the undamaged reinforcing bar (Figure 6a). A smaller change was visible in the case of specimen Z with an impaired reinforcing bar (Figure 6b) and a marked change was apparent in the case of specimen P with an interrupted reinforcing bar (Figure 6c).

Mean values of length change d and depth change h in view of boundaries representing the reinforcing bar upper surface are summarized in Table 2.

Table 2. Mean values of GPR measurement results in a longitudinal direction.

Measurement Position	Length of Change d [mm]	Change of Depth h [mm]
N–B2	-	1
Z–B2	**55**	**3**
P–B2	**47** [1]	**5** [1]

[1] The interrupted reinforcement was captured in the record.

From Table 2 it is evident that in the case of specimen N, with an undamaged reinforcing bar, changes were not observed. In the case of specimen Z with an impaired reinforcing bar in its center, the necking (3 mm) of the reinforcing bar upper surface in the length of 55 mm was apparent. The biggest changes corresponded to specimen P with a reinforcing bar interrupted in its center. In this case, we could see the amplitude of the lower value and the necking (5 mm) of the reinforcing bar upper surface in a length of 47 mm in its center due to the missing part of the bar.

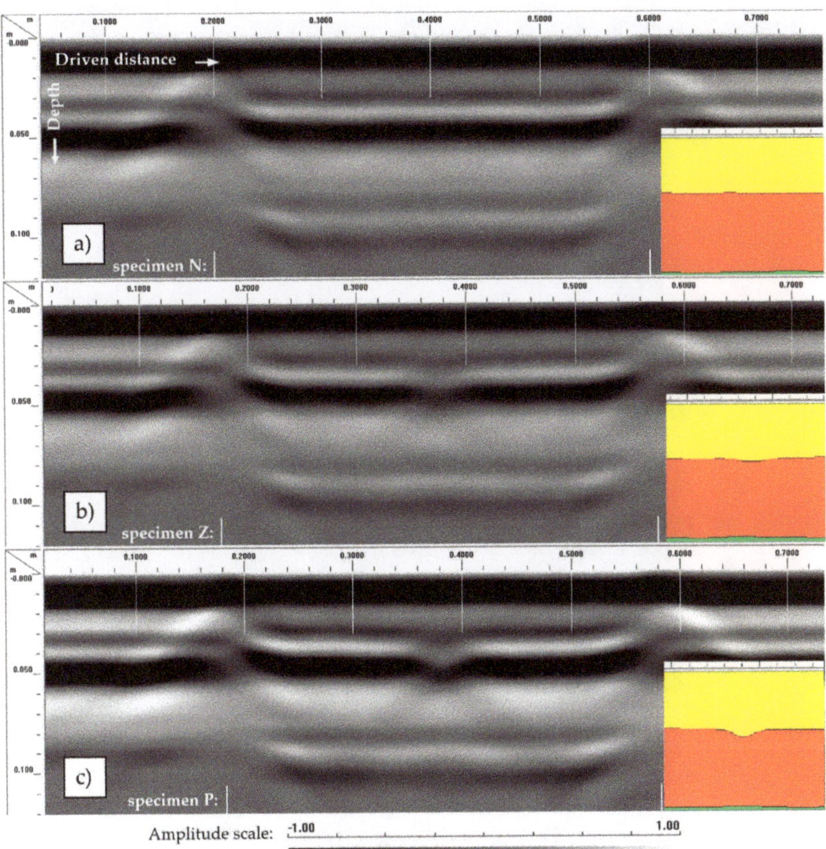

Figure 6. Radargrams from GPR measurements performed in a longitudinal direction: (**a**) across the specimen N centre crossing *N*-B1a; *N*-B2; *N*-B1b; (**b**) across the specimen Z center crossing *Z*-B1a; *Z*-B2 a *Z*-B1b; (**c**) across the specimen P center crossing *P*-B1a; *P*-B2; *P*-B1b.

4. Impact-Echo Method

The second applied method was impact-echo. It is based on the propagation of stress waves, which are generated by a mechanical impulse; the scheme is shown in Figure 7. A short-duration mechanical impact, produced by tapping a hammer against the surface of concrete, produces low-frequency stress waves (from 1 to 60 kHz) that propagate into the structure and are reflected by flaws and/or external surfaces [14,19]. Reflected waves are recorded on the surface by a sensor in the form of a voltage signal. The resulting voltage versus time plot (time-domain realization) is digitized and fed into the memory of a computer, which subsequently performs the frequency analysis. A time realization and the corresponding frequency spectrum were the results of this test.

Any implementation of such tests requires a reference standard and its frequency spectrum to be set down. An undamaged reinforcing bar of the same diameter along the whole length of the beam N was the reference standard in our case. If the test specimen material properties differ from those of the reference standard (i.e., impaired reinforcing bar represented by the specimen Z and interrupted reinforcing bar represented by the specimen P), the resonance frequency will be shifted against that of the reference standard.

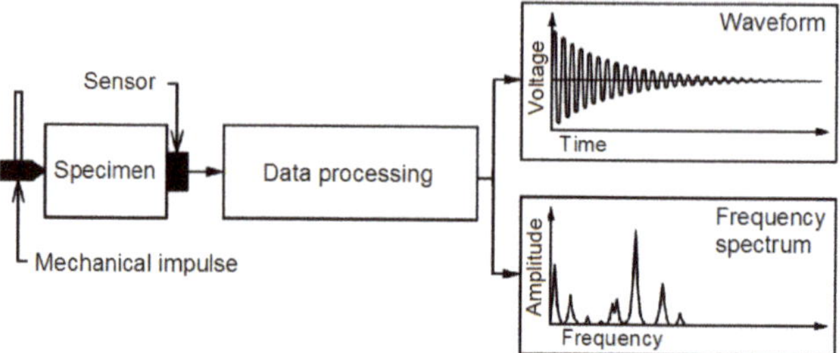

Figure 7. Scheme of the impact-echo method [19].

4.1. Data Acquisition

A mechanical impulse, provided by a special hammer, was applied at the exciter points E. A piezoelectric sensor S, operating in a frequency range from 100 Hz to 50 kHz, was used to pick up the specimen response, see Figure 8. The THPS3_25 HandyScope3 measuring unit and a special signal-analysis software package were used for sampling and further processing of captured signals.

Seven mechanical impulse-to-sensor configurations were measured, see Figure 8a. The results were verified by repeated measurement (three times). Various impulse intensity and response frequency spectra were also analyzed from the viewpoint of nonlinear effects [18,19].

The results obtained from three configurations (Figure 8b) are presented below.

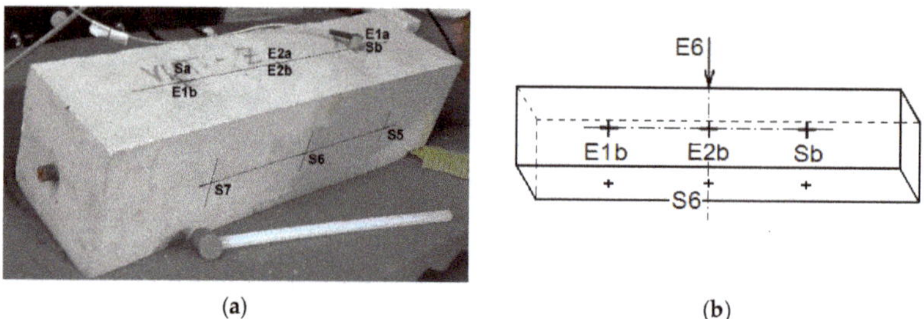

(a)

(b)

Figure 8. Mechanical impulse (E) to sensor (S) configurations: (**a**) test beam with a drawing of all configurations; (**b**) three presented configurations.

4.2. Results and Discussion

Measurement results are presented in the form of response frequency spectra. Figure 9 shows a frequency spectrum obtained from all specimen types (N, Z, P) corresponding to the 3rd configuration of E1b impulse-to-Sb sensor location, see Figure 8b. The predominant frequencies occurred in the frequency range of 4700 to 4850 Hz, as shown in Figure 9. Figure 9A shows the whole frequency range and Figure 9B shows details of the predominant frequency range.

The predominant frequency of the value of 4773 Hz (graph B) corresponded to specimen N, with an undamaged reinforcing bar. The response frequency spectrum Z corresponded to a specimen with a bar locally impaired in its central part. The peak frequency of 4802 Hz signified a shift of 29 Hz to the right in comparison with the N specimen. The response frequency spectrum P corresponded to a specimen with a reinforcing bar interrupted in its central part. The predominant frequency shifted to a lower value, namely 4762 Hz, which corresponded to a shift of 11 Hz in comparison with the N specimen.

Figure 10A analogically compares measurement results from the 4th configuration of E2b impulse-to-Sb sensor location. Figure 10B illustrates a detail of the frequency range from 2120 Hz to 2200 Hz. In this case, the dominant frequency value of 2151 Hz pertains to the N specimen with an undamaged reinforcing bar. The difference is evident in the case of specimen Z with a reinforcing bar locally impaired in its central part; two peaks were evident in this response frequency spectrum. The first peak of 2136 Hz occurred to the left of the predominant frequency of specimen N. The second peak occurred to the right of the predominant frequency of specimen N and its value equaled 2165 Hz. The frequency spectrum P corresponded to the specimen with a reinforcing bar interrupted in its central part. In this case, the predominant frequency shifted to value 2137 Hz on the left similar to specimen P (Figure 9B). Figure 10C illustrates a detail of the frequency range from 4700 Hz to 4 50 Hz, which corresponded to the predominant frequency range in the case of the 3rd configuration (Figure 9). The peak frequency of 4776 Hz pertained to the N specimen, with an undamaged reinforcing bar. The response frequency spectrum Z corresponded to a specimen with a reinforcing bar locally impaired in its central part. Its frequency peak occurred at the value of 4802 Hz, which meant a shift of 26 Hz to the right, as in the case of the Z specimen (Figure 9B), in comparison with the N specimen. The frequency spectrum P corresponded to a specimen with a reinforcing bar interrupted in its central part. In this case, the predominant frequency shifted to the value of 4760 Hz to the left also similar to specimen P (Figure 9B).

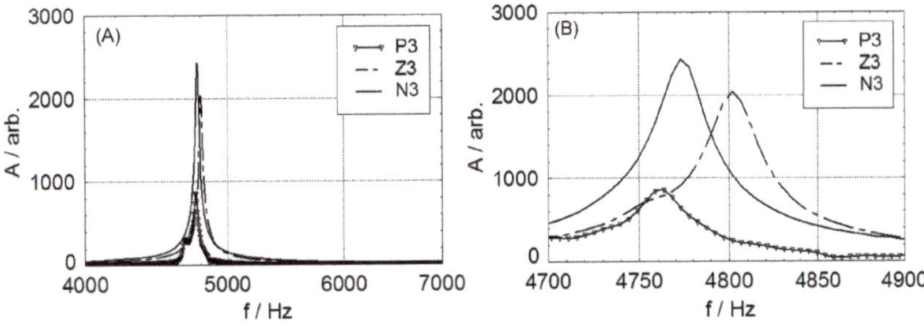

Figure 9. Response frequency spectrum representing measurements of 3rd configuration E1b impulse-to-Sb sensor location: (**A**) whole frequency range; (**B**) details of the predominant frequency range.

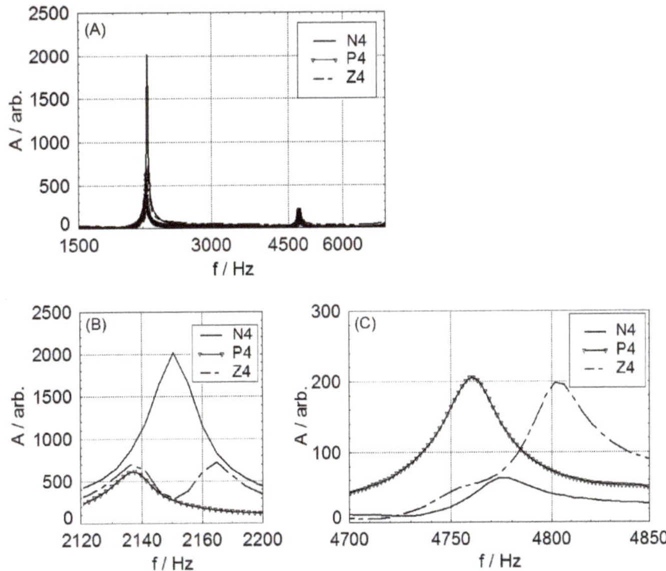

Figure 10. Response frequency spectrum representing measurements of the 4th configuration: E2b impulse-to-Sb sensor location: (**A**) whole frequency range; (**B**) detail of the frequency range from 2120 Hz to 2200 Hz; (**C**) detail of the frequency range from 4700 Hz to 4850 Hz.

Figure 11 shows the response frequency spectrum for the 6th configuration of the E6 impulse-to-S6 sensor location. Figure 11A shows the whole frequency range and Figure 11B shows details of the predominant frequency range. The predominant frequencies proved similar values as shown in Figure 10B.

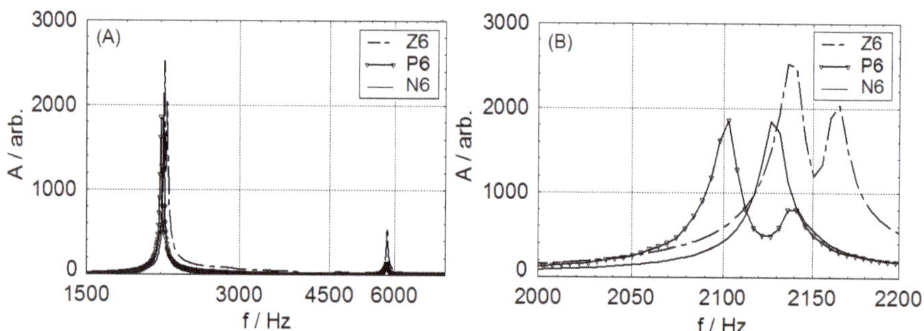

Figure 11. Response frequency spectrum representing measurements of the 6th configuration: E6 impulse-to-S6 sensor location: (**A**) whole frequency range; (**B**) details of the predominant frequency range [16].

The dominant frequency value of 2129 Hz pertained to specimen N, with an undamaged reinforcing bar. Frequency spectrum Z6 corresponded to a specimen with a reinforcing bar locally impaired in its central part. It was seen that the predominant frequency with two peaks was similar to the 4th orientation (Figure 10B); however, both peaks occurred to the right of the predominant frequency of specimen N. The first peak value equaled 2138 Hz and the second peak value equaled 2164 Hz. The frequency spectrum P corresponded to the specimen with a reinforcing bar interrupted in its central part. The predominant frequency shifted to the value of 2102 Hz to the left of the predominant frequency of specimen N, similar to Figures 9 and 10B.

To verify measurement results, reproducibility measurements were repeated three times. Mean values of dominant frequencies are summarized in Table 3.

Table 3. Mean values of specimen dominant frequencies from all measurements.

Configuration		Dominant Frequencies [Hz]		
Order No.	Location *E-to-S*	Specimen *N*	Specimen *Z*	Specimen *P*
1st	*E1a-to-Sa*	4774	4751; 4802	4764
2nd	*E2a-to-Sa*	2151	2138; 2164	2138
		4772	4802	4760
3th	*E1b-to-Sb*	4771	4800	4762
4th	*E2b-to-Sb*	2150	2137; 2164	2137
		4773	4801	4761
5th	*E5-to-S5*	4730	4754	4674
6th	*E6-to-S6*	2129	2138; 2164	2102
7th	*E7-to-S7*	4730	4754	4674

A mechanical impulse of various intensities was used and response frequency spectra were also analyzed from the viewpoint of nonlinear effects. Higher impulse intensity normally produces a shift of dominant frequency in the case of a presence of cracks and other defects in the specimen structure. All specimens proved good reproducibility of measurement results and the presence of nonlinear effects, as the indicators of concrete structure damage, were not observed.

5. Metal Magnetic Memory Method

The metal magnetic memory (MMM) method is an NDT method based on sensing the magnetic field strength *Hp* in the form of residual magnetization on the surface of the material. It uses one or more probes, each sensing the magnetic field in three perpendicular planes, see Figure 12. The residual magnetization of a material is affected by the production process, machining, cutting, welding, bending, heat treatment, cooling, operational stress, corrosion of the material, etc. A very important factor influencing the magnetic memory of a material is the degradation process caused by the operational stress and conditions to which the structure is exposed.

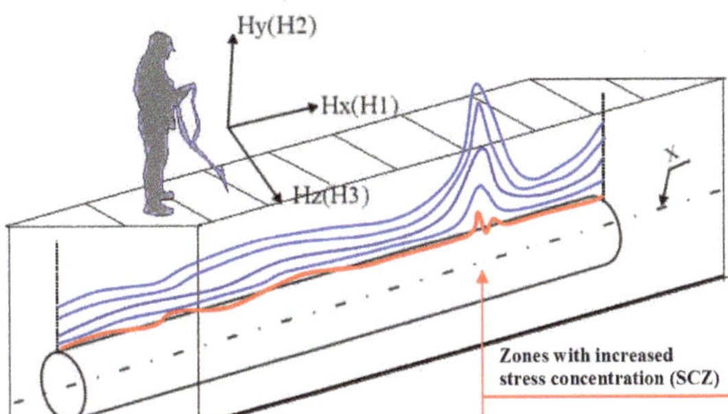

Figure 12. Scheme of measurement by the metal magnetic memory (MMM) method above a steel pipe in the ground in the longitudinal direction.

The MMM method can detect these effects from a magnetogram, i.e., the dependence of the magnetic field strength on the distance of the probe from the beginning of the measurement.

A special cart, on which the probes are placed, is moved above the surface of the measured area (e.g., in a pipe or beam axis) and the wheel of the device measures the driven distance. The signal from each individual channel, together with the distance, is recorded in digital form in the device's memory and displayed graphically either directly as *Hp* or in the form of gradient *dHp/dx*.

The recorded data are presented in the form of a graph (so-called magnetogram). Based on the evaluation, we can then determine zones with increased stress concentration (SCZ), where there is an increased probability of changes or defects in the material structure. The stress concentration is proportional to the measured magnitude of the magnetic field strength gradient around a given position.

A significant advantage of this NDT method is the measurement speed and its high sensitivity.

5.1. Data Acquisition

The measurement technique used is shown in Figure 13 and includes the following parts:

- Tester of stress concentration with evaluation unit TSC-3M-12;
- Scanning device type 11-6W with a probe comprising 2 triaxial sensors;
- Cart with the device measuring the driven distance.

The measuring range for magnetic field strength *Hp* was from −2000 A/m to +2000 A/m. The used measurement step was every 1 mm.

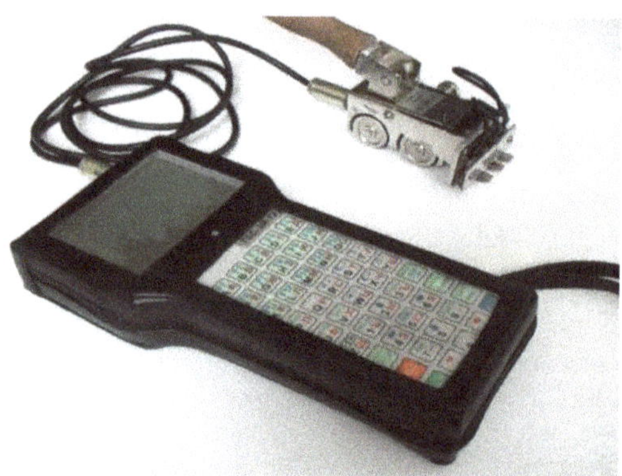

Figure 13. Used device for MMM measurement with an evaluation unit.

5.2. Results and Discussion

The concrete beams were arranged in a line on the ground. Firstly, the indicative contactless measurement was applied at the distance of 50 mm from the surface of the beams, the contact measurement followed. The measurement was performed on specimen surfaces in the central part of the specimens along the length of 200 mm. The measurement was repeated four times to verify the reproducibility of the measurement results. A magnetogram in Figure 14 illustrates the results of the continuous contact scanning of specimen N with an undamaged reinforcing bar.

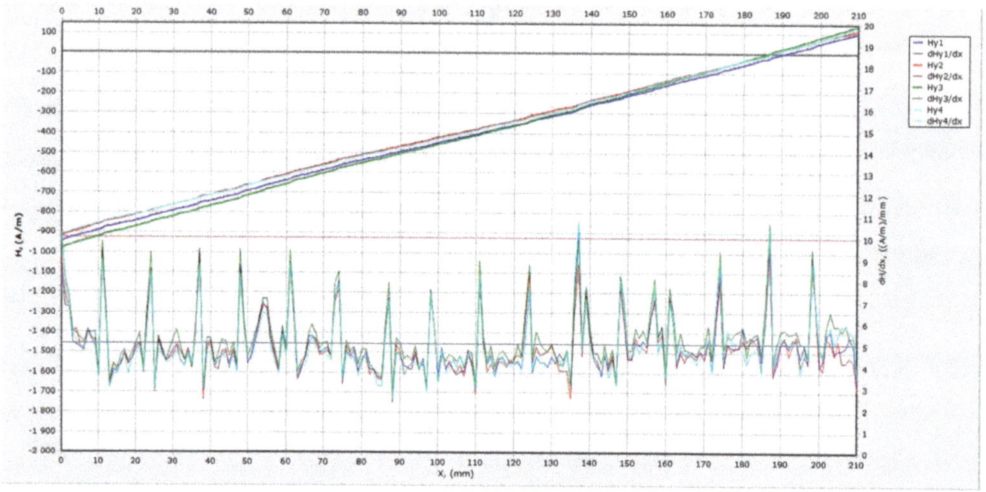

Figure 14. Magnetogram from scanning with contact probe, specimen N with undamaged reinforcing bar.

The magnetogram graphically showed the field strength *Hy* and the field gradient *dHy/dx* as a function of distance *x*. The uniform distribution of the field strength *Hy* in the magnetogram confirmed that the sample structure was uniform along its length.

Figure 15 presents results of the continuous contact scanning of specimen Z with a locally impaired reinforcing bar.

The different distribution of field strength H_y compared to the magnetogram of specimen N was visible. The distribution was not uniform, in this case. The zone of the increased field strength gradient (area marked in a red rectangle) corresponded to the locally impaired part of the reinforcing bar.

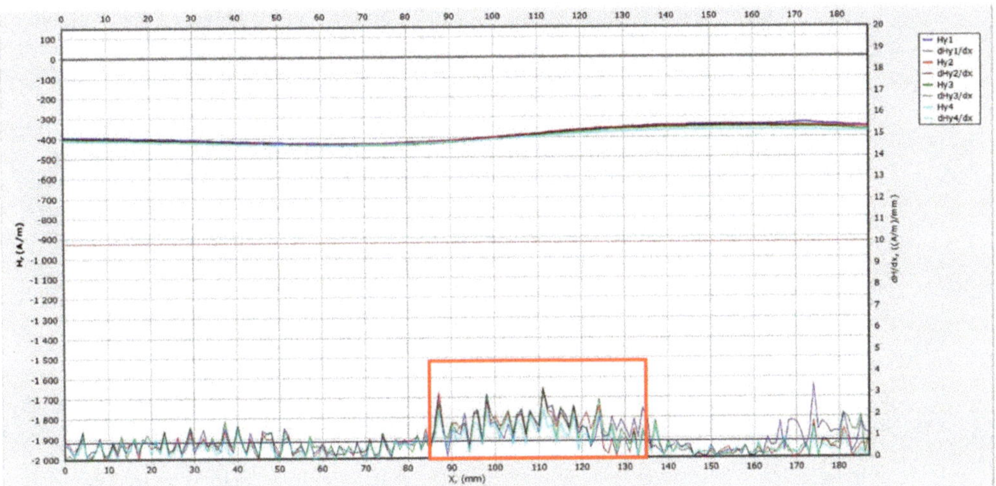

Figure 15. Magnetogram from scanning with contact probe, specimen Z with locally impaired reinforcing bar.

Figure 16 presents a magnetogram from the measurement of specimen P with an interrupted reinforcing bar.

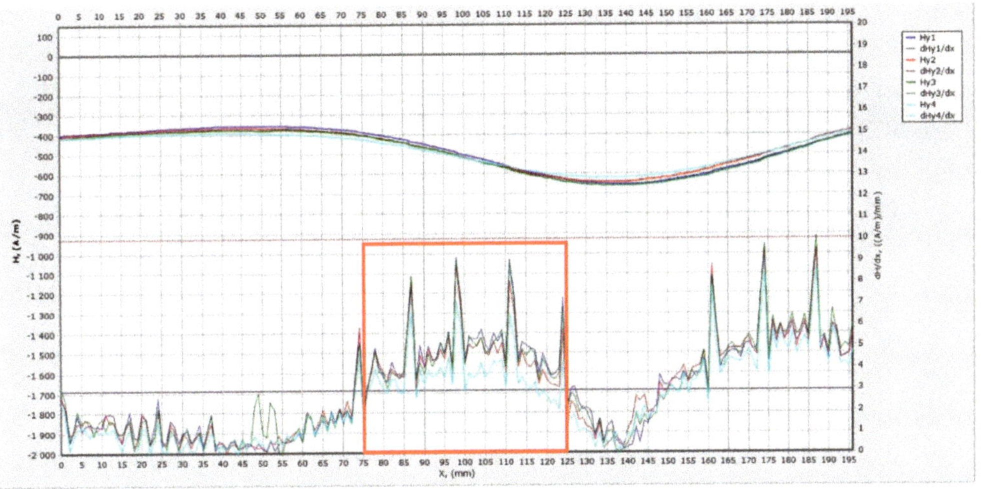

Figure 16. Magnetogram from scanning with contact probe, specimen P with interrupted reinforcing bar.

The magnetogram showed in shape analogous distribution of the field strength similar to the magnetogram in Figure 15 (corresponding to specimen Z) with a larger range of values. The zone of increased field strength gradient (area marked in a red rectangle) corresponded to the missing part of the reinforcing bar. The adjoining zone with an increased gradient must have been caused by another reason, possibly the mechanical preparation of bars which needs to be further studied.

6. Conclusions

The presented paper describes the application of three non-destructive methods for the evaluation of different reinforcing bar conditions in concrete beams. These specimens were prepared in laboratory conditions and the only difference between them is in the central part of the reinforcing bars (diameter of 10 mm, diameter of 5 mm, and a missing piece of bar). The applied methods were: the ground-penetrating radar, the impact-echo method, and the metal magnetic memory method.

The ground-penetrating radar was applied in transversal and in longitudinal directions. The measurement results in both directions proved the correlation of evaluated parameters (changes of the observed depth of the reinforcing bar from the concrete surface and direct wave attenuation) with a reinforcing bar variant. It was confirmed that a 2.6 GHz antenna is able to measure the position of the reinforcement placed at a depth of about 50 mm with millimeter accuracy.

The results of the impact-echo method proved that this method is sensitive to the reinforcement conditions. The shifts of the dominant frequencies obtained with fast Fourier transform from the response signal correspond with the reinforcing bar changes in the concrete beams. It was shown that the impaired or interrupted reinforcing bars have an influence on the dominant frequency. However, these differences are difficult to interpret and without a comparison to a reference value measured in the same positions, an evaluation would not be possible.

The measured parameters of the metal magnetic method proved correlation with concrete beams with different reinforcing bars. The measurement with a contact probe shows that the uniform specimen structure along the whole length produced a uniform distribution of magnetic field strength Hy. In the case of impaired and interrupted reinforcing bars, the change appeared in Hy and became much clearer on its gradient. It turned out that it will be necessary to distinguish in the records other influences to which this method is sensitive. The measurement is similar to GPR in terms of time and comfort of measurement.

The results of these measurements showed how they can help in assessing conditions of steel reinforcement in concrete structures and have proved to constitute a starting point for specifying their use in the next step involving testing of concrete specimens with prestressed reinforcement, either directly embedded in the concrete or placed in a protecting duct.

GPR will be used to determine the position of structural reinforcement (stirrups, etc.) and ducts for conducting prestressed reinforcement in the form of strands.

The impact-echo method allows measurements directly on the prestressing reinforcement or using holes in the concrete and waveguides connected directly to the reinforcement (at different distances from each other). The use of waveguides to transmit the excitation pulse directly to the reinforcement and to sense the response on the reinforcement will allow a substantial increase in the distance between the exciter and the sensor and refine the measurement results. A combination of the impact-echo method with nonlinear ultrasonic spectroscopy will be used.

The MMM method will be tested to evaluate the state of prestressing reinforcement in the longitudinal direction when measured on the surface of concrete specimens. For the purposes of proper evaluation, it will be necessary to test variously damaged and modified samples of reinforcement and their location in various types of protective ducts, especially made of steel.

Author Contributions: Conceptualization and methodology, K.P.; GPR measurement and analysis, R.M. and J.S.; impact-echo measurement and analysis, M.M. and M.K.; MMM method measurement and analysis V.S. All authors have read and agreed to the published version of the manuscript.

Funding: This paper was produced with the state support of the Technological Agency of the Czech Republic within the program TRANSPORT2020+, project No. CK01000108: New approaches in diagnosis of reinforcement of pre-stressed/post-tensioned concrete bridge beams.

Institutional Review Board Statement: Not applicable.

Informed Consent Statement: Not applicable.

Data Availability Statement: No new data were created or analyzed in this study. Data sharing is not applicable to this article.

Conflicts of Interest: The authors declare no conflict of interest.

References

1. Zollini, S.; Alicandro, M.; Dominici, D.; Quaresima, R.; Giallonardo, M. UAV Photogrammetry for Concrete Bridge Inspection Using Object-Based Image Analysis (OBIA). *Remote Sens.* **2020**, *12*, 3180. [CrossRef]
2. Pieraccini, M.; Miccinesi, L. Ground-Based Radar Interferometry: A Bibliographic Review. *Remote Sens.* **2019**, *11*, 1029. [CrossRef]
3. Tang, Y.; Li, L.; Wang, C.; Chen, M.; Feng, W.; Zou, X.; Huang, K. Real-time detection of surface deformation and strain in recycled aggregate concrete-filled steel tubular columns via four-ocular vision. *Robot. Comput. Int. Manuf.* **2019**, *59*, 36–46. [CrossRef]
4. Tosti, F.; Ferrante, C. Using Ground Penetrating Radar Methods to Investigate Reinforced Concrete Structures. *Surv. Geophys.* **2020**, *41*, 485–530. [CrossRef]
5. Lai, W.W.; Derobert, X.; Annan, P. A review of Ground Penetrating Radar application in civil engineering: A 30-year journey from locating and testing to imaging and diagnosis. *NDT E Int.* **2018**, *96*, 58–78. [CrossRef]
6. He, X.Q.; Zhu, Z.Q.; Liu, Q.Y.; Lu, G.Y. Review of GPR rebar detection. In Proceedings of the Progress in Electromagnetics Research Symposium, Beijing, China, 23–27 March 2009; pp. 804–813.
7. Stryk, J.; Matula, R.; Pospisil, K. Possibilities of ground penetrating radar usage within acceptance tests of rigid pavements. *J. Appl. Geophys.* **2013**, *97*, 11–26. [CrossRef]
8. Zhou, F.; Chen, Z.; Liu, H.; Cui, J.; Spencer, B.F.; Fang, G. Simultaneous Estimation of Rebar Diameter and Cover Thickness by a GPR-EMI Dual Sensor. *Sensors* **2018**, *18*, 2969. [CrossRef] [PubMed]
9. Chang, C.W.; Lin, C.H.; Lien, H.S. Measurement radius of reinforcing steel bar in concrete using digital image GPR. *Constr. Build. Mater.* **2009**, *23*, 1057–1063. [CrossRef]
10. Zaki, A.; Johari, M.A.M.; Aminuddin Wan Hussin, W.M.; Jusman, Y. Experimental Assessment of Rebar Corrosion in Concrete Slab Using Ground Penetrating Radar (GPR). *Int. J. Corros.* **2018**, *2018*, 5389829. [CrossRef]
11. Sbartaï, Z.M.; Laurens, S.; Rhazi, J.; Balayssac, J.P.; Arliguie, G. Using radar direct wave for concrete condition assessment: Correlation with electrical rezistivity. *J. Appl. Geophys.* **2007**, *62*, 361–374. [CrossRef]
12. Cassidy, N.J.; Eddies, R.; Dods, S. Void detection beneath reinforced concrete sections: The practical application of ground-penetrating radar and ultrasonic techniques. *J. Appl. Geophys.* **2011**, *74*, 263–276. [CrossRef]
13. Janku, M.; Cikrle, P.; Grosek, J.; Anton, O.; Stryk, J. Comparison of infrared thermography, ground-penetrating radar and ultrasonic pulse echo for detecting delaminations in concrete bridges. *Constr. Build. Mater.* **2019**, *225*, 1098–1111. [CrossRef]
14. Hong, S.U.; Lee, Y.T.; Kim, S.H.; Na, J.H. Estimation of Thickness of Concrete Slab Members Using Impact Echo Method. *Key Eng. Mater.* **2014**, *605*, 139. [CrossRef]
15. Zhang, J.K.; Weizhong, Y.; Cui, D.M. Concrete Condition Assessment Using Impact-Echo Method and Extreme Leatning Machines. *Sensors* **2016**, *16*, 447. [CrossRef] [PubMed]
16. Manychova, M.; Fuciman, O.; Pazdera, L. An Initial Investigation on the Potential Applicability of Non-Destructive Methods to Assessing Joint Condition in Prefabricated Structures. *Solid State Phenom.* **2017**, *258*, 489–492. [CrossRef]
17. Antonaci, P.; Bruno, C.L.; Gliozzi, A.S.; Scalerandi, M. Monitoring evolution of compressive damage in concrete with linear and nonlinear ultrasonic methods. *Cem. Concr. Res.* **2010**, *40*, 1106–1113. [CrossRef]
18. Lesnicki, K.J.; Kim, J.-Y.; Kurtis, K.E. Characterization of ASR damage in concrete using nonlinear impact resonance acoustic spectroscopy technique. *NDT E Int.* **2011**, *44*, 721–727. [CrossRef]
19. Korenska, M.; Manychova, M.; Pazdera, L. Experimental Study of the Nonlinear Effects Generated in a Concrete Structure with Damaged Integrity. *Russ. J. Nondestruct. Test.* **2013**, *49*, 530–537. [CrossRef]
20. Dubov, A.A. New requirements to methods and means of materials' stress-strain state diagnostics. *MIR Izmereni* **2012**, *6*, 38–42.
21. Dubov, A.A. Fundamentals difference of the metal magnetic memory method from other known magnetic NDT methods. Results and prospects of the method development. *Territ. NDT* **2016**, *2*, 64–68.
22. Gorkunov, E.S.; Efimov, A.G.; Shubochkin, A.E.; Artemyev, B.V. On the question of magnetic NDT application to determining the stress-strain state of metal structures. *World NDT* **2016**, *3*, 52–55.
23. Svoboda, V.; Gajdos, L.; Janovac, J. Verification and localization of defects (cracks) at boiler waterwall tubes using magnetic memory method. In Proceedings of the 3rd International Conference on Diagnostics of Structures and Components Using Magnetic Memory Method, Prague, Czech Republic, 22–23 May 2019; pp. 42–48.

Technical Note

Optimising the Complex Refractive Index Model for Estimating the Permittivity of Heterogeneous Concrete Models

Hossain Zadhoush [1,*], Antonios Giannopoulos [1] and Iraklis Giannakis [2]

[1] School of Engineering, Institute for Infrastructure and Environment, The University of Edinburgh, Edinburgh EH8 9YL, UK; A.Giannopoulos@ed.ac.uk
[2] School of Geosciences, University of Aberdeen, Aberdeen AB24 3UE, UK; Iraklis.Giannakis@abdn.ac.uk
* Correspondence: h.zadhoush@ed.ac.uk

Abstract: Estimating the permittivity of heterogeneous mixtures based on the permittivity of their components is of high importance with many applications in ground penetrating radar (GPR) and in electrodynamics-based sensing in general. Complex Refractive Index Model (CRIM) is the most mainstream approach for estimating the bulk permittivity of heterogeneous materials and has been widely applied for GPR applications. The popularity of CRIM is primarily based on its simplicity while its accuracy has never been rigorously tested. In the current study, an optimised shape factor is derived that is fine-tuned for modelling the dielectric properties of concrete. The bulk permittivity of concrete is expressed with respect to its components i.e., aggregate particles, cement particles, air-voids and volumetric water fraction. Different combinations of the above materials are accurately modelled using the Finite-Difference Time-Domain (FDTD) method. The numerically estimated bulk permittivity is then used to fine-tune the shape factor of the CRIM model. Then, using laboratory measurements it is shown that the revised CRIM model over-performs the default shape factor and provides with more accurate estimations of the bulk permittivity of concrete.

Keywords: GPR; FDTD; antenna; time-zero; permittivity measurement; gprMax; NDT; concrete; CRIM model; shape factor; GPU

Citation: Zadhoush, H.; Giannopoulos, A.; Giannakis, I. Optimising the Complex Refractive Index Model for Estimating the Permittivity of Heterogeneous Concrete Models. *Remote Sens.* **2021**, *13*, 723. https://doi.org/10.3390/rs13040723

Academic Editor: Mercedes Solla
Received: 17 December 2020
Accepted: 10 February 2021
Published: 16 February 2021

Publisher's Note: MDPI stays neutral with regard to jurisdictional claims in published maps and institutional affiliations.

1. Introduction

Ground penetrating radar (GPR) is a non-destructive geophysical technique which has a wide range of applications [1–3] and it is extensively used for civil engineering. It is used for surveying buried utilities [4,5], roads [6,7], tunnels [8], bridges [9–11] and concrete [12,13]. It is also used for detecting landmines [14], concrete steel bars [15], moisture clusters [16] and air voids in asphalt [17,18]. GPR is an electromagnetic investigative tool which it has been around for many years but GPR modelling is rapidly becoming increasingly useful and the quality of GPR models is becoming more realistic [1].

One of the most important applications of GPR is monitoring and condition assessment of concrete structures. Investigations that usually take place for GPR applications involve shallow-depth buried targets or ones that are located close to the surface. For instance, such applications are detecting the location of rebars, air voids, moisture content and cracks [19]. For concrete construction, many mixtures exist with a variety of content combinations dependant on the application. These mixtures have various material percentages for aggregate, cement and sand which can result in a different dielectric constant of the resulting concrete product. Concrete is a heterogeneous material and the calculation of the dielectric constant has been the interest of many researchers [20–22]. When analysing GPR data, having the wrong estimation of the dielectric properties will result in an incorrect interpretation of key parameters that will be extracted from them. These parameters are the GPR wave velocity in the mixture which is used to convert its two-way travel time in depth, and also, the moisture content. So, if the dielectric constant is not estimated correctly, the GPR analysis of the data will be wrong and problematic. Therefore, in order to investigate the electrical properties of concrete, the dielectric constant should be calculated accurately [23,24]. There are many methodologies that have been used in the past for this purpose which raises an issue concerning which method is the most effective for estimating the bulk permittivity of concrete mixtures.

Over the years many methods have been developed to estimate the bulk permittivity of heterogeneous materials [25,26]. The most common methods [27] are based on the Complex Refractive Index Model (CRIM) model [28], the Rayleigh model [29], the Böttcher model [30], the Brown model [31], the Wagner model [32], the Bruggeman model [33] and the Topp model [34]. These models calculate the dielectric constant of the mixture with respect to the dielectric properties and the volumetric fractions of its components. From the aforementioned models, CRIM has been established as the mainstream methodology in the GPR community mainly due to its simplicity and its straightforward implementation.

In this study, the CRIM mixing model was investigated in detail focusing on the key geometric parameter of the model. A methodology is presented in order to investigate the shape factor further and evaluate different estimates of it. This allows to determine the best shape factor and present a fine-tuned value for concrete structures. Our approach is based on numerical-synthetic experiments executed using gprMax [35,36] an open source electromagnetic solver using finite difference time domain method (FDTD) [37,38]. FDTD is robust, accurate, flexible, computationally efficient and uses time domain discretisation which is ideal for GPR scenarios [39–41]. As computational resources have improved and become more accessible, an increase in the knowledge and effectiveness of GPR modelling has been observed [42]. One of the advantages of GPR numerical modelling is that it is able to produce models that are close to reality and support research effort when restrictions exist to execute it physically. Numerical modelling has been widely used for designing various models and optimising complex antennas [43–45]. FDTD has been widely used to simulate different antenna models such as bowties [46–48], dipoles [49–52] and horn antennas [53,54]. In this study, the antenna used for the simulations was a model-equivalent of a GSSI 1.5 GHz centre frequency antenna structure available for experimentation [45,55].

Numerous realistic concrete models were simulated using an automatic framework that generates different distribution of aggregate particles, cement, air-voids and moisture content. The chloride content within the concrete is negligible thus the effects are not noticeable and were not considered in the numerical experiments [56,57]. The moisture content is a very important aspect and has been shown that greatly affects the overall dielectric properties of the concrete mixture [58,59]. It has been reported that moisture content greater than 5% has an important influence on the transmitted signal travelling through the concrete [60]. Therefore, in the current study, different mixtures with various degrees of moisture content have been numerically simulated and tested. In another framework, different mixtures with constituent variations have been tested and have shown minor effects on the permittivity in comparison with the moisture content, air-voids, cement and aggregate particles [61].

Using the aforementioned numerical framework, a coherent set of synthetic examples was generated. The synthetic set was subsequently used to fine-tune and optimise the shape factor such as the CRIM-based bulk permittivity to match the actual one. Using numerical experiments allowed to have full control on the volumetric fractions of the concrete's components. Thus, every term in the CRIM formula can be accurately implemented. The resulting optimised shape factor using the suggested scheme was evaluated in both synthetic and laboratory experiments indicating the validity and robustness of the revised CRIM model.

2. Methodology

CRIM (1) is one of the most used models by GPR practitioners for predicting the bulk permittivity of complex materials like concrete.

$$\varepsilon_b = \left(V_{ag}\varepsilon_{ag}^{\alpha} + V_a\varepsilon_a^{\alpha} + V_c\varepsilon_c^{\alpha} + V_w\varepsilon_w^{\alpha} \right)^{1/\alpha} \tag{1}$$

where ε_b is the bulk permittivity of the concrete mixture, V_{ag} and ε_{ag} are the volume and relative permittivity of aggregate, V_a and ε_a are the volumetric fraction of air-voids and relative permittivity of free space, V_c and ε_c are the volumetric fraction and relative permittivity of cement, V_m and ε_m are the volumetric fraction and relative permittivity of the moisture content and α is the shape factor.

The most common shape factor of CRIM model is usually $\alpha = 0.5$. This study investigates the CRIM formula in order to find an optimised shape factor for this specific problem. To be able to take this matter further, a training set needs to be built and to have control over this, numerical modelling was used.

Numerical modelling is a great tool when it comes to optimisation. In order to find the optimum shape factor, we used synthetic data to define the shape factor. After creating and simulating heterogeneous concrete models, a reflection from a Perfect Electric Conductor (PEC) target was used to predict the velocity and calculate the bulk permittivity. In order to achieve this, GPR data should be accurately picked regarding the time-zero position.

2.1. Time-zero

GPR applications requires great accuracy and precision. Such cases are when trying to locate gas pipes [62], for landmine detection [63] or concrete inspection [64] where it is necessary to position the time-zero accurately. Defining the exact location of time-zero on a GPR trace is still an open issue with no specific conclusive solution and usually addressed by taking into account the GPR manufacturers recommendations [65].

In Yelf's paper [65], a number of positions have been discussed as possible candidates for picking the time-zero for a GPR trace, such as the positive peak, negative peak, mid-amplitude point, and the first break position of the direct wave wavelet. Additionally, for the arrival time of the target's response, there are a number of possible positions that can be considered for calculating the two-way travel time. All of these cases have been considered and compared resulting in a better time-zero position. In this paper, the time-zero was positioned on the first peak of the direct wave and two way travel time was calculated from the time arrival average where the three reflected peaks of the target occur [66]. Figure 1 presents a GPR trace with a reflection from a PEC target. This time-zero picking methodology has been successfully evaluated using numerous numerical scenarios indicating the validity of the current approach.

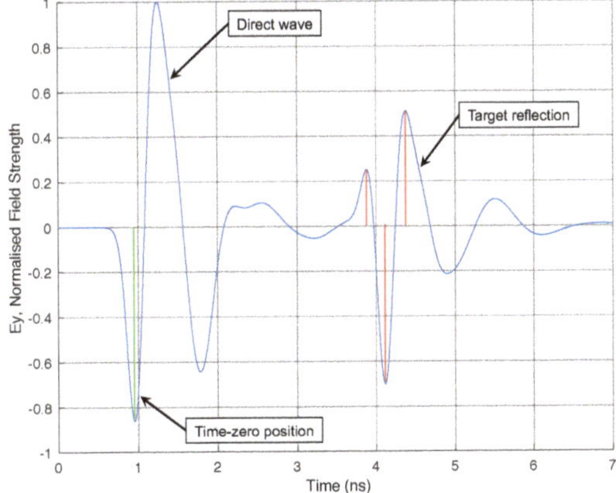

Figure 1. Time-zero is positioned on the first peak of the direct wave (green line). The two way travel time is calculated from the time-zero position to the average time of the three peaks (Perfect Electric Conductor (PEC) target response—red lines).

2.2. Concrete Modelling

The initial investigation of the CRIM model was performed based on a numerical homogeneous two dimensional (2D) model with a simple theoretical excitation source. The simulation showed promising results for the permittivity estimation of the concrete. In other words, the real and the estimated permittivities using CRIM were close. This provided confidence to upgrade the framework and design models which are much closer to reality. To increase the complexity of the model and be more realistic, a three dimensional (3D) concrete model was created including a realistic GSSI-like antenna.

Aggregates come in different distribution types, shapes and forms. In reality, aggregates do not overlap or collide with each other therefore, they were modelled in a non-overlapping environment. The disadvantage of this type of packing is that it is very time-consuming as it requires a long CPU run-time. This is because the aggregates need to find a position that does not collide or overlap with another aggregate. Simulating such models can take days or even weeks of run-time depending on the volume of spheres and the degree of compactness. The simulation results for both overlapping and non-overlapping aggregates were obtained and compared. The results were very similar and did not affect the outcomes of this study. Therefore, the numerical modelling was continued with overlapping aggregates as it required less computational run-time.

Another challenge for designing a realistic concrete model was the aggregate shape (random polygon shapes—rock). Numerical results showed a similar output despite the shape of the aggregate therefore, for this study, the type of aggregate used can be discarded. The rock aggregates come with some disadvantages. Firstly, it is very hard to calculate the volume of each aggregate hence not able to model a medium with a specific aggregate percentage. Secondly, much higher computational run-times are needed to produce the rock models thus it is very time-consuming when dealing with a large number of aggregates. Consequently, although the rock aggregate model displays a more realistic concrete mixture, taking into account the limitations and the fact that the results are similar to the sphere aggregates, in this study, the aggregates were kept as spheres and were randomly distributed with different diameters.

After establishing the aggregate shape the next step was to implement additional mixture content that exist in a concrete slab. As mentioned, concrete is a mixture of cement, aggregates, air-voids and moisture content. To add to the complexity of the model, air-voids and moisture particles are randomly distributed in the model in order to achieve a more realistic concrete structure. By changing the moisture or air-void percentages we can achieve different medium permittivities.

There are a limited number of studies that use realistic models due to their complexity and lack of available computational resources. The open-source GPU engine for gprMax has greatly accelerated the simulations. The simulations that are conducted using the GPU based gprMax solver are up to 30 times faster in comparison with the traditional CPU based one. CPU uses a few cores and is generally used for simple tasks whereas GPU works with thousands of efficient cores with a parallel architecture [36].

Figure 2 illustrates a heterogeneous concrete model with a size of 30 cm $\times$ 20 cm $\times$ 36 cm and a spatial discretisation of 1 mm. Aggregate, cement, air-voids and moisture content with different percentages were randomly distributed in order to achieve different concrete mixtures. A PEC plate was placed at the bottom of the model that resulted in a strong reflection which would be used later on to derive the bulk permittivity of the investigated medium. Each mixture was simulated and the reflected signals were processed to find the GPR wave propagation of velocity and therefore calculate the bulk permittivity. The GSSI-like 1.5 GHz centre frequency antenna structure was coupled to the concrete surface. This transducer consist of a transmitter/receiver (Tx/Rx) bowtie (copper) pair, printed circuit boards—PCBs (glass fiber), electromagnetic absorber (carbon-loaded foam), shield (PEC) and a red case (polypropylene) [67].

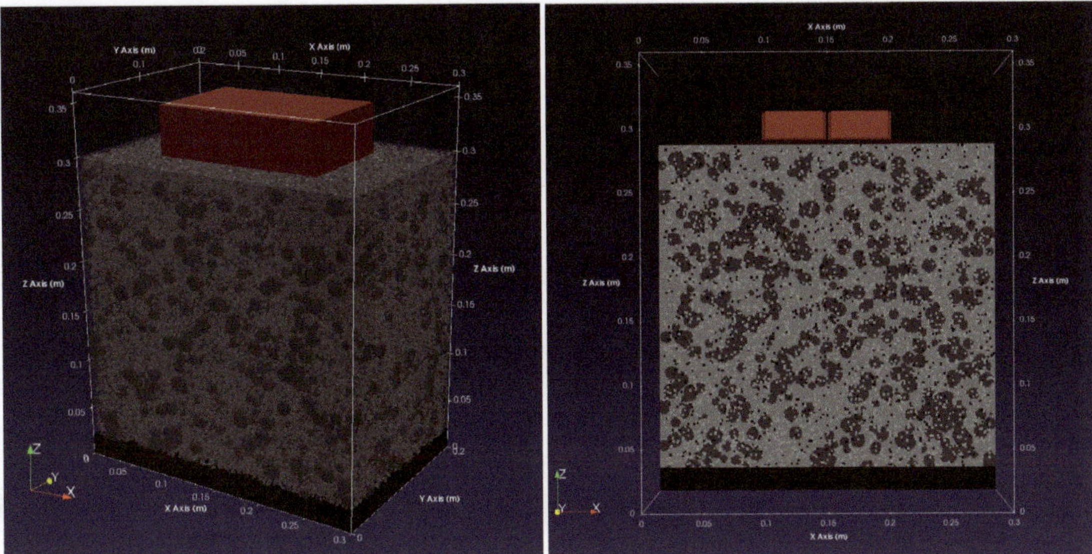

(**a**) Heterogeneous 3D concrete model   (**b**) Slice of heterogeneous concrete model

Figure 2. Heterogeneous concrete model is plotted using Paraview [68]. The model employs a 1.5 GHz centre frequency GSSI-like antenna structure on top of the concrete. A PEC plate is placed below the concrete model in order to obtain a perfect reflection. The geometry dimension of the models is 30 cm × 20 cm × 36 cm. (**a**) 3D view of the concrete model. (**b**) A slice of the 3D model which allows for a better understanding of the material distribution.

Producing a realistic concrete model requires the material mixture to follow a rational percentage range. The permittivity of the materials used in the concrete mixture combinations with their corresponding percentage ranges are shown in Table 1. Notice that the imaginary is omitted from the simulations since in the current study we try to infer the permittivity from the bulk velocity which is not affected by electromagnetic losses. Numerous concrete mixtures were synthetically generated based on the percentages shown in Table 1. Various concretes with different aggregate, cement, air and water fraction percentages were investigated numerically in order to calculate their resulting bulk permittivity. Each concrete mixture was simulated multiple times to find the average permittivity as presented in Figure 3. The estimated bulk permittivity from the numerical modelling will be used in section 2.3 to calibrate the shape factor of the CRIM model and generate a better formula with a more accurate shape factor that can accurately predict the bulk permittivity of concrete mixtures based on its aggregate, cement, water and air fraction.

Table 1. The range of components used to generate the training data. Notice that the water is assumed to be bound and therefore its relaxation frequency is shifted to lower frequencies which results to a lower permittivity value for the frequency range of interest [69]. The percentages of the components are for on service concrete.

Material	Mixture Percentage	Permittivity (ε)
Aggregate	60–75%	7
Cement	7–15%	3
Air	1–8%	1
Water	14–21%	37.54 [69]

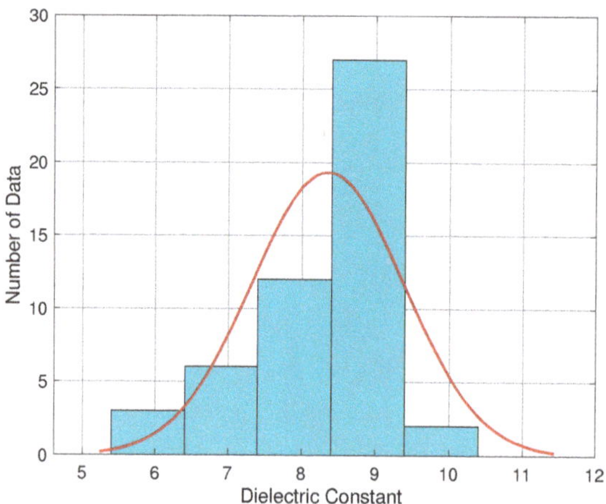

Figure 3. A random concrete model is simulated a number of times (×50) with different particle distributions in order to find its average bulk permittivity based on the reflection from the PEC reflector. This process neglects abnormal permittivity and allows the output to be more precise.

The PEC plate at the bottom of the model had a thickness of 2 cm. The GPR signals propagating through the medium did not transmit through the PEC and all of the energy from the impinging signals on the plate was reflected hence the thickness is not important. In reality, aggregates come in various shapes and sizes. To keep the numerical modelling as realistic as possible, different aggregate sizes were used. The size of the aggregates distributed in the mixture had a radius range from 4 to 8 mm [69,70] for coarse aggregates. As the aggregate percentage reached 60%, the radius was automatically decreased to a range of 1 to 2 mm [69,71] in order to simulate smaller or fine aggregates into the model and reach the specified percentage. This made the resulting shape factor tuned for an average size of particles and not for a specific one. Moisture and air-void particles were randomly distributed around the aggregates according to the selected percentage. The cement material was used as the background material. In other words, it filled up the remaining spaces of the concrete mixture. In regards to the water permittivity, there are two types of water. Bound and unbound (free) water. Bound water is a thin layer of water or moisture which surrounds mineral surfaces such as soil and concrete. Water has a strong electrical polarity hence it bounds very easily with other surfaces [72]. This has a high impact on the permittivity of the material which the water molecules bound with. The dielectric constant of bound water in comparison with free water ($\varepsilon_r = 81$) is much smaller. In a recent study, the permittivity of the liquid phase was fitted in a non-linear CRIM and permittivity of $\varepsilon_r = 37.54$ was calculated [69]. Figure 4 illustrates a set of concrete mixtures generated using the aforementioned procedure. The volumetric percentages of the components for each model are given in Table 2.

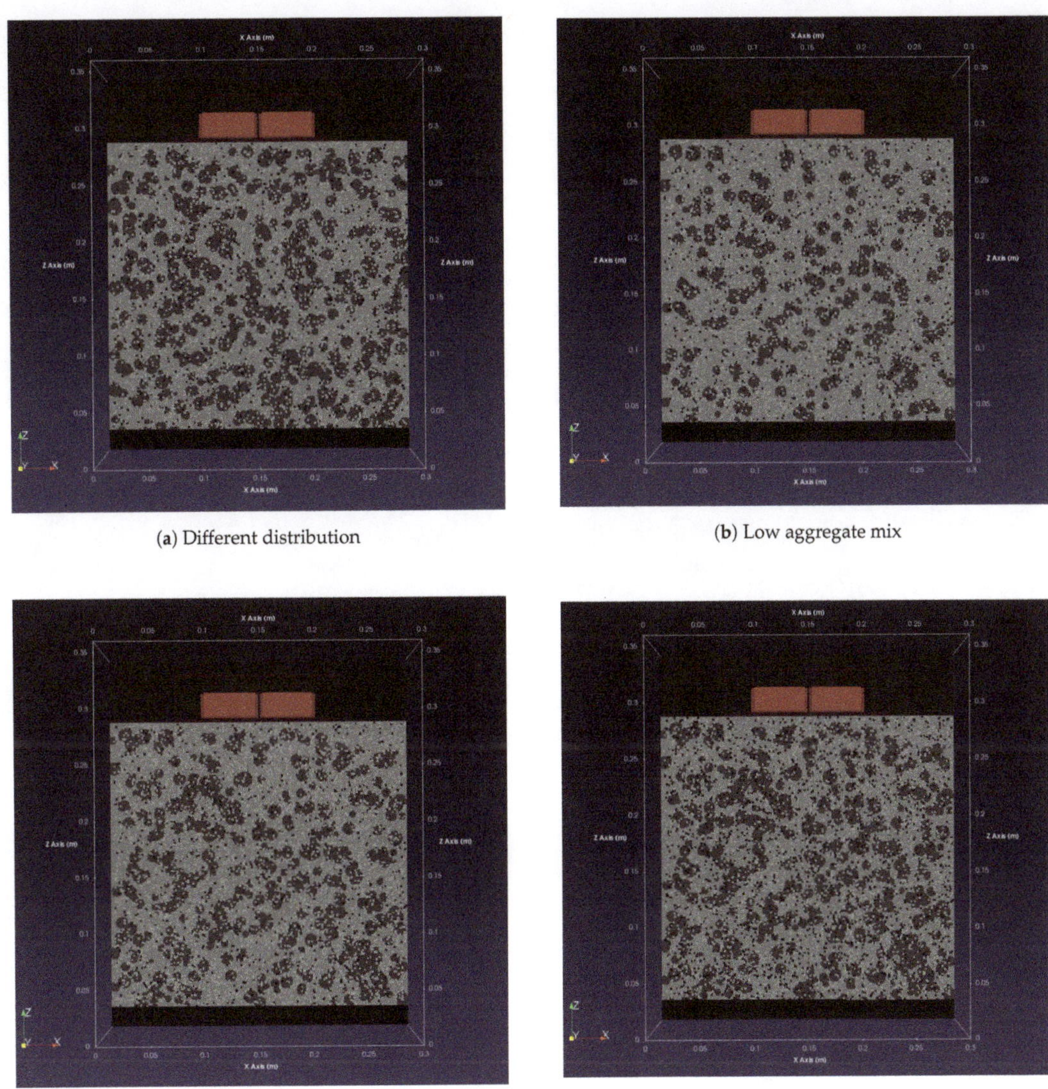

(a) Different distribution

(b) Low aggregate mix

(c) High moisture mix

(d) High air-void mix

Figure 4. Heterogeneous concrete mix with different aggregates, air-voids and moisture content. (a) presents the same mixture content as (b) but with different distribution. (b) shows a mixture with low aggregate content. (c) illustrates high moisture content concrete resulting in a high permittivity. Finally, (d) indicates high air-void content allowing the GPR signal to travel with a higher velocity. 64 representative concrete mixtures were selected from the training pool and each one was simulated 50 times resulting in 3200 simulations. The red box corresponds to the numerical equivalent of the GSSI 1.5 GHz antenna structure. The volumetric percentages of the components for each model are shown in Table 2.

Table 2. The volumetric fraction of the concrete's components of the models shown in Figure 4. Some of these percentages are not within the ranges shown in table 1 in order to illustrate the modelling capabilities of the current framework for extreme cases.

Model	Aggregate	Cement	Air-Voids	Moisture Content
a	65%	15%	5%	15%
b	45%	27%	11%	17%
c	60%	14%	5%	21%
d	65%	10%	15%	10%

2.3. Optimisation and Comparison

A simple method that is used to estimate the bulk relative effective permittivity for complex mixtures such as heterogeneous concrete models is the CRIM model as presented in Equation (2)

$$\varepsilon_{mix} = \left(\sum_{i=1}^{N} f_i \varepsilon_i^{\alpha} \right)^{1/\alpha} \tag{2}$$

where ε_{mix} is the bulk permittivity of a mixture, f_i is the volume fraction of *i*th material, ε_i is the permittivity of the *i*th material, N is the number of phases and α is a constant that is usually set up to $\alpha = 1/2$. For concrete, Equation (2) becomes

$$\varepsilon_r^{\alpha} = V_{ag}\varepsilon_{ag}^{\alpha} + V_c\varepsilon_c^{\alpha} + V_a\varepsilon_a^{\alpha} + V_w\varepsilon_w^{\alpha} \tag{3}$$

where in (3):

$$
\begin{aligned}
\alpha &= \text{geometric parameter} \\
\varepsilon_r &= \text{relative bulk permittivity} \\
V_{ag} &= \text{aggregate volume} \\
V_c &= \text{cement volume} \\
V_a &= \text{air-void volume} \\
V_w &= \text{water volume} \\
\varepsilon_{ag} &= \text{relative permittivity of aggregate (solid phase–matrix)} \\
\varepsilon_c &= \text{relative permittivity of cement (solid phase–matrix)} \\
\varepsilon_a &= \text{relative permittivity of air-void (gaseous phase–air)} \\
\varepsilon_w &= \text{relative permittivity of water or moisture content (liquid phase–water)}
\end{aligned}
$$

It has been reported that the most common value used is $\alpha = 0.5$ [73,74]. In some studies, the value of $\alpha = 0.46$ is substituted and other studies have shown that $\alpha = 0.66$ is more satisfactory for the research conducted [75]. Other work presented the shape factor to be $\alpha = 0.65$ [76,77].

This paper is focused on finding the optimum α for modelling concrete mixtures. In order for the shape factor investigation to take place, all the material permittivities and volumetric properties were implemented in to the CRIM formula. In comparison with the simulated results (derived in the previous section), the error difference is plotted in Figure 5a with respect to α. The error with respect to the shape factor was calculated using 64 representative and realistic mixing models from the available training pool. As shown in Figure 5a, the error has minimised for a shape factor of $\alpha = 0.13$.

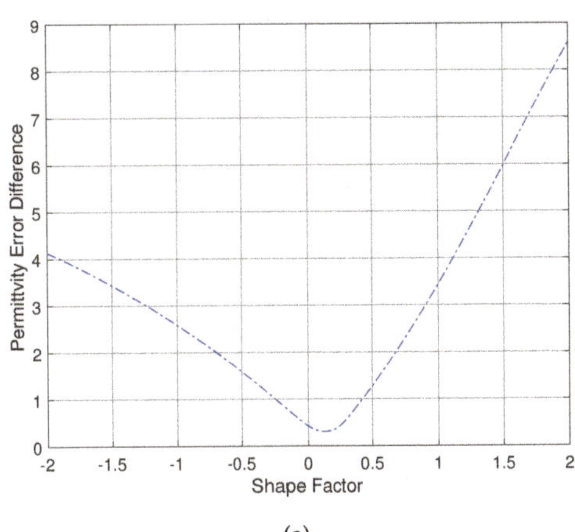
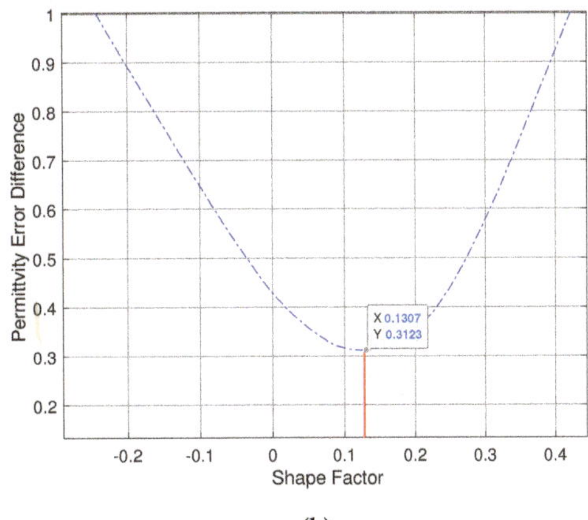

(a)   (b)

Figure 5. The error betweenthe estimated permittivity using Finite-Difference Time-Domain (FDTD) and Complex Refractive Index Model (CRIM) models (**a**). Sub-figure (**b**), zooms in to better visualise the resulting shape factor.

To compare the results of this research with other mixing models such as the Rayleigh [29] and the Böttcher [30] models, the same calculation process was done.

The Rayleigh mixing model is presented as:

$$\frac{\varepsilon_{bulk} - \varepsilon_b}{\varepsilon_{bulk} + 2\varepsilon_b} = V_{sb}\frac{\varepsilon_s - \varepsilon_b}{\varepsilon_s + 2\varepsilon_b} + V_a\frac{\varepsilon_a - \varepsilon_b}{\varepsilon_a + 2\varepsilon_b} + V_{sw}\frac{\varepsilon_{sw} - \varepsilon_b}{\varepsilon_{sw} + 2\varepsilon_b} \tag{4}$$

The Böttcher mixing model is presented as:

$$\frac{\varepsilon_{bulk} - \varepsilon_b}{3\varepsilon_{bulk}} = V_{sb}\frac{\varepsilon_s - \varepsilon_b}{\varepsilon_s + 2\varepsilon_{bulk}} + V_a\frac{\varepsilon_a - \varepsilon_b}{\varepsilon_a + 2\varepsilon_{bulk}} + V_{sw}\frac{\varepsilon_{sw} - \varepsilon_b}{\varepsilon_{sw} + 2\varepsilon_{bulk}} \tag{5}$$

where in (4) and (5):

$$
\begin{aligned}
\varepsilon_{bulk} &= \text{bulk permittivity} \\
\varepsilon_b &= \text{dielectric constant of binder} \\
\varepsilon_s &= \text{dielectric constant of the solid phase (matrix)} \\
\varepsilon_a &= \text{dielectric constant of the gaseous phase (air)} \\
\varepsilon_{sw} &= \text{dielectric constant of the liquid phase (water)} \\
V_{sb} &= \text{bulk volume of aggregate} \\
V_a &= \text{volume of air} \\
V_{sw} &= \text{volume of water}
\end{aligned}
$$

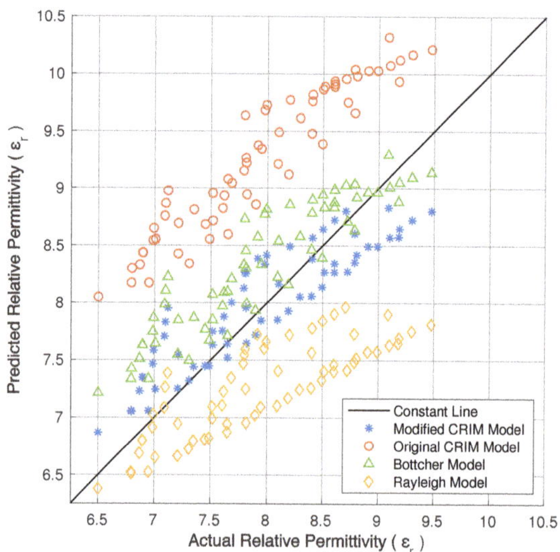

Figure 6. Comparison between different mixing models and the new modified CRIM model. The actual relative permittivity is based on the FDTD algorithm and the predicted relative permittivity is from the mixing models.

Equations (4) and (5) were used to calculate the bulk permittivity of the Rayleigh and the Böttcher models, respectively. By inserting the dielectric constant and the volume of each material, the formula calculates a specific bulk permittivity according to the given parameters. The Rayleigh mixing model has one output hence the permittivity calculation was simple. On the other hand, the Böttcher mixing models output was mathematically more complex. To overcome this time-consuming calculation, an automated procedure was programmed in MATLAB. A symbolic variable was created followed by a variable precision arithmetic (VPA) operation in order to derive four potential solutions. The positive solution was chosen as the bulk electric permittivity while the negative solutions were omitted.

Figure 6 illustrates the estimated permittivty using FDTD and the predicted permittivity using the aforementioned mixing models. It is apparent that the revised geometric factor performs better in comparison with the traditional CRIM model ($\alpha = 0.5$) and the other mixing models. This supports the premise that the revised CRIM using a shape factor of $\alpha = 0.13$ is a reliable mixing formula for predicting the bulk permittivity of concrete based on its aggregate, cement, water and air fraction content .

3. Laboratory Experiments

The revised shape factor was tested in laboratory experiments using a commercial horn antenna with 1 GHz central frequency. The experimental configuration is shown in Figure 7. The horn antenna was placed ≈40 cm above a concrete surface which consisted of 18 homogeneous and well-matured concrete blocks tightly packed. The concrete blocks were placed in a 6 × 3 grid format. The dimensions of each concrete block is 40 cm × 20 cm × 10 cm. The bulk permittivity of the concrete surface was evaluated based on equation [7]

$$\varepsilon = \left(\frac{1 + A_1/A_m}{1 - A_1/A_m} \right)^2 \tag{6}$$

where A_1 is the amplitude of the reflection from the concrete surface and A_m is the amplitude of the reflection when the concrete surface is covered with with a PEC plate [7]. A zero-offset correction was applied to each scan in an effort to remove static components that might compromise the accuracy of the measurements. The bulk permittivity of the concrete blocks was estimated using (6) $\varepsilon_c \approx 7.8$. Subsequently, the gaps between the concrete blocks were gradually widened from 0–2-cm with 5 mm step in an effort to artificially increase the air-voids in a controlled manner and see the overall effects to the bulk permittivity. By increasing the gaps larger than 2 cm, it would act as voids rather than cracks hence the limited to 2 cm. Additionally, as the concrete blocks were homogeneous, the water fraction was consistent throughout the concrete block. Doing this, the bulk permittivity of the mixture concrete/air-voids could be evaluated with respect to the artificially created air-voids.

Since the permittivity of the concrete blocks was known ($\varepsilon_c \approx 7.8$), the bulk permittivity of the concrete/air-voids mixture could also be estimated using the CRIM model.

$$\varepsilon_b = ((1 - V_a)\varepsilon_c^\alpha + V_a\varepsilon_a^\alpha)^{1/\alpha} \tag{7}$$

where ε_b is the CRIM-based bulk permittivity of the mixture concrete/air-voids, V_a is volumetric fraction of the artificially created air-voids, $\varepsilon_c \approx 7.8$ is the relative bulk permittivity of the concrete blocks, $\varepsilon_a = 1$ is the relative permittivity of free space and α is the shape factor. The revised shape factor $\alpha = 0.13$ and the default-one $\alpha = 0.5$ were used for the current example. The results are shown in Figure 8. The measured relative permittivity using (6) with respect to the artificially created air-gaps is illustrated with dots. The relative bulk permittivity estimated using CRIM (7) with $\alpha = 0.5$ and $\alpha = 0.13$ are illustrated with dotted and solid lines, respectively. It is apparent that the CRIM model using the optimised shape factor $\alpha = 0.13$ matched the measured bulk permittivity better and clearly over-performed the default $\alpha = 0.5$.

Figure 7. The experimental framework used to validate the revised shape factor. A horn antenna with 1 GHz central frequency is placed on top of a surface consisted of concrete blocks. The gaps between the concrete blocks are gradually increased in an effort to increase the overall volumetric fraction of air.

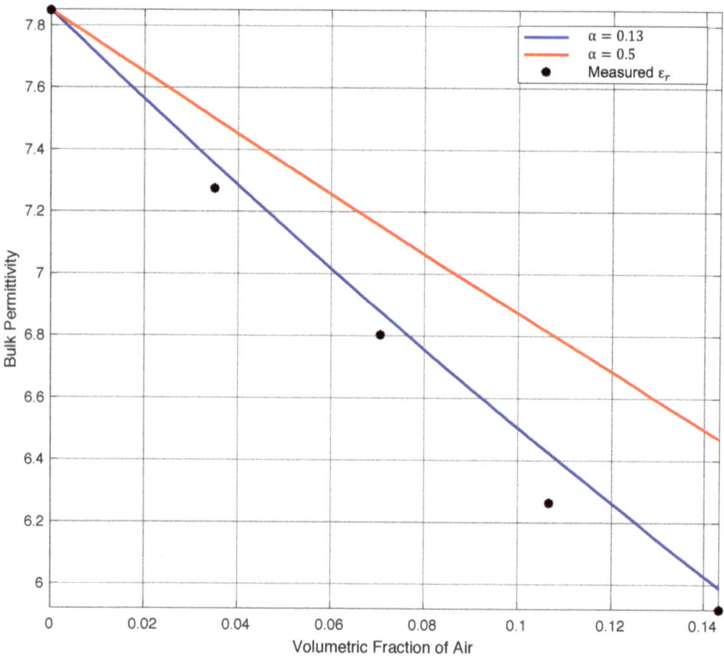

Figure 8. The measured and the calculated bulk permittivity using $\alpha = 0.5$ and $\alpha = 0.13$. It is apparent that the revised shape factor $\alpha = 0.13$ over-performs the default $\alpha = 0.5$.

4. Conclusions

An optimised shape factor for the CRIM mixing model has been obtained using realistic synthetic GPR models and verified by numerical experiments. The shape factor is fine-tuned for concrete applications and it is used to estimate the bulk permittivity of concrete based on its individual components (aggregates, cement, air voids and water fraction). The optimisation was done based on numerical simulations for a wide range of concrete samples with different properties and compositions. The concrete models were designed in 3D with a 1.5 GHz centre frequency GSSI-like antenna structure on the surface. The revised CRIM formula was in good agreement with the simulation results and clearly performed better compared to the default CRIM model. Subsequently, the modified CRIM was applied in a laboratory experiment in order to predict the effects of air voids to the bulk permittivity of concrete. The results using the numerically-derived shape factor were in good agreement with the laboratory measurements. Through numerical and laboratory measurements we have supported the premise that the proposed modified CRIM is a more reliable method for predicting the dielectric properties of concrete based on its components.

Author Contributions: Conceptualisation, H.Z., A.G. and I.G.; methodology, H.Z. and I.G.; execution, H.Z.; writing-review and editing, H.Z.; Proofreading, I.G. and A.G.; Supervision, A.G.; Project administration, A.G. All authors have read and agreed to the published version of the manuscript.

Funding: This research received no external funding.

Institutional Review Board Statement: Not applicable.

Informed Consent Statement: Not applicable.

Data Availability Statement: The data used in this research are available from the authors upon reasonable requests.

Conflicts of Interest: The authors declare that there are no conflicts of interest.

References

1. Daniels, D.J. *Ground Penetrating Radar*, 2nd ed.; The University of Michigan: London, UK, 2005.
2. Jol, H.M. *Ground pEnetrating Radar Theory and Applications*; Elsevier: Amsterdam, The Netherlands, 2008.
3. Annan, A. Ground-penetrating radar. In *Near-Surface Geophysics*; Society of Exploration Geophysicists: Tulsa, Oklahoma, 2005; pp. 357–438.
4. El-Mahallawy, M.S.; Hashim, M. Material classification of underground utilities from GPR images using DCT-based SVM approach. *IEEE Geosci. Remote Sens. Lett.* **2013**, *10*, 1542–1546. [CrossRef]
5. Gabryś, M.; Kryszyn, K.; Ortyl, Ł. GPR surveying method as a tool for geodetic verification of GESUT database of utilities in the light of BSI PAS128. *Rep. Geod. Geoinform.* **2019**, *107*, 49–59.
6. Kang, M.S.; Kim, N.; Im, S.B.; Lee, J.J.; An, Y.K. 3D GPR Image-based UcNet for Enhancing Underground Cavity Detectability. *Remote Sens.* **2019**, *11*, 2545. [CrossRef]
7. Saarenketo, T.; Scullion, T. Road evaluation with ground penetrating radar. *J. Appl. Geophys.* **2000**, *43*, 119–138. [CrossRef]
8. Cardarelli, E.; Marrone, C.; Orlando, L. Evaluation of tunnel stability using integrated geophysical methods. *J. Appl. Geophys.* **2003**, *52*, 93–102. [CrossRef]
9. Hugenschmidt, J. Concrete bridge inspection with a mobile GPR system. *Constr. Build. Mater.* **2002**, *16*, 147–154. [CrossRef]
10. Hugenschmidt, J.; Mastrangelo, R. GPR inspection of concrete bridges. *Cem. Concr. Compos.* **2006**, *28*, 384–392. [CrossRef]
11. Diamanti, N.; Annan, A.P.; Redman, J.D. Concrete bridge deck deterioration assessment using ground penetrating radar (GPR). *J. Environ. Eng. Geophys.* **2017**, *22*, 121–132. [CrossRef]
12. Maierhofer, C. Nondestructive evaluation of concrete infrastructure with ground penetrating radar. *J. Mater. Civ. Eng.* **2003**, *15*, 287–297. [CrossRef]
13. Xie, X.; Li, P.; Qin, H.; Liu, L.; Nobes, D.C. GPR identification of voids inside concrete based on the support vector machine algorithm. *J. Geophys. Eng.* **2013**, *10*, 034002. [CrossRef]
14. Giannakis, I.; Giannopoulos, A.; Davidson, N. Realistic modelling of ground penetrating radar for landmine detection using FDTD. In Proceedings of the 15th International Conference on Ground Penetrating Radar, Brussels, Belgium, 30 June–4 July 2014; pp. 954–959.
15. Chang, C.W.; Lin, C.H.; Lien, H.S. Measurement radius of reinforcing steel bar in concrete using digital image GPR. *Constr. Build. Mater.* **2009**, *23*, 1057–1063. [CrossRef]
16. Mai, T.C.; Razafindratsima, S.; Sbartaï, Z.M.; Demontoux, F.; Bos, F. Non-destructive evaluation of moisture content of wood material at GPR frequency. *Constr. Build. Mater.* **2015**, *77*, 213–217. [CrossRef]
17. Hoegh, K.; Khazanovich, L.; Dai, S.; Yu, T. Evaluating asphalt concrete air void variation via GPR antenna array data. *Case Stud. Nondestr. Test. Eval.* **2015**, *3*, 27–33. [CrossRef]
18. Shang, J.Q. Effects of asphalt pavement properties on complex permittivity. *Int. J. Pavement Eng.* **2002**, *3*, 217–226. [CrossRef]
19. Pérez-Gracia, V.; García, F.G.; Abad, I.R. GPR evaluation of the damage found in the reinforced concrete base of a block of flats: A case study. *NDT e Int.* **2008**, *41*, 341–353. [CrossRef]
20. Klysz, G.; Balayssac, J.; Ferrières, X. Evaluation of dielectric properties of concrete by a numerical FDTD model of a GPR coupled antenna—parametric study. *NDT e Int.* **2008**, *41*, 621–631. [CrossRef]
21. Tsui, F.; Matthews, S. Analytical modelling of the dielectric properties of concrete for subsurface radar applications. *Constr. Build. Mater.* **1997**, *11*, 149–161. [CrossRef]
22. Bourdi, T.; Rhazi, J.E.; Boone, F.; Ballivy, G. Application of Jonscher model for the characterization of the dielectric permittivity of concrete. *J. Phys. D Appl. Phys.* **2008**, *41*, 205410. [CrossRef]
23. Bourdi, T.; Rhazi, J.E.; Boone, F.; Ballivy, G. Modelling dielectric-constant values of concrete: An aid to shielding effectiveness prediction and ground-penetrating radar wave technique interpretation. *J. Phys. D Appl. Phys.* **2012**, *45*, 405401. [CrossRef]
24. Redman, J.D.; Annan, A.P.; Diamanti, N. Measurement of bulk electrical properties using GPR with a variable reflector. *J. Environ. Eng. Geophys.* **2018**, *23*, 489–496. [CrossRef]
25. Böttcher, C.J.F. *Theory Electric Polarisation*; Elsevier Publishing Company: Amsterdam, The Netherlands, 1952.
26. Sihvola, A.H. *Electromagnetic Mixing Formulas and Applications*; IET Digital Library: Padstow, UK, 1999.
27. Al-Qadi, I.L.; Leng, Z.; Lahouar, S.; Baek, J. In-place hot-mix asphalt density estimation using ground-penetrating radar. *Transp. Res. Rec.* **2010**, *2152*, 19–27. [CrossRef]
28. Birchak, J.R.; Gardner, C.G.; Hipp, J.E.; Victor, J.M. High dielectric constant microwave probes for sensing soil moisture. *Proc. IEEE* **1974**, *62*, 93–98. [CrossRef]
29. Rayleigh, L. LVI. On the influence of obstacles arranged in rectangular order upon the properties of a medium. *Lond. Edinb. Dublin Philos. Mag. J. Sci.* **1892**, *34*, 481–502. [CrossRef]
30. Böttcher, C.J.F.; van Belle, O.C.; Bordewijk, P.; Rip, A. *Theor. electr. Polarization*; Elsevier Science Ltd: Amsterdam, The Netherlands, 1978.
31. Brown, W.F.; Franz, W.; Forsbergh, P. Dielectrics/Dielektrika. *Handbuch der Physik*; Springer: Heidelberg, Germany, 1956.
32. Wagner, A. Physik, 40, 817 (1913). *Arch. Elektroteeh* **1914**, *3*, 83.
33. Bruggeman, V.D. Berechnung verschiedener physikalischer Konstanten von heterogenen Substanzen. I. Dielektrizitätskonstanten und Leitfähigkeiten der Mischkörper aus isotropen Substanzen. *Annalen der physik* **1935**, *416*, 636–664. [CrossRef]

34. Topp, G.C.; Davis, J.L.; Annan, A.P. Electromagnetic determination of soil water content: Measurements in coaxial transmission lines. *Water Resour. Res.* **1980**, *16*, 574–582. [CrossRef]
35. Warren, C.; Giannopoulos, A.; Giannakis, I. gprMax: Open source software to simulate electromagnetic wave propagation for Ground Penetrating Radar. *Comput. Phys. Commun.* **2016**, *209*, 163–170. [CrossRef]
36. Warren, C.; Giannopoulos, A.; Gray, A.; Giannakis, I.; Patterson, A.; Wetter, L.; Hamrah, A. A CUDA-based GPU engine for gprMax: open source FDTD electromagnetic simulation software. *Comput. Phys. Commun.* **2019**, *237*, 208–218. [CrossRef]
37. Taflove, A.; Hagness, S.C. *Computational Electromagnetics: The Finite-Difference Time-Domain Method*, 3rd ed.; Artech House Publishers: London, UK, 2005.
38. Yee, K. Numerical solution of initial boundary value problems involving Maxwell's equations in isotropic media. *IEEE Trans. Antennas Propag.* **1966**, *14*, 302–307.
39. Sadiku, M.N. *Numerical Techniques in Electromagnetics*; CRC press: Boca Raton, FL, USA, 2000.
40. Cassidy, N.J. A review of practical numerical modelling methods for the advanced interpretation of ground-penetrating radar in near-surface environments. *Near Sur. Geophys.* **2007**, *5*, 5–21. [CrossRef]
41. Cassidy, N.J.; Millington, T.M. The application of finite-difference time-domain modelling for the assessment of GPR in magnetically lossy materials. *J. Appl. Geophys.* **2009**, *67*, 296–308. [CrossRef]
42. Giannakis, I.; Giannopoulos, A.; Warren, C. A realistic FDTD numerical modeling framework of ground penetrating radar for landmine detection. *IEEE J. Sel. Top. Appl. Earth Obs. Remote Sens.* **2016**, *9*, 37–51. [CrossRef]
43. Lee, K.H.; Chen, C.C.; Teixeira, F.L.; Lee, R. Modeling and investigation of a geometrically complex UWB GPR antenna using FDTD. *IEEE Trans. Antennas Propag.* **2004**, *52*, 1983–1991. [CrossRef]
44. Uduwawala, D. Modeling and investigation of planar parabolic dipoles for GPR applications: A comparison with bow-tie using FDTD. *J. Electomagn. Waves Appl.* **2006**, *20*, 227–236. [CrossRef]
45. Giannakis, I.; Giannopoulos, A.; Warren, C. Realistic FDTD GPR antenna models optimized using a novel linear/nonlinear Full-Waveform Inversion. *IEEE Trans. Geosci. Remote Sens.* **2019**, *57*, 1768–1778. [CrossRef]
46. Bourgeois, J.M.; Smith, G.S. A fully three-dimensional simulation of a ground-penetrating radar: FDTD theory compared with experiment. *IEEE Trans. Geosci. Remote Sens.* **1996**, *34*, 36–44. [CrossRef]
47. Klysz, G.; Balayssac, J.; Laurens, S.; Ferrieres, X. Numerical FDTD simulation of the direct wave propagation of a GPR coupled antenna. In Proceedings of the Tenth International Conference on Grounds Penetrating Radar, Delft, The Netherlands, 21–24 June 2004; pp. 45–48.
48. Caratelli, D.; Yarovoy, A.; Ligthart, L.P. Accurate FDTD modelling of resistively-loaded bow-tie antennas for GPR applications. In Proceedings of the 2009 3rd European Conference on Antennas and Propagation, Berlin, Germany, 23–27 March 2009; pp. 2115–2118.
49. Bourgeois, J.; Smith, G. A complete electromagnetic simulation of a ground penetrating radar for mine detection: Theory and experiment. In Proceedings of the IEEE Antennas and Propagation Society International Symposium 1997, Montreal, QC, Canada, 13–18 July 1997; pp. 986–989.
50. Radzevicius, S.J.; Chen, C.C.; Peters Jr, L.; Daniels, J.J. Near-field dipole radiation dynamics through FDTD modeling. *J. Appl. Geophys.* **2003**, *52*, 75–91. [CrossRef]
51. Diamanti, N.; Annan, A.P. Characterizing the energy distribution around GPR antennas. *J. Appl. Geophys.* **2013**, *99*, 83–90. [CrossRef]
52. Lampe, B.; Holliger, K. Resistively loaded antennas for ground-penetrating radar: A modeling approach. *Geophysics* **2005**, *70*, K23–K32. [CrossRef]
53. Venkatarayalu, N.V.; Chen, C.C.; Teixeira, F.L.; Lee, R. Numerical modeling of ultrawide-band dielectric horn antennas using FDTD. *IEEE Trans. Antennas Propag.* **2004**, *52*, 1318–1323. [CrossRef]
54. Turk, A.S.; Sahinkaya, D.A.; Sezgin, M.; Nazli, H. Investigation of convenient antenna designs for ultra-wide band GPR systems. In Proceedings of the 2007 4th International Workshop on, Advanced Ground Penetrating Radar, Aula Magna Partenope, Italy, 27–29 June 2007; pp. 192–196.
55. Warren, C.; Giannopoulos, A. Creating finite-difference time-domain models of commercial ground-penetrating radar antennas using Taguchi's optimization method. *Geophysics* **2011**, *76*, G37–G47. [CrossRef]
56. Robert, A. Dielectric permittivity of concrete between 50 MHz and 1 GHz and GPR measurements for building materials evaluation. *J. Appl. Geophys.* **1998**, *40*, 89–94. [CrossRef]
57. Al-Saleh, S.A. Analysis of total chloride content in concrete. *Case Stu. Constr. Mater.* **2015**, *3*, 78–82. [CrossRef]
58. Laurens, S.; Balayssac, J.; Rhazi, J.; Klysz, G.; Arliguie, G. Non-destructive evaluation of concrete moisture by GPR: experimental study and direct modeling. *Mater. Struct.* **2005**, *38*, 827–832. [CrossRef]
59. Wu, Z.; Wong, H.; Buenfeld, N. Transport properties of concrete after drying-wetting regimes to elucidate the effects of moisture content, hysteresis and microcracking. *Cem. Concr. Res.* **2017**, *98*, 136–154. [CrossRef]
60. Shaw, M. The permittivity and conductivity of concretes at ground-penetrating radar frequencies. *Adv. Cem. Res.* **1998**, *10*, 187–194. [CrossRef]
61. Soutsos, M.; Bungey, J.; Millard, S.; Shaw, M.; Patterson, A. Dielectric properties of concrete and their influence on radar testing. *NDT e Int.* **2001**, *34*, 419–425. [CrossRef]

62. Terrasse, G.; Nicolas, J.M.; Trouvé, E.; Drouet, É. Automatic localization of gas pipes from GPR imagery. In Proceedings of the 2016 24th European Signal Processing Conference (EUSIPCO), Budapest, Hungary, 29 August–2 September 2016; pp. 2395–2399.

63. Daniels, D.J. A review of GPR for landmine detection. *Sens. Imaging Int. J.* **2006**, *7*, 90–123. [CrossRef]

64. Brown, E. Density of asphalt concrete-how much is needed? In Proceedings of the 69th Annual Meeting of the TransportationResearch Board, Washington, DC, USA, 8–9 January 1990.

65. Yelf, R. Where is true time zero? In Proceedings of the Tenth International Conference on Grounds Penetrating Radar, Delft, The Netherlands, 21–24 June 2004; pp. 279–282.

66. Zadhoush, H. Numerical modelling of ground penetrating radar for optimisation of the time-zero adjustment and complex refractive index model. Ph.D. Thesis, The University of Edinburgh, Edinburgh, UK, 2020.

67. Warren, C.; Giannopoulos, A. Investigation of the directivity of a commercial ground-penetrating radar antenna using a finite-difference time-domain antenna model. In Proceedings of the 2012 14th International Conference on Ground Penetrating Radar (GPR), Shanghai, China, 4–8 June 2012; pp. 226–231.

68. Ahrens, J.; Geveci, B.; Law, C. Paraview: An end-user tool for large data visualization. *Vis. Handb.* **2005**, *717*, 8.

69. Lachowicz, J.; Rucka, M. A concept of heterogeneous numerical model of concrete for GPR simulations. In Proceedings of the 2017 9th International Workshop on Advanced Ground Penetrating Radar (IWAGPR), Edinburgh, UK, 28–30 June 2017; pp. 1–4.

70. Grassl, P.; Wong, H.S.; Buenfeld, N.R. Influence of aggregate size and volume fraction on shrinkage induced micro-cracking of concrete and mortar. *Cem. Concr. Res.* **2010**, *40*, 85–93. [CrossRef]

71. Kemper, W.; Rosenau, R. Aggregate stability and size distribution. *Methods Soil Anal. Part I Phys. Mineral. Methods* **1986**, *5*, 425–442.

72. Jury, W.A.; Horton, R. *Soil Physics*; John Wiley & Sons: Hoboken, NJ, USA, 2004.

73. Ukaegbu, I.K.; Gamage, K.A.; Aspinall, M.D. Nonintrusive Depth Estimation of Buried Radioactive Wastes Using Ground Penetrating Radar and a Gamma Ray Detector. *Remote Sens.* **2019**, *11*, 141. [CrossRef]

74. Roth, K.; Schulin, R.; Flühler, H.; Attinger, W. Calibration of time domain reflectometry for water content measurement using a composite dielectric approach. *Water Resour. Res.* **1990**, *26*, 2267–2273. [CrossRef]

75. Gardner, C.; Dean, T.; Cooper, J. Soil water content measurement with a high-frequency capacitance sensor. *J. Agric. Eng. Res.* **1998**, *71*, 395–403. [CrossRef]

76. Dobson, M.C.; Ulaby, F.T.; Hallikainen, M.T.; El-Rayes, M.A. Microwave dielectric behavior of wet soil-Part II: Dielectric mixing models. *IEEE Trans. Geosci. Remote Sens.* **1985**, *GE-23*, 35–46. [CrossRef]

77. Peplinski, N.R.; Ulaby, F.T.; Dobson, M.C. Dielectric properties of soils in the 0.3-1.3-GHz range. *IEEE Trans. Geosci. Remote Sens.* **1995**, *33*, 803–807. [CrossRef]

MDPI

St. Alban-Anlage 66

4052 Basel

Switzerland

Tel. +41 61 683 77 34

Fax +41 61 302 89 18

www.mdpi.com

Remote Sensing Editorial Office

E-mail: remotesensing@mdpi.com

www.mdpi.com/journal/remotesensing